Power Electronics and Power Systems

Volume 94

Series Editors

M. A. Pai
Alex M. Stankovic

For further volumes:
http://www.springer.com/series/6403

Joe H. Chow
Editor

Power System Coherency and Model Reduction

Springer

Editor
Joe H. Chow
Department of Electrical, Computer,
and Systems Engineering
Rensselaer Polytechnic Institute
Troy, NY
USA

ISSN 2196-3185 ISSN 2196-3193 (electronic)
ISBN 978-1-4614-1802-3 ISBN 978-1-4614-1803-0 (eBook)
DOI 10.1007/978-1-4614-1803-0
Springer New York Heidelberg Dordrecht London

Library of Congress Control Number: 2013938034

Printed on acid-free paper

Springer is part of Springer Science+Business Media (www.springer.com)

To
Robin Podmore, William Price, and
the Singular Perturbations team: Petar
Kokotović, Peter Sauer, James Winkelman,
John Allemong, Boza Avramović, and
George Peponides

Preface

In the simulation of large power system dynamics for stability analysis on a digital computer, there is always a tug of war between capturing as much detail as possible in the simulation and completing the simulation in a reasonable amount of time so that the results can be useful. Dynamic equivalencing, that is, obtaining a reduced-order power system model to capture the relevant dynamics, has been an active research area since the early 1970s led by the work of Robin Podmore. It is interesting to note that the original dynamic equivalencing procedure of Podmore is still being used in the industry today, although some of the algorithms in the procedure have been updated.

In the late 1970s, Petar Kokotović and a few of his students investigated the applications of singular perturbations and time scales for power system analysis. In a somewhat surprising development, they were able to show a relationship linking weak connections between coherent areas and the slow interarea modes. This relationship becomes an analytical basis for large power system analysis. These results were published in a 1982 Springer monograph [1]. The monograph was out-of-print a few years after it was published, but now is again available in electronic form from the Springer website.

After the 1982 monograph went out-of-print, I thought about writing a new monograph with some incremental updates, but that idea did not seem to be productive. Most recently, however, with many new model reduction results, a monograph summarizing the old monograph and some early results, and encompassing several new results, including synchrophasor measurements, becomes much more appealing. Thus, the objectives of this new monograph include the following:

1. A summary of the original coherency and aggregation procedure and the slow coherency theory—This description is intended to help a reader to quickly catch on to the coherency idea, especially a student interested in pursuing research in this area, or a practicing engineer planning to use the dynamic equivalencing (DYNRED) program (Chaps. 2, 3, and 7).
2. Recent applications of linear model reduction methods for power systems—The linear model reduction methods available now are more rigorous than the modal equivalents originally proposed by John Undrill and William Price.

3. Other methods for obtaining nonlinear reduced power system models, such as artificial neural networks—They can be quite useful as these methods are response based and not model based.
4. Application of synchrophasor measurements—Instead of developing reduced-order models using power system model data, it is possible to use synchrophasor measurements to develop simple reduced-order models to capture across a power transfer interface.
5. Tracing interarea modes—Illustrations of interarea modes in a power network will help to extend various notions, such as mode shapes and damping control design.
6. Availability of dynamic equivalencing software—Some of the slow coherency and aggregation algorithms presented in this monograph and built on the MATLAB based Power System Toolbox [2] are now available as part of toolbox, which can be downloaded for free from the author's website [3].

I hope that a reader will find the updated materials and extended coverage of this new monograph useful in understanding power system coherency, interarea modes, and model reduction.

I would like to thank Professor M. A. Pai, the editor of this Springer monograph series, for his encouragement in the development of the monograph, as well as being an author, and Ms. Ania Levison at Springer for coordinating the publication. Dr. Barbara Lewis at the RPI Communications Center helped me in polishing the text. I would like to thank the contributors: Robin Podmore, Vijay Vittal, Feng Ma, Luis Rouco, Luis Pagola, Ignacio Peréz-Arriaga, George Verghese, Shanshan Liu, Peter Sauer, Dimitrios Chaniotis, Lei Wang, Kip Morrison, and Yuwa Chompoobutrgool, for writing some excellent chapters, and my former students, Ranjit Date, Ricardo Galarza, Aranya Chakrabortty, and Luigi Vanfretti. I am also grateful to Felipe Wilches and Amy Chow for drawing some of the figures.

My power system coherency and model reduction research has been supported by industry and several government funding agencies. The most recent ones are the Power System Research Consortium at Rensselaer Polytechnic Institute, funded by AEP, FirstEnergy, ISO-NE, NYISO, and PJM, the Engineering Research Center Program of the US National Science Foundation, and the US Department of Energy under NSF Award Number EEC-1041877.

This monograph is dedicated to Robin and Bill, two of the early pioneers of power system model reduction, and the slow coherency research team at University of Illinois, Urbana–Champaign, and General Electric Company, Schenectady, New York.

Finally, I would like to thank my wife Doris and my daughters Amy and Tammy for their love and patience.

Troy, NY, December 2012 Joe H. Chow

References

1. J.H. Chow, G. Peponides, P.V. Kokotović, B. Avramović, J.R. Winkelman, *Time-Scale Modeling of Dynamic Networks with Applications to Power Systems* (Springer, New York, 1982)
2. J.H. Chow, K.W. Cheung, A toolbox for power system dynamics and control engineering education. IEEE Trans. Power Syst. **7**, 1559–1564 (1992)
3. www.ecse.rpi.edu/homepages/chowj

Contents

Chapter 1
Introduction

Joe H. Chow

Abstract This introductory chapter gives a brief overview of power system coherency and model reduction literature. This survey focuses on both the early results and some more recent developments, and organizes power system model reduction techniques into two broad categories. One category of methods is to use coherency and aggregation methods to obtain reduced models in the form of nonlinear power system models. The other category is to treat the external system or the less relevant part of the system as an input–output model and obtain a lower order linear or nonlinear model based on the input–output properties. This chapter also provides a synopsis of the remaining chapters in this monograph.

1.1 Introduction

In the simulation of power system dynamics for stability analysis on a digital computer, a prudent approach is to develop the most comprehensive power system models so that the relevant dynamics can be accurately simulated given the computing resources and a desired simulation completion time.

One of the decisions that a power system engineer has to make is the geographic extent of the power system data set. Although the purpose of a stability investigation is to determine the dynamic response of generators and control systems in a study region due to disturbances inside the region, because of the interconnected nature of large power systems, these disturbances will impact the neighboring and other areas, the so-called external system, which in turn will impact the study region (Fig. 1.1).[1] For example, a short-circuit fault cleared by a line trip will redistribute the pre-fault

J. H. Chow (✉)
Rensselaer Polytechnic Institute, Troy, New York, USA
e-mail: chowj@rpi.edu

[1] This discussion can readily be extended to multiple external systems, as well as some buffer or boundary systems between the study system and the external system.

J. H. Chow (ed.), *Power System Coherency and Model Reduction*,
Power Electronics and Power Systems 94, DOI: 10.1007/978-1-4614-1803-0_1,

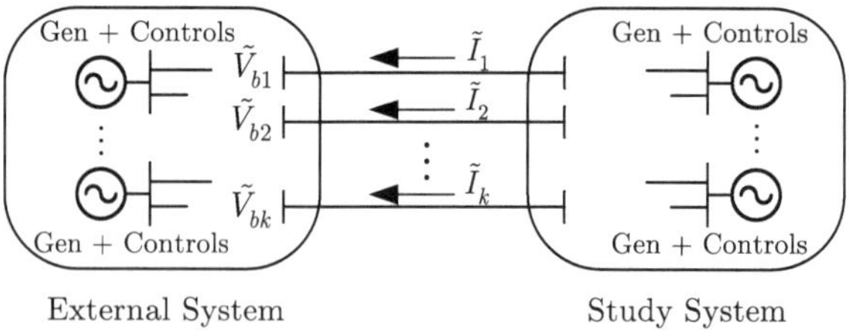

Fig. 1.1 Separation of a power system into a study system and an external system

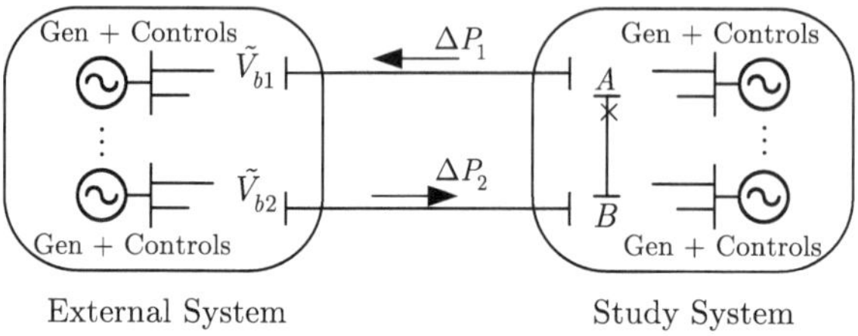

Fig. 1.2 Loop flow through the external system

flow on the tripped line to other paths, some of which may circulate through the external system. This situation is illustrated in Fig. 1.2, in which tripping the line from Bus A to Bus B causes a portion of the pre-fault line flow to be exported into the external system at one boundary bus and to return via another boundary bus. These uncontrolled loop flows need to be modeled accurately in order for the stability simulation to be valid. On the other hand, disturbances in the study region will most likely not excite significantly the internal dynamics of the external system. Thus a less detailed model of the external system can be used. As a result, for system studies with a strict turnaround time requirement, such as contingency analysis, or limited computing resources, power engineers commonly construct low-order models with the study region modeled in full detail and a reduced external system. For practical applications in large power grids, the low-order models, after reducing the size of the external system, may still have upwards of 5,000 buses and 1,000 generators.

To facilitate the discussions in the rest of this monograph, we express the power system model in the nonlinear differential-algebraic equation form for the study system as

$$\dot{x}_s = f_s(x_s, V_s, u_s, t), \quad V_s = g_s(x_s, x_e, u_s, u_e, t) \tag{1.1}$$

and the external system as

$$\dot{x}_e = f_e(x_e, V_e, u_e, t), \quad V_e = g_e(x_s, x_e, u_s, u_e, t) \tag{1.2}$$

where x denotes the state vector, V is a vector of the real and imaginary parts of the voltage phasors $\tilde{V}$, u is the control, t denotes time, f represents the dynamic equations such as the swing equation and excitation system models, g contains the algebraic network equations, and the subscripts s and e denote "study" and "external", respectively. Note that f_s and f_e depend only on the bus voltage phasors within the study area and the external system, respectively. The bus voltage phasors are, in general, functions of all the states of the interconnected study and external systems.

Thus the objective of the model reduction task is to obtain a model

$$\dot{x}_s = f_s(x_s, V_s, u_s, t), \quad V_s = g_s(x_s, \bar{x}_e, u_s, \bar{u}_e, t) \tag{1.3}$$

$$\dot{\bar{x}}_e = \bar{f}_e(\bar{x}_e, \bar{V}_e, \bar{u}_e, t), \quad \bar{V}_e = \bar{g}_e(x_s, \bar{x}_e, u_s, \bar{u}_e, t) \tag{1.4}$$

in which the external system has been reduced. Here $\bar{x}_e$, $\bar{V}_e$, and $\bar{u}_e$ are the variables, $\bar{f}_e$ is the dynamic equation, and $\bar{g}_e$ is the network equation for the reduced external system. Often the boundary bus voltages $\tilde{V}_{bi}$, $i = 1, ..., k$, are kept in the reduced external system, so that the interface currents $\tilde{I}_i$, $i = 1, ..., k$, between the study system and the reduced external system can still be computed individually. The reduced external system may also be a linear model.

Many model reduction techniques for general applications have been proposed and advanced techniques are still being developed. We will limit the model reduction overview here to those methods that have been applied to power systems. To provide some structure to the overview, we group the methods under coherency and aggregation, linear and nonlinear input–output methods, and the relatively new phasor measurement-based method.

1.2 Coherency

The practical method most commonly used by power utilities to derive reduced models of large power systems is based on the concept of coherency and aggregation. This method uses the inherent properties, such as line admittances and loading, and machine inertias, in a practical power system to derive a reduced nonlinear external system that retains the relevant dynamics (1.4). The method consists of two main steps: (1) identifying coherent groups of machines, and (2) aggregating each coherent group of machines into a single equivalent machine.

1.2.1 Time Simulation

Coherency means that some machines exhibit similar rotor angle swings after a disturbance. Podmore, deMello, and Germond proposed using linear time simulation to identify the coherent groups of machines [1–3]. The advantage is that linear time simulation can be computed much more quickly than a full nonlinear simulation. System disturbances can be modeled as equivalent injections so that the admittance matrix does not need to be rebuilt and refactorized. From the time response of the synchronous machines, a grouping algorithm with a specific tolerance value for coherency can be used to identify the coherent machines.

1.2.2 Modal Coherency

Following Podmore's coherency result and Undrill's modal equivalent result (see Sect. 1.4), Schlueter and coworkers [4] proposed a modal-coherency approach in identifying coherent groups that can approximately capture the modal frequencies. The key idea is the use of a zero-mean, independent, and identically distributed disturbance to compute a rms coherency measure. By letting the time period of simulation approach infinity, an analytical formula without numerical integration can be developed. The success of the method was partly attributed to the rms measure being determined by the synchronizing torque coefficients, which depends on transmission line stiffness and generator inertias.

1.2.3 Slow Coherency

Shortly after the work of [2, 3], a group of singular perturbations researchers from the University of Illinois, Urbana-Champaign and the General Electric Company, Schenectday, New York, investigated the use of singular perturbations for power system model reduction. One of the significant results is establishing the connection between slow coherency and weak connections in power systems [5, 6]. Slow coherency is coherency arising from the slower interarea modes, which are oscillatory modes due to groups of machines oscillating against each other across power transfer interfaces. These interarea modes, if negatively damped, can lead to system separation and extensive loss of load [7].

From the slow coherency theory, the eigen-subspace of the interarea modes can be used to identify the slow coherent machines. To apply this grouping method, only the slowest electromechanical modes and their mode shapes represented by eigenvectors need to be computed. For large power systems, sparsity-based computation methods for calculating only the slow eigenvalues and their corresponding eigenvectors are preferred. An early attempt using the Lanzcos method was documented in [8]. More successful methods include the inverse iteration method [9], the S-matrix method [10], and the Arnoldi method [11].

1.2.4 Weak-Link Methods

The grouping algorithm proposed in [5] requires the computation of eigenvalues, which can be time consuming for large power systems, despite the availability of sparsity-based partial eigenvalue and eigenvector computation routines as mentioned earlier. It may be desirable to develop methods to identify the coherent areas without eigensubspace computation. The weak-connection and slow-coherency relationship in [5] points to exploring weak connections to find the coherent groups. In [12], a clustering algorithm based on the state matrix derived from the synchronizing torque coefficients and the machine inertias has been proposed to find the weakly connected machine clusters. A reduced incidence matrix is constructed by setting rows of off-diagonal entries of the state matrix whose sum is less than a threshold to zero. The connectivity that remains in the incidence matrix defines the slow coherent groups of machines.

The results in [13] extend the search of weak links to form weakly coherent areas and strongly coherent areas. The identification procedure starts by iteratively computing a coupling factor derived from the synchronizing torque coefficients, through a sequential search of the machines. Then relative changes in the coupling factors define the weakly coherent groups, similar to the slow coherent groups from the grouping algorithm [5]. In addition, relative changes in the second variation of the coupling factors can be used to determine the strongly coherent areas. The technique has been demonstrated on a 50-machine model of the northern India power grid.

1.3 Aggregation

The second part of the coherency approach is to aggregate each coherent group of machines into a single equivalent machine, followed by eliminating load buses in the external system that are not needed.

1.3.1 Generator Aggregation

The technique proposed by Podmore and Germond [3] is to connect all the coherent generators to a common bus. All the generators are then aggregated into a single generator. This equivalent generator construction preserves the power flow in the power system as well as the power system model and data structure. Furthermore, exciter and governing capability of the equivalenced generator can be obtained from an aggregate frequency response approximation.

1.3.2 Singular Perturbation Models

An advantage of the singular perturbations method is that it generates asymptotic series expansion terms to improve the slow subsystem [14]. For power system model reduction, it is possible to improve the reduced model by aggregating the coherent generators at the generator internal node, rather than at the generator bus as in the Podmore aggregation [15, 16]. The inertia aggregation method and the Podmore method both stiffen the power network by adding connections with infinite admittances. This stiffening causes the reduced model to have higher interarea mode frequency. Another term can be taken from the asymptotic series expansion to reduce the stiffening effect using only parameters from within the coherent area [15, 16].

1.4 Linear Equivalent Input–Output Models

One of the premises of reducing the external system is that it is not perturbed significantly by a disturbance in the study area. As a result, the external system can be represented by a linear model. To derive the linear model, one can detach the external system from the study system, in which the tielines to the study system are represented by current injections, as shown in Fig. 1.3.

The input–output model of the external system can be represented by

$$\dot{\bar{x}}_e = \bar{f}_e(\bar{x}_e, \bar{V}_e, u_e, \bar{I}, t), \quad \bar{V}_e = \bar{g}_e(\bar{x}_e, u_e, \bar{I}, t), \quad \bar{V}_b = \bar{g}_{be}(\bar{x}_e, u_e, \bar{I}, t) \tag{1.5}$$

in which the tieline current injections $\bar{I}$ are the input to the model and the boundary bus voltages $\bar{V}_b$ are the output. The dynamic equation $\bar{f}_e$ is driven by the bus voltages $\bar{V}_e$ and the input $\bar{I}$. Note that the network equation $\bar{g}_e$ is no longer dependent on the study network variables. In particular, the current injection $\bar{I}$ is an independent variable, and no longer a function of both the study and external systems.

Note that it is also possible to use a formulation with the boundary bus voltages as input variables and the currents injected into the study system as the output variables [17]. This formulation is equivalent to (1.5) if the boundary buses are extended into the study system.

One of the first model reduction ideas is to simply linearize the external model (1.5) as

$$\Delta\dot{\bar{x}}_e = A\Delta\bar{x}_e + B\Delta I, \quad \Delta\bar{V}_b = C\Delta\bar{x}_e + D\Delta I \tag{1.6}$$

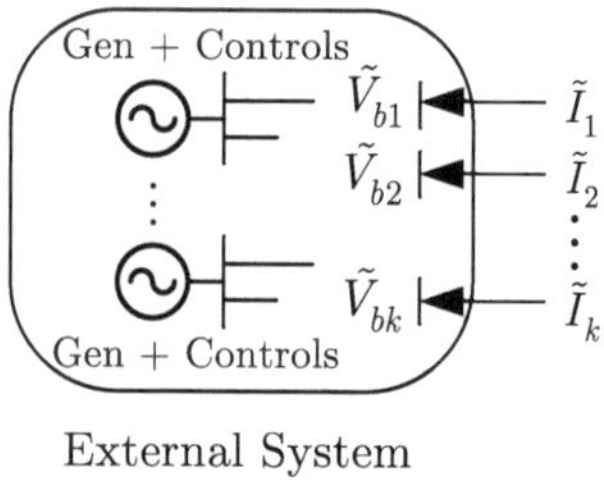

Fig. 1.3 Input–output model of the external system

at the pre-fault power flow condition. The network equation and any feedback control have been included in (1.6) without showing them explicitly. Expressions for linearized models can be found in [18]. Alternatively, linearized models can be obtained via a numerical derivative process [19].

Representing the external system with a linear model would decrease the computation needed for simulating a disturbance in the study system. The linear model will be able to account for the loop flow shown in Fig. 1.2. In addition, a linear network is solved in the external system. To achieve further computation reduction, various approaches have been proposed for reducing the linear model of the external system. A recent survey of linear model reduction approaches can be found in [20].

1.4.1 Modal Truncation

In the early 1970s, Undrill, Price, and others developed a modal truncation approach [17, 21–23] following such work as Davison [24] by determining the modes to be retained in the reduced external systems. In this approach, (1.6) is represented by a reduced model in the modal form

$$\Delta\dot{\bar{x}}_m = A_m \Delta\bar{x}_m + B_m \Delta I, \quad \Delta\bar{V}_b = C_m \Delta\bar{x}_m + D\Delta I \tag{1.7}$$

in which only the dominant modes, including the electromechanical modes, are retained. The dimension of $\Delta\bar{x}_m$ is less than that of $\Delta\bar{x}_e$.[2] Furthermore, if the state matrix A_m is expressed as a diagonal matrix, in which 2×2 diagonal blocks are used for complex eigenvalues, the computation needs during simulation can be further reduced. Special computer code needs to be developed to interface the reduced linear model of the external system to the study system in a nonlinear power system simulation program.

One of the concerns with the modal truncation approach is that it does not preserve steady-state values [20]. This shortfall can be overcome using the singular perturbations technique [14]. However, the modes kept by the reduced model, the so-called slow subsystem, will no longer be identical to the modes of the full external system model.

1.4.2 Selective Modal Analysis

The modal truncation method [24] and the singular perturbations method [14] eliminate the modes with fast decaying transients and as such are less important.

---

[2] Note that (1.7) can be shown to be equivalent to the formulation in [17] using the study system boundary buses as inputs and current injected into the study system as the outputs, and maintaining a linearized model of the external system power network.

The selective modal analysis technique designates the modes with high modal participating factors as relevant modes which will be retained, and those modes with lower modal participating factors as less relevant modes which will be eliminated [25, 26]. From a state-space model form, the relevant reduced model can be computed iteratively. The reduced models are also suitable for designing damping controllers, such as power system stabilizers.

1.4.3 Krylov and Balanced Model Reduction Methods

Another modal elimination approach is based on balanced truncation methods, which eliminate modes that are less controllable or observable [27, 28]. With balanced truncation, the reduced linear model of the external system will have the model

$$\Delta\dot{\bar{x}}_{\text{br}} = A_{\text{br}}\Delta\bar{x}_{\text{br}} + B_{\text{br}}\Delta I, \quad \Delta\bar{V}_b = C_{\text{br}}\Delta\bar{x}_{\text{br}} + D\Delta I \tag{1.8}$$

where the reduced matrices $(A_{\text{br}}, B_{\text{br}}, C_{\text{br}})$ are obtained using controllability and observability Gramians. Such methods do not keep the external system modes exactly but provide a better frequency response approximation of the input–output model compared to the modal truncation method. These balanced truncation methods are supported by efficient computation algorithms [29].

A third class of linear model reduction methods is based on the Krylov method [30], in which the Markov parameters of the linear input–output model (1.6) are preserved up to a certain index. This matching part of the controllability-observability subspace is also known as moment matching. The Krylov is less computationally expensive than the balanced truncation method. The method does not provide any error bounds but, in general, seems to work well.

1.5 Nonlinear Equivalent Input–Output Models

Besides reduced linear models, an alternative approach for capturing the nonlinear dynamics of the input–output model (1.5) shown in Fig. 1.3 is to use reduced nonlinear models. In general, such nonlinear model reduction methods have to be tailored to the characteristics of the physical processes. Power systems have many different types of nonlinearities, such as sinusoidal functions, deadbands, saturation functions, and limits, to name a few. The coherency and aggregation method discussed earlier attempts to preserve these nonlinearities. A few methods have been proposed for developing reduced nonlinear external systems, including the two methods described next.

1.5.1 Singular Perturbations Methods

Once slow coherency has been identified for a power system, one can obtain the quasi-steady-state approximation from singular perturbations as presented in [31]. In this approach, the intraarea modes within each coherent area are assumed to be fast and have settled to their quasi-steady-state values. As a result, the differential equations describing the intra-area modes are solved as algebraic equations. This approach would require the development of special computer code to perform the quasi-steady solution in a conventional power system simulation program [31].

1.5.2 Computational Intelligence Methods

Computation intelligence methods have been used to capture nonlinear model dynamics [32]. One such non-model based method is the artificial neural network (ANN) approach [32]. In an ANN, neurons represented by selected nonlinear functions are arranged in layers connected by weights. These weights are trained from input–output data using a variety of tuning methods. Upon convergence, the method provides a reduced nonlinear model. The ANN can be used to represent all or part of an external system.

1.6 Measurement-Based Reduced Models

Most of the model reduction techniques developed until recently are all based on power system load flow and dynamic data sets. With the advent of synchrophasor measurement technology [33], time-synchronized voltage and current phasor measurements across wide areas can be obtained. These synchrophasor measurements, obtained from phasor measurement units (PMUs), are particularly useful for extracting interarea oscillatory modes and their mode shapes [34]. Such a capability opens up the possibility of using synchrophasor measurements to construct simple interarea models. The results in [35] show that by monitoring voltage phasors at both ends of a power transfer path, it is possible to develop a simple two-area model to emulate the oscillatory modes and establish an energy function. Interestingly, the results were derived from examining the voltage magnitude oscillation of the interarea mode along the transfer path such that the effective impedance connecting the two areas can be computed. This property has been overlooked in coherency and other power system dynamics literature. Research is ongoing to extend this interarea model estimation method to develop simple models for multiple interarea modes.

1.7 Applications

In this section, we discuss some of the impacts of the research on power system model reduction. This list is by no means exhaustive.

1.7.1 Dynamic Model Reduction Programs

Model reduction programs are available for practicing power engineers to reduce data sets of upwards of 30,000 buses to a more manageable 5,000–10,000 buses, which can be handled by power system analysis programs with functions such as transient stability simulation, voltage stability analysis, and optimal power flow. The main steps of the model reduction programs still follow the original coherency and aggregation approach [2] proposed more than 30 years ago.

1.7.2 Interarea Mode Analysis and Damping Control Design

The development of coherency has contributed to the understanding of interarea mode oscillations. Interarea mode damping is an operational concern for systems with heavily loaded long distance transmission lines. Recommendations for interarea mode damping enhancement with power system stabilizers can be found in [36] and for flexible AC transmission systems in [37, 38]. An interarea damping controller applied to a Thyristor-Controlled Series Compensator (TCSC) is critical in the operation of the Brazil North–South Intertie [39]. With the deployment of synchrophasor measurement systems [33], the center of angle or speed of a coherent area, can be measured precisely. Hence, the interarea modes can be computed accurately in real-time and be used to improve the performance of wide-area damping control.

1.7.3 Islanding

System separation or islanding is often a last but necessary resort to prevent a cascading blackout by preserving viable islands of generation and load. This is a difficult task because it needs to balance a number of factors, such as the load and generation balance in each island and the subsequent generator swings after separation. An adaptive out-of-step relay-based islanding strategy using synchrophasors and coherency on a radial-like system (US Florida-Georgia interface) is described in [40]. Another islanding strategy [41, 42] for a more complex situation is to identify the tielines connecting multiple slow coherent groups as cutsets. Such islands have the advantage of strong internal connections such that after islanding, the sum of the synchronizing torques and hence the potential energy [43] in the islands are high.

1.7.4 Dynamic Security Analysis

Power system dynamic security analysis concerns transient, voltage, and small-signal stability [44] and normally gives a yes-or-no answer for a given set of contingencies. An important issue that is not well studied is how interarea modes travel through the various coherent areas for each of the contingencies and the voltage fluctuations on the transfer paths. A practical example of such an application is that the power import on a HVDC system is limited by the voltage stability of a transfer path in a neighboring area; that is, a bipolar fault on the HVDC system would cause voltage collapse on that transfer path. Such wide-area stability issues will become more important with higher power transfer levels between operating regions. Some investigations of the interarea modes in power networks can be found in [45].

1.8 Chapter Guide

In this monograph, we selectively cover three of the model reduction concepts and approaches discussed earlier in this chapter.

The first topic is on coherency and aggregation and is covered in the following chapters:

1. Chapter 2 describes the coherency and aggregation ideas.
2. Chapter 3 describes the slow coherency concept and algorithms.
3. Chapter 4 describes a method to obtain equivalent nonlinear exciter models.
4. Chapter 7 describes the practical application of the dynamic model reduction program (DYNRED) on a large power system.

The second topic is on input–output models for external systems. Chapter 5 describes an approach using ANN to model the external system as a nonmodel-based nonlinear system. The ANN is successfully trained to show an improvement in the time response of the reduced-order model. There are also two chapters on linear reduced-order models:

1. Chapter 6 is on the Krylov method and the balanced truncation method.
2. Chapter 9 describes the selective modal analysis method.

In Chap. 9, the use of selective modal analysis technique for control design is also discussed.

The third topic is on using synchrophasor measurements to develop simple interarea models. Chapter 8 describes the Interarea Model Estimation method and uses measured PMU data from several WECC disturbances to develop two-area models for two transfer paths in the WECC system.

Many of the model reduction methods are based on analyzing interarea modes. Although coherency depends only on the mode shapes of the generators, there are many other interesting and important issues related to interarea modes. The last chapter, Chap. 10, in this monograph discusses the use of synchrophasor data to track the propagation of interarea mode oscillations resulting from large disturbances.

1.9 Conclusions

In this chapter, we have provided a brief survey of some of the fundamental approaches and results in equivalencing and model reduction of dynamic power system models. Methods using similar concepts or techniques are grouped together to provide a reader with a better perspective of the field. In the reminder of this monograph, various authors will share their many years of research and development results in this important field.

References

1. R.W. deMello, R. Podmore, K.N. Stanton, Coherency-based dynamic equivalents: Applications in transient stability studies. *PICA Conference Proceedings* (1975) , pp. 23–31
2. R. Podmore, Identification of coherent generators for dynamic equivalents. IEEE. Trans. Power Apparatus Syst. **PAS–97**(4), 1344–1354 (1978)
3. A.J. Germond, R. Podmore, Dynamic aggregation of generating unit models. IEEE Trans. Power Apparatus Syst. **PAS–97**(4), 1060–1069 (1978)
4. J. Lawler, R.A. Schlueter, P. Rusche, D.L. Hackett, Modal-Coherent Equivalents Derived from an RMS Coherency Measure. IEEE Trans. Power Apparatus Syst. **PAS–99**(4), 1415–1425 (1980)
5. J.H. Chow, G. Peponides, P.V. Kokotović, B. Avramović, J.R. Winkelman, *Time-Scale Modeling of Dynamic Networks with Applications to Power Systems* (Springer-Verlag, New York, 1982)
6. J.H. Chow, J.R. Winkelman, M.A. Pai, P.W. Sauer, Singular perturbation analysis of large scale power systems. J. Electr. Power Energy Syst. **12**, 117–126 (1990)
7. C.W. Taylor, D.C. Erickson, Recording and analyzing the July 2 cascading outage. IEEE Comput. Appl. Power. **10**(1), 26–30 (1997)
8. J.H. Chow, J. Cullum, R.A. Willoughby, A sparsity-based technique for identifying slow-coherent areas in large power systems. EEE Trans. Power Apparatus Syst. **PAS–103**, 463–473 (1983)
9. N. Martins, Efficient eigenvalue and frequency response methods applied to power system small-signal stability studies. IEEE Trans. Power Syst. **1**, 217–225 (1986)
10. N. Uchida, T. Nagao, A new eigen-analysis methcd of steady-state stability studies for large power systems: S matrix method. IEEE Trans. Power Syst. **2**, 706–714 (1988)
11. L. Wang, A. Semlyen, Applications of sparse eigenvalue techniques to the small signal stabiity analysis of large power systems. IEEE Trans. Power Syst. **5**, 635–642 (1990)
12. J. Zaborszky, K.-W. Whang, G.M. Huang, L.-J. Chiang, and S.-Y. Lin, A clustered dynamical model for a class of linear autonomous systems using simple enumerative sorting, IEEE Trans on Circuits and Systems, vol. CAS-29, 747–758, (1982).
13. R. Nath, S.S. Lamba, K.S.P. Rao, Coherency based system decomposition into study and external areas using weak coupling. IEEE Trans. Power Apparatus Syst. **PAS–104**, 1443–1449 (1985)

14. P.V. Kokotović, H. Khalil, J. O'Reilly, *Singular Perturbation Methods in Control: Analysis and Design* (Academic Press, London, 1986)
15. R.A. Date, J.H. Chow, Aggregation properties of linearized two-times-scale power networks. IEEE Trans. Circuits Syst. **38**, 720–730 (1991)
16. J.H. Chow, New algorithms for slow coherency aggregation of large power systems, in *Systems and Control Theory for Power Systems, IMA Volumes in Mathematics and its Applications*, vol. 64, ed. by J.H. Chow, R.J. Thomas, P.V. Kokotović (Springer-Verlag, New York, 1994)
17. J.M. Undrill, A.E. Turner, Construction of power system electromechanical equivalents by modal analysis. IEEE Trans. Power Apparatus Syst. **PAS–90**, 2049–2059 (1971)
18. G. Rogers, *Power System Oscillations* (Kluwer Academic, Dordrecht, 2000)
19. J.H. Chow, K.W. Cheung, A toolbox for power system dynamics and control engineering education. IEEE Trans. Power Syst. **7**, 1559–1564 (1992)
20. S.D. Dukić, A.T. Sarić, Dynamic model reduction: An overview of available techniques with application to power systems, Serbian. J. Electr. Eng. **9**(2), 131–169 (2012)
21. J.M. Undrill, J.A. Casazza, E.M. Gulachenski, L.K. Kirchmayer, Electromechanical equivalents for use in power system stability studies. IEEE Trans. Power Apparatus Syst. **PAS–90**, 2060–2071 (1971)
22. W.W. Price, E.M. Gulachenski, P. Kundur, F.J. Lange, G.C. Loehr, B.A. Roth, R.F. Silva, Testing of the modal dynamic equivalents technique. IEEE Trans. Power Apparatus Syst. **PAS–97**, 1366–1372 (1978)
23. W.W. Price, B.A. Roth, B.A. Roth, Large-scale implementation of modal dynamic equivalents. IEEE Trans. Power Apparatus Syst. **PAS–100**, 3811–3817 (1981)
24. E.J. Davison, A method for simplifying dynamic systems. IEEE Trans. Autom. Control **AC–11**, 93–101 (1966)
25. I.J. Pérez-Arriaga, G.C. Verghese, F.C. Schweppe, Selective modal analysis with applications to electric power systems. part I: Heuristic introduction. part II: The dynamic stability problem. IEEE Trans. Power Apparatus Syst. **PAS–101**, 3117–3134 (1982)
26. F.L. Pagola, L. Rouco, I.J. Pérez-Arriaga, Analysis and control of small signal stability in electric power systems by selective modal analysis, in *Eigenanalysis and Frequency Domain Methods for System Dynamic Performance*. (IEEE Publication 90TH0292-3-PWR, 1990) , pp. 77–96
27. B.C. Moore, Principal component analysis in linear systems: Controllability, observability, and model reduction. IEEE Trans. Autom. Control **AC–26**, 17–32 (1981)
28. K. Glover, All optimal Hankel-norm approximations of linear multivariable systems and their L^∞ norms. International Journal of Control **39**, 1115–1193 (1984)
29. P. Benner, V. Mehrmann, D.C. Sorensen, *Dimension Reduction of Large-Scale Systems, Lecture Notes in Computational Sciences and Engineering*, vol. 45 (Springer, Berlin, 2005)
30. A.C. Antoulas, *Approximation of Large-Scale Dynamical Systems* (SIAM, Philadelphia, 2005)
31. J.R. Winkelman, J.H. Chow, B.C. Bowler, B. Avramovic, P.V. Kokotović, An analysis of inter-area dynamics of multi-machine systems. IEEE Trans. Power Apparatus Syst. **PAS–100**, 754–763 (1981)
32. S. Haykin, *Neural Networks and Learning Machines*, 3rd edn. (Prentice Hall, Englewood Cliffs NJ, 2008)
33. A.G. Phadke, J.S. Thorp, *Synchronized Phasor Measurements and their Applications* (Springer, New York, 2008)
34. D.J. Trudnowski, Estimating electromechanical mode shape from synchrophasor measurements. IEEE Trans. Power Syst. **23**(3), 1188–1195 (2008)
35. J.H. Chow, A. Chakrabortty, L. Vanfretti, M. Arcak, Estimation of radial power system transfer path dynamic parameters using synchronized phasor data. IEEE Trans. Power Syst. **23**(2), 564–571 (May 2008)
36. A. Murdoch, G. Boukarim, *Performance Criteria and Tuning Techniques, Chapter 3 in IEEE Tutorial Course - Power System Stabilization via Excitation Control* (Tampa, Florida, 2007)
37. E.V. Larsen, J.H. Chow, SVC control design concepts for system dynamic performance, in *IEEE Power Engineering Society Publication 87TH0187-5-PWR Application of Static Var Systems for System Dynamic Performance*, 1987

38. E.V. Larsen, J.J. Sanchez-Gasca, J.H. Chow, Concepts for design of FACTS controllers to damp power swings. IEEE Trans. Power Syst. **10**, 948–956 (1995)
39. C. Gama, L. Änguist, G. Ingeström, M. Noroozian, Commissioning and operative experience of TCSC for damping power oscillation in the Brazilian north-south interconnection. Paper 14–104, CIGRE Session 2000
40. V. Centeno, A.G. Phadke, A. Edris, J. Benton, M. Gaugi, G. Michel, An adaptive out-of-step relay. IEEE Trans Power Syst. **26**, 334–343 (1997)
41. H. You, V. Vittal, X. Wang, Slow cherency-based islanding. IEEE Trans. Power Syst. **19**, 483–491 (2004)
42. G. Xu, V. Vittal, A. Anatoliy, J.E. Thalman, Controlled islanding demonstrations in WECC system. IEEE Trans. Power Deliv. **12**, 61–71 (2011)
43. A.-A. Fouad, V. Vittal, *Power System Transient Stability Analysis using the Transient Energy Function Method* (Prentice-Hall, Englewood Cliffs NJ, 1992)
44. P. Kundur, *Power System Stability and Control* (McGraw-Hill, New York, 1994)
45. L. Vanfretti, Phasor measurement based state estimation of electric power systems and linearized analysis of power system network oscillations, PhD thesis, Rensselaer Polytechnic Institute, 2009

Chapter 2
Coherency in Power Systems

Robin Podmore

Abstract There has been a continuing need over the past several decades to model larger and larger interconnection wide models. Models of the complete interconnections with up to 50,000 buses are regularly used for system planning studies. These models typically go down to 115 and 69 kV levels, but ignore underlying 35 kV sub-transmission networks. With the growing deployment of plug-in vehicles, distributed generation and smart load controls, along with the need to perform realistic system restoration drills there is a need to model interconnections down to the feeder breaker level. Restoration drills also require modeling of power plant auxiliaries and emergency generator systems, especially for nuclear units. It is conceivable that the size of interconnection wide models could grow by another order of magnitude. The EPRI DYNRED (Dynamic Reduction) computer program reduces a large-scale system model into a smaller equivalent model for use in transient stability studies. The program has been used since the 1970s to build equivalent models of the Eastern U.S. and Western U.S. interconnected power systems. The DYNRED program accepts a normal transient stability database as input, and develops an equivalent that is a fraction of the size of the full power system representation, while adequately retaining the dynamic characteristics of the full system. The reduction process requires only a fraction of the time needed for a transient stability simulation.

2.1 Introduction

2.1.1 Background

Electric utilities expend a considerable effort performing transient stability studies to evaluate a power system's ability to withstand large disturbances; such a disturbance might be caused by a faulted transmission line, which might lead to a loss of

---

R. Podmore (✉)
IncSys, Bellevue, Washington, USA
e-mail: robin@incsys.com

J. H. Chow (ed.), *Power System Coherency and Model Reduction*,
Power Electronics and Power Systems 94, DOI: 10.1007/978-1-4614-1803-0_2,

synchronism and possible customer interruptions. Many transient stability solutions are routinely required for both long-range planning studies, short-range planning studies, near-term operations planning studies and real-time on-line studies. Long-range studies may require solutions involving various combinations of proposed transmission and generator configurations. Short-range planning may involve solutions for coordination of protection schemes and design of control systems. Near term operational planning studies may be used to determine secure transfer limits. Real-time on-line transient stability studies can support system operator situation awareness and decision making.

Since most major disturbances can propagate through tie-lines to neighboring systems, it is important for transient stability studies to represent not only the power system in question, but also the neighboring utilities. The representation of the neighboring utilities can be difficult because most power systems in the United States are extensively interconnected. For example, to study the dynamic behavior of a power system in California, it is often necessary to represent systems as far north as British Columbia, as far south as Arizona and New Mexico, and as far east as Colorado. As a result, transient stability studies still require large computers and use a significant amount of computer time even though Moore's law continues to provide improvements in price and performance of computers.

2.1.2 EPRI Coherency-Based Dynamic Equivalents Project

The foundation for Coherency-Based Dynamic Equivalents was laid by two EPRI sponsored projects performed by Systems Control Inc., Palo Alto in the mid-1970s [1–4]. Generators that are closely coupled in an electrical sense tend to swing together in groups during disturbances, and this characteristic behavior can be exploited to reduce the size of the power system model. The characteristic of generators swinging together is referred to as coherency and is illustrated in Fig. 2.1, which shows the swing curves for a coherent group of 21 generators when a fault occurs in the Western U.S. system [3]. In the coherency-based method, each coherent group of generators is replaced by a single equivalent generator.

The EPRI Dynamic Equivalents projects accomplished the following:

- Highly efficient methods for computing dynamic equivalents, without requiring a transient stability simulation of the full system, were developed.
- The coherency-based equivalencing techniques were extended to encompass more comprehensive and detailed models of generating units and loads.
- The EPRI DYNRED program was written for forming coherency-based dynamic equivalents.
- The methods and DYNRED program were tested and successfully validated on two large-scale models: one representing the Eastern U.S. interconnected systems and one representing the Western U.S. interconnected systems.

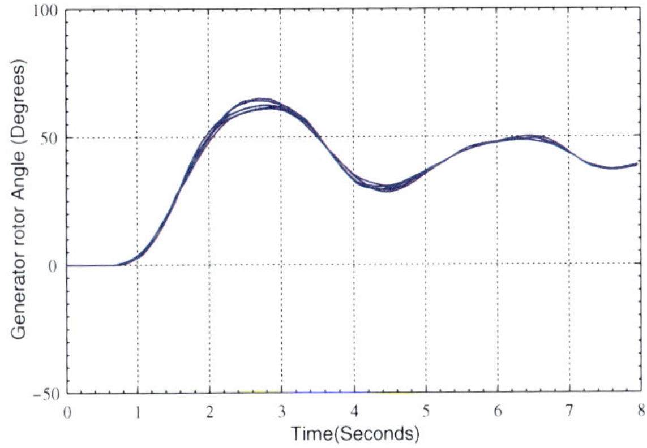

Fig. 2.1 Swing curves for a group of 21 generators in Arizona area of WECC system

The dynamic equivalents were typically two to six times more efficient than the original system model in terms of computer running time. Also, the results which are obtained by using the dynamic equivalent are not significantly different than those obtained from using the original system model. The dynamic equivalencing procedure has been successfully applied and validated using models of the Eastern U.S. [5] and Western U.S. interconnected systems.

2.1.3 Industry Applications and Benefits

The DYNRED program is structured to interface conveniently with existing transient stability programs and data files. It may therefore be implemented by utilities with a modest manpower effort. The various benefits that would result directly from a utility's implementation and application of the programs are summarized below:

- Reduction of engineering effort in preparing the system representation. The alternate manual methods for obtaining system equivalents for transient stability studies are tedious, time consuming, and expensive in terms of experienced engineering effort. These are replaced by a dynamic equivalencing program package that forms dynamic equivalents for large-scale systems in a routine manner comparable to the formation of steady-state equivalents for load flow studies.
- Less dependence on engineering judgment when forming dynamic equivalents. Some judgment would be helpful in deriving the maximum benefits, but the situation of two engineers obtaining different equivalents, both apparently valid, would no longer occur.
- Reduction of computational costs for transient stability studies. The DYNRED program can achieve significant reductions in size of the system model without any notable loss of accuracy.

- More effective planning studies. Engineering skill can be concentrated on system planning with less distraction by problems associated with the boundaries of system representation. Also, the reduced system representation provides greater insight into the system behavior and better opportunity for testing alternative plans.
- Ability to perform on-line transient stability studies for larger system models, with the same amount of computing power.

2.1.4 Problem Formulation

For the purpose of analysis, the power system is divided into two parts:

- A study area in which the system behavior is of direct interest and where all faults and configuration changes are assumed to occur.
- An external area where detailed information on the system responses is not required. The network configuration and the status of the generating units in the external system are assumed to be unaffected by the fault.

The study area may be defined in various ways and its buses need not be contiguous. For example, two alternative definitions for the study area are described below:

- A generating station and its local transmission system could be defined as the study area in order to analyze the stability of the station with respect to the rest of the system.
- The high voltage bulk transmission system and major generating units could be defined as the study area in order to assess the regional stability of an interconnected system.

The following requirements were used to guide the development of the DYNRED program:

- The method should achieve reductions in the size of the transient stability model without introducing significant differences between the equivalent and full system simulations.
- The equivalents should be efficiently computed. Preferably, the computer effort for calculating an equivalent should be much less than the effort for a transient stability run on the original system.
- The method and associated software should be developed in a form which allows it to be readily implemented in a utility planning or operating department with a modest manpower effort. To this end, the equivalent must be composed of normal power system components and is therefore compatible with existing transient stability programs.

2.1.5 Overview of Procedure

The overall procedure for forming coherency-based dynamic equivalents for a defined study area can be divided into four basic steps:

1. Identification of groups of coherent generators that are valid for faults in the study area.
2. Reduction of generator buses.
3. Reduction of load buses.
4. Dynamic aggregation of generating unit models.

The procedure is illustrated by the 39-bus network shown in Fig. 2.2. Consider that the study area is defined as shown. Also, assume for the sake of illustration, that we are interested in a fault on Bus 29 and that units 2, 3, 4, 5, 6, and 7 have been determined to be coherent for this fault. In the generator bus reduction, all the generator terminal buses are replaced by an equivalent bus, as shown in Fig. 2.3. Subsequently, the load buses are eliminated to obtain the network in Fig. 2.4.

At this point, the size of the network has been reduced but the number of generating units is unchanged. The dynamic aggregation procedure replaces each group of parallel generating units with a single equivalent generating unit, as shown in Fig. 2.5, except that several equivalent generating units may be sometimes used to avoid combining units which have markedly different characteristics, e.g., steam and hydro units.

The problem of reducing the network, which is described by algebraic equations, is decoupled from the problem of reducing the dynamic order of the system through

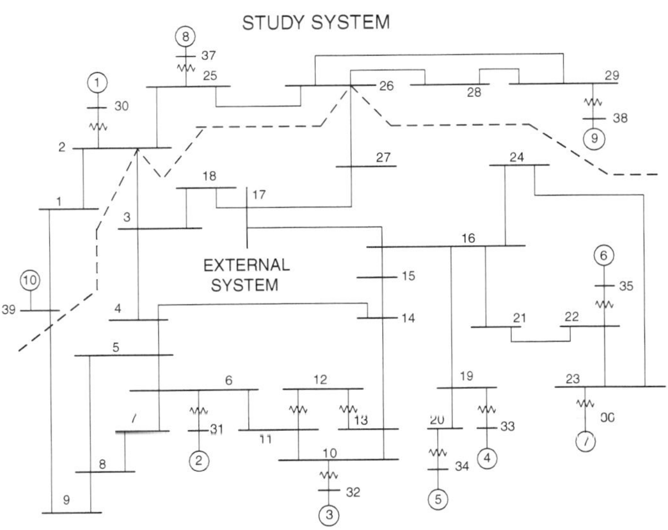

Fig. 2.2 Definition of study area in sample 39-bus system

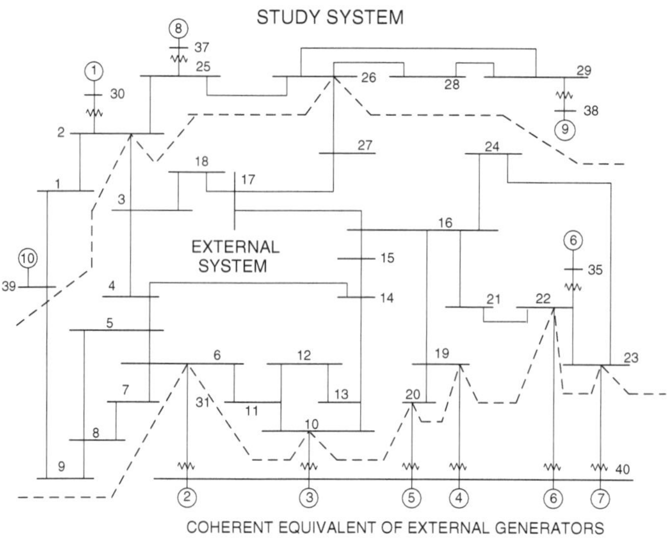

Fig. 2.3 System configuration after coherency-based reduction of generator buses

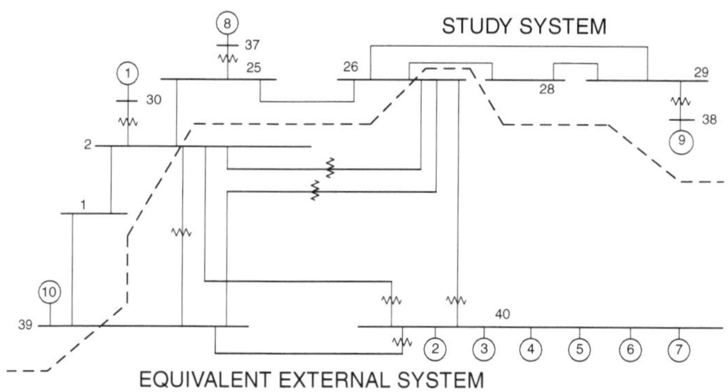

Fig. 2.4 System configuration after gaussian elimination of load buses

aggregation of the generating unit models. This decoupling effect is a key to the development of a procedure which is efficient and generally applicable to a wide range of system models.

The four steps in the procedure are described in the following sections.

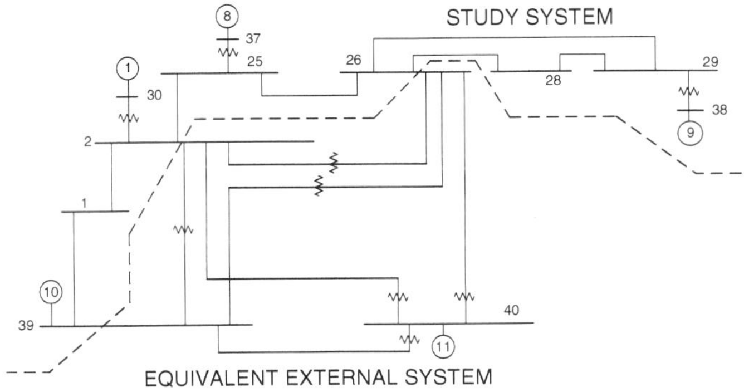

Fig. 2.5 System configuration after dynamic aggregation of generating unit models

2.2 Identification of Coherent Generators

Two generator buses are defined as coherent if their angular difference is constant within a certain tolerance over a certain time interval. The coherency of both generator internal and terminal buses is of interest. The coherency of the generator terminal buses forms the basis for the network reduction step. The coherency of the generator internal bus is assumed in the dynamic aggregation step. Coherent groups of generators can be defined for a single fault or multiple faults occurring within an area. In the remainder of this section a procedure for identifying coherency for a single fault is described briefly. The extension to considering multiple faults is straight-forward and is discussed toward the end of the section.

The method for identifying coherent generators is based upon the formation of a simplified model of the power system which uses the following assumptions:

- The coherent groups of generators are independent of the size of the disturbance. Therefore, coherency can be determined by considering a linearized system model.
- The coherent groups are independent of the amount of detail in the generating unit models. Therefore, a classical synchronous machine model is considered and the excitation and turbine-governor systems are ignored.
- The effect of a fault may be reproduced by considering the unfaulted network and pulsing the mechanical powers to achieve the same accelerating powers that would have existed in the faulted network.

The first assumption may be confirmed by considering a fault on a certain bus, and observing that the coherency behavior of the generators is not significantly changed as the fault clearing time is increased. The second assumption is based upon the observation that although the amount of detail in the generating unit models has a significant effect upon the swing curves, particularly the damping, it does not radically affect the more basic characteristics such as the natural frequencies and mode shapes. The third assumption recognizes that the generator accelerating powers are approximately constant during faults with typical clearing times.

Based upon these assumptions, the coherent groups of generators can be detected by generating power angle curves using a linear time step simulation of the decoupled power angle equations. By using the numerically robust trapezoidal integration rule [6], step sizes of 0.1 s can be used. The solution is non-iterative, so the overall solution time can be 10–50 times faster than a full transient stability solutions. More sophisticated methods can be developed, but it is hard to improve on the performance of the linear simulation method.

The mechanical equations for the motion of a synchronous generator are

$$M_i \frac{\mathrm{d}\Delta\omega_i}{\mathrm{d}t} = \Delta P_{Mi} - \Delta P_{Gi} - D_i \Delta\omega_i \tag{2.1}$$

$$\frac{\mathrm{d}\Delta\delta_i}{\mathrm{d}t} = 2\pi f_o \Delta\omega_i \tag{2.2}$$

where

- i subscript for generator i
- Δ indicates that this variable represents a deviation from a specific steady-state operating point
- M_i inertia constant (pu)
- $\Delta\omega_i$ speed deviation (pu)
- $\Delta\delta_i$ rotor angle deviation (radians)
- D_i damping constant (pu)
- f_o synchronous frequency (Hz)
- ΔP_{Mi} change in mechanical input power (pu)
- ΔP_{Gi} change in electrical output power (pu)

The linearized loadflow expression relating the change in active and reactive power generation and load to the generator and load bus voltages and angles are given by

$$\begin{bmatrix} \Delta P_G \\ \Delta P_L \\ \Delta Q_G \\ \Delta Q_L \end{bmatrix} = \begin{bmatrix} \partial P_G/\partial\delta & \partial P_G/\partial\theta & \partial P_G/\partial E & \partial P_G/\partial V \\ \partial P_L/\partial\delta & \partial P_L/\partial\theta & \partial P_L/\partial E & \partial P_L/\partial V \\ \partial Q_G/\partial\delta & \partial Q_G/\partial\theta & \partial Q_G/\partial E & \partial Q_G/\partial V \\ \partial Q_L/\partial\delta & \partial Q_L/\partial\theta & \partial Q_L/\partial E & \partial Q_L/\partial V \end{bmatrix} \begin{bmatrix} \Delta\delta \\ \Delta\theta \\ \Delta E \\ \Delta V \end{bmatrix} \tag{2.3}$$

The variables used in (2.3) are defined as

- P_G, Q_G real and reactive power injections (pu) at the internal generator buses
- P_L, Q_L real and reactive power loads (pu) at the load buses
- E, δ voltage magnitudes (pu) and angles (radians) at generator internal buses
- V, θ voltage magnitudes (pu) and angles (radians) at load buses

The voltage dependence of the load powers is included in the $\partial P_L/\partial V$ and $\partial Q_L/\partial V$ terms, and the changes in the power residuals ΔP_L and ΔQ_L are normally zero. But they may be assigned certain values in order to model a disturbance such as bus load shedding.

Equation (2.3) may be simplified by accounting for the decoupling which exists between the real and reactive power flows for a transmission system with high X/R ratios [7]. The real power flows are largely dependent upon the voltage angles, and as a first order approximation, the effect of variations in load bus voltage magnitude may be neglected by setting the terms $\partial P_G/\partial V$ and $\partial P_L/\partial V$ to zero. The voltage behind the generator transient reactance is constant, thus $\Delta E = 0$.

Equation (2.3) therefore simplifies to

$$\begin{bmatrix} \Delta P_G \\ \Delta P_L \end{bmatrix} = \begin{bmatrix} \partial P_G/\partial \delta & \partial P_G/\partial \theta \\ \partial P_L/\partial \delta & \partial P_L/\partial \theta \end{bmatrix} \begin{bmatrix} \Delta \delta \\ \Delta \theta \end{bmatrix} \tag{2.4}$$

For convenience, (2.4) is rewritten with the following notation

$$\begin{bmatrix} \Delta P_G \\ \Delta P_L \end{bmatrix} = \begin{bmatrix} H_{GG} & H_{GL} \\ H_{LG} & H_{LL} \end{bmatrix} \begin{bmatrix} \Delta \delta \\ \Delta \theta \end{bmatrix} \tag{2.5}$$

The partial derivatives in (2.4) are most precisely calculated using the voltages and angles at the pre-fault steady-state operating point. As a further simplification, it is possible to assume a flat voltage profile with unit voltage magnitudes and zero voltage angles.

The simplified linear model can be applied to model various power system disturbances. The electrical power output of the generating units during a fault is computed by solving the faulted network equations with the generator transient voltages fixed at the pre-fault values. For typical clearing times the relative rotor angles do not change by more than a few degrees and the generator accelerating powers are approximately constant during the fault. This characteristic response allows the effects of the faulted network condition to be closely duplicated by considering an unfaulted network and increasing the generator mechanical powers by the accelerating power for a time equal to the fault clearing time.

A bus load-dropping disturbance can be modeled by introducing step changes in the ΔP_L and ΔQ_L variables for the selected bus at the appropriate time. The dropping of a generating unit can also be modeled by introducing step changes in the ΔP_L and ΔQ_L variables for the generating unit terminal bus. The generator internal bus and transient reactance is not included in the formation of the Jacobian matrix in this case.

The modeling of line outages is illustrated by considering the sample system shown in Fig. 2.6. Figure 2.6a shows the pre-fault state. Figure 2.6b shows a hypothetical network situation in which the line from Bus 2 to Bus 3 has been removed but injections equal to the original line flows have been inserted so that the bus voltages are not changed. This hypothetical steady-state operating condition is used as the base point for linearizing the system equations. The effect of line removal is then modeled by introducing load changes so that the equivalent injections at Buses 2 and 3 are set to zero as shown in Fig. 2.6c.

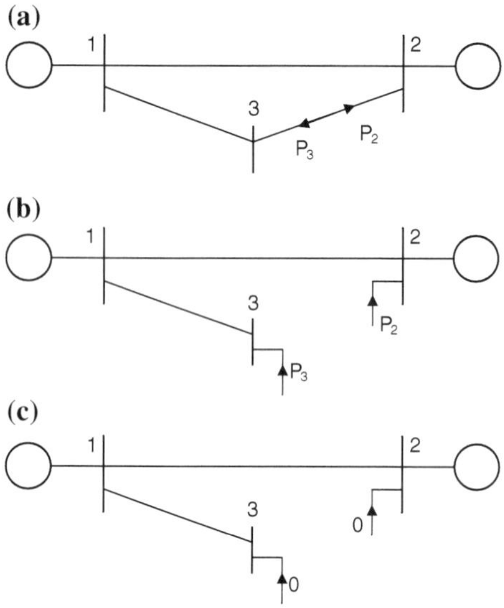

Fig. 2.6 Illustrating treatment of line outages. **a** Prefault network. **b** Hypothetical network. **c** Postfault network

The trapezoidal integration technique [6] is applied to obtain a time domain solution of the linearized swing equations. The trapezoidal integration algorithm is ideal for this application because a highly accurate solution is not required and because the method is numerically stable, large step sizes can be used. Also, a direct non-iterative solution is obtained at each time step since the system of equations is linear. These two factors combine to allow a highly efficient algorithm to be developed. The application of the trapezoidal integration rule to (1) and (2) yields

$$\begin{aligned} M_i \Delta\omega_i(t) &= M_i \Delta\omega_i(t-\Delta t) - (\Delta t/2) D_i(\Delta\omega_i(t) + \Delta\omega_i(t-\Delta t)) \\ &\quad - (\Delta t/2)(\Delta P_{Gi}(t) + \Delta P_{Gi}(t-\Delta t)) + \Delta t \Delta P_{Mi}(t) \end{aligned} \tag{2.6}$$

$$\Delta\delta_i(t) = \Delta\delta_i(t-\Delta t) + (\Delta t/2) 2\pi f_o(\Delta\omega_i(t) + \Delta\omega_i(t-\Delta t)) \tag{2.7}$$

The variables $\Delta\omega$, $\Delta\delta$ and ΔP_G are assumed to vary linearly over the interval $t - \Delta t$ to t. In order to facilitate the modeling of disturbances, the input variables $\Delta P_M(t)$ and $\Delta P_L(t)$ are assumed to vary in a stepwise fashion. The value of $\Delta P_M(t)$ is defined to be constant from $t - \Delta t$ to t, and a similar definition applies to $\Delta P_L(t)$. Equations (2.6) and (2.7) are solved simultaneously with (2.5) at each integration time step to calculate the variables at time t in terms of the variables at time $t - \Delta t$.

The overall algorithm has the following steps.

A Linear Time Simulation Algorithm

Step 1. Initialize $\Delta\delta(0)$, $\Delta\omega(0)$, and $\Delta P_G(0)$.
Step 2. Increment the time from $t - \Delta t$ to t.
Step 3. Set $\Delta P_M(t)$ and $\Delta P_L(t)$ according to the disturbance being modeled.
Step 4. Calculate the following variables for each generating unit using

$$A_i(t - \Delta t) = \left(1 - \frac{D_i \Delta t}{2M_i}\right) \Delta\omega_i(t - \Delta t) - \frac{\Delta t}{2M_i} \Delta P_{Gi}(t - \Delta t) + \frac{\Delta t}{M_i} \Delta P_{Mi}(t) \tag{2.8}$$

$$B_i(t - \Delta t) = \pi f_o \Delta t \Delta\omega_i(t - \Delta t) + \Delta\delta_i(t - \Delta t) \tag{2.9}$$

$$C_i(t - \Delta t) = \frac{2M_i}{\Delta t} A_i(t - \Delta t) - \frac{2M_i}{\Delta t^2 \pi f_o} \left(1 + \frac{D_i \Delta t}{2M_i}\right) B_i(t - \Delta t) \tag{2.10}$$

Step 5. Solve the following matrix equation

$$\begin{bmatrix} \Delta C(t - \Delta t) \\ \Delta P_L(t) \end{bmatrix} = \begin{bmatrix} H'_{GG} & H_{GL} \\ H_{LG} & H_{LL} \end{bmatrix} \begin{bmatrix} \Delta\delta(t) \\ \Delta\theta(t) \end{bmatrix} \tag{2.11}$$

for the new voltage angles $\Delta\theta(t)$ and $\Delta\delta(t)$, where H'_{GG} is the matrix H_{GG} with

$$\frac{2M_i}{\Delta t^2 \pi f_o} \left(1 + \frac{D_i \Delta t}{2M_i}\right) \tag{2.12}$$

added to the diagonal elements.
Step 6. Calculate the new generator electric power using

$$\Delta P_{Gi}(t) = \frac{2M_i}{\Delta t} A_i(t - \Delta t) - \frac{2M_i}{\Delta t^2 \pi f_o} \left(1 + \frac{D_i \Delta t}{2M_i}\right) (\Delta\delta_i(t) - B_i(t - \Delta t)) \tag{2.13}$$

Step 7. Calculate the new generator speeds using (2.9) rearranged as

$$\Delta\omega_i(t) = \frac{\Delta\delta_i(t) - B_i(t - \Delta t)}{\pi f_o \Delta t} \tag{2.14}$$

Step 8. Stop if time t exceeds the specified value; otherwise, return to Step 2.

The matrix in Step 5 is constant and has the same sparse symmetric structure as the power-angle Jacobian matrix. The diagonal terms for the internal generator buses are the only ones which distinguish it from the power-angle Jacobian. The triangular factors for the matrix in Step 5 are precomputed and are used at each integration time interval. The bulk of the computation effort for the algorithm consists of: one triangular factorization of a real sparse matrix with the same structure as the power-angle Jacobian, and one repeat solution of this matrix at each integration time step.

The natural frequencies of the rotor angle oscillations typically lie in the range 0.25–2.0 Hz. An integration time step of 0.1 s is quite accurate enough for this range, and in some cases 0.2 s is also satisfactory.

2.2.1 Generator Clustering Algorithm

A clustering algorithm is used to process the approximate swing curves which are produced by the linear simulation and thereby determine the coherent generator groups. The clustering algorithm can actually be applied to either generator internal voltage swing curves or terminal voltage swing curves. In the procedure that was found most convenient, the clustering algorithm is initially applied to the terminal voltage angles. The groups of coherent terminal buses so obtained are saved for use by the network reduction program. An additional check on the coherency of the generator internal bus voltages is then made to determine if all the generators within each group should be dynamically aggregated; generally this is the case.

The clustering algorithm minimizes the number of data curve comparisons by recognizing that the coherency of the generators is a transitive process; i.e., if Unit A is coherent with Unit C, and Unit B is coherent with Unit C, then it follows that Units A and B are coherent. A reference generator is defined in each group, and other generators are always compared against this reference unit in order to determine whether they should fall in the same group. The first generating unit is arbitrarily defined as the reference unit for group one. The remainder of the generating units are evaluated in turn with two alternative consequences; either the unit is combined with an existing group or the unit does not combine with any existing group, and a new group is created with the unit defined as the reference.

The criterion used for determining whether a unit should be added to an existing group is

$$\Delta\theta_i(t) - \Delta\theta_r(t) < \varepsilon \tag{2.15}$$

for all the samples of t, where

ε the specified tolerance in degrees
i the index for generator being clustered
r the index for the reference generator of the group under consideration.

2.3 Reduction of Generator Buses

This section describes the generator bus reduction procedure as a series of operations on the physical network elements. This physical interpretation of the procedure is used as a basis for developing an efficient algorithm. It is also useful for obtaining

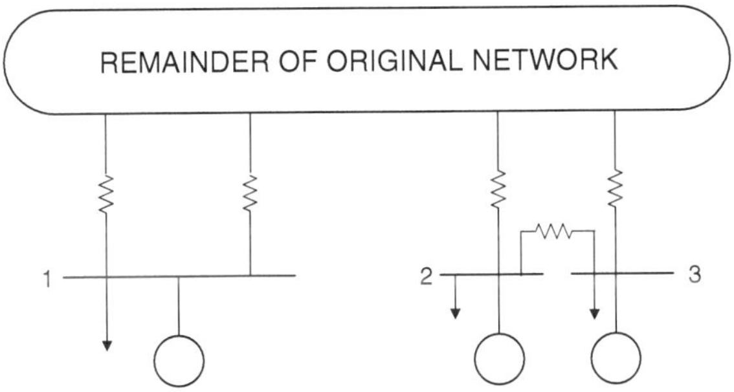

Fig. 2.7 Configuration of coherent generators in original network

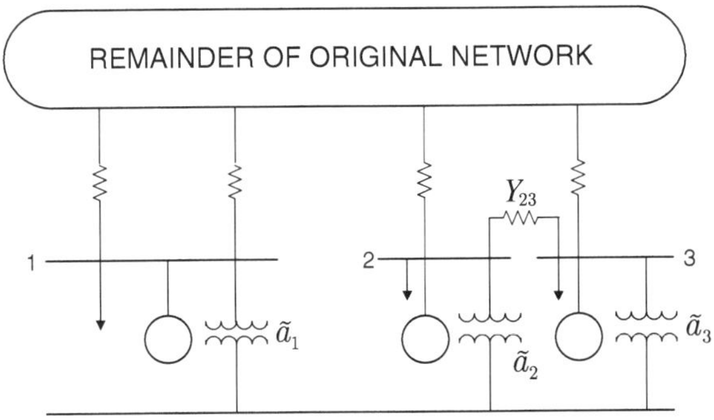

Fig. 2.8 Coherent generator buses are connected to an equivalent bus through ideal transformers with complex ratio

insight into how the equivalent model should characteristically perform in a transient stability simulation.

The procedure consists of four basic steps. Each of the steps is described in turn below. The example network in Fig. 2.7 is used to illustrate the procedure; consider that the generator terminal buses 1, 2, and 3 have been identified as coherent and are to be replaced by a single equivalent Bus 4.

Step 1: The voltage V_t in the equivalent bus is defined; either an average voltage of the group or the voltage of an individual bus is selected. Each terminal bus is connected through an ideal transformer with complex turns ratio to the equivalent bus. The turns ratio is directed as shown in Fig. 2.8 and is calculated as: $\tilde{a}_k = V_k / V_t$, where V_k is the voltage on Bus k.

Under coherent conditions, the ratio $\tilde{a}_k$ is constant for each bus in the group and no circulating power flows through any of the phase shifters. It follows that the introduction of the phase shifters will have no effect on the response of the network voltages and currents under these conditions. The situation described is analogous to the well-known method for simplifying dc networks, where a short circuit is introduced between two nodes that always have the same dc voltage (Fig. 2.9).

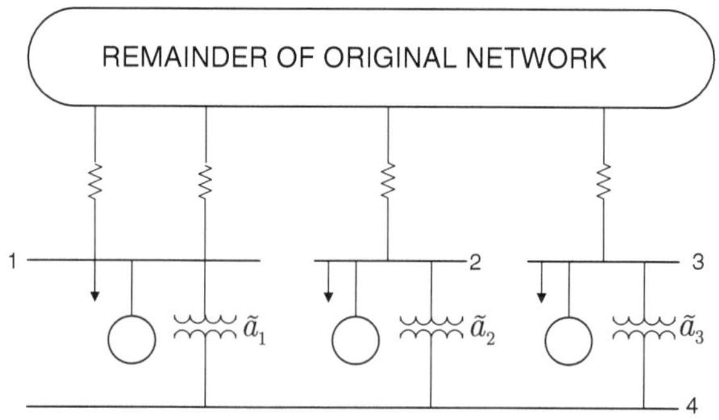

Fig. 2.9 Branch between coherent buses 2 and 3 is replaced by equivalent shunt admittance on buses 2 and 3

Step 2: The generator terminal buses will generally be connected radially through a step-up transformer to the rest of the network. However, in some cases the low voltage bus may have been eliminated by combining the transformer reactance with the generator internal reactance. In this circumstance, several non-radial buses may be included within the coherent group and a common branch may connect them (e.g., the branch between Buses 2 and 3 in Fig. 2.6b). The function of the second step is to detect this situation and to remove the intragroup branch by replacing it by equivalent shunt admittances. To explain this, consider the current flow in the branch between Buses 2 and 3

$$I_{23} = (V_2 - V_3)Y_{23} \tag{2.16}$$

Because V_2/V_3 is constant, the current can be expressed as a linear function of either V_2 or V_3. The effect of the branch can thus be replaced by a shunt admittance $(1 - V_3/V_2)Y_{23}$ at Bus 2 and a shunt admittance $(1 - V_2/V_3)Y_{23}$ at Bus 3.

Step 3: The generation, load, and shunt admittances on the coherent buses are transferred and summed on the equivalent bus as illustrated in Fig. 2.10. The generation and load are not modified by the transfer. The shunt admittance is scaled to account for the off-nominal tap ratio of the ideal transformer. If a nonlinear load representation is used, then the constant MVA, constant current, and constant impedance load components are transferred individually and kept separate.

Step 4: The original coherent buses are eliminated by series combination of the original branch and the ideal transformer (see Fig. 2.11). If several original branches connect to the eliminated bus (as for Bus 1), the ideal transformer is combined with each of them.

The reduction procedure only affects those branches connected to buses in the coherent group. The balanced load flow condition on the original buses is preserved, and a balanced load flow condition is created for the equivalent bus.

The connection of the generator terminal buses to the equivalent bus through the ideal transformers introduces infinitely strong synchronizing ties between them. For this reason, it is preferable that the coherency-based reduction be performed on the

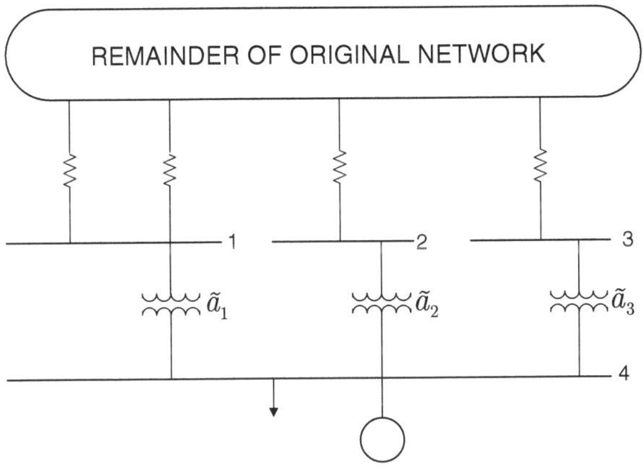

Fig. 2.10 Generation, loads, and shunt admittances on original buses are transferred to the equivalent bus

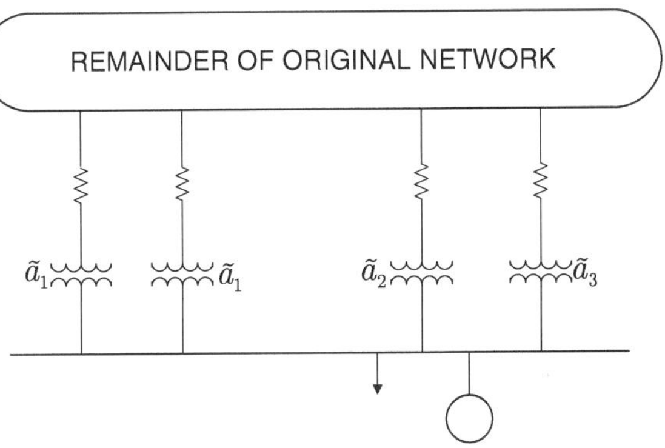

Fig. 2.11 Original generator terminal buses are Eliminated by Series Combination of ideal transformers with original branches

generator low voltage buses in order that the effect of this artificial tie, as seen from the rest of the network, be minimized by the relatively high transformer reactances.

2.4 Reduction of Load Buses

In this section we shall first consider the reduction of buses which have constant impedance loads. In this case, a linear relation exists between the load voltage and current, and an exact reduction method is possible. Later, an approximate method for reducing buses with more general nonlinear load characteristics is described.

The Ward-Hale [8] or Gaussian elimination method has been applied extensively to reduce constant impedance load buses for purposes of load flow and transient stability. Although the Gaussian elimination method effectively reduces the number of buses, there is no guarantee that the number of lines will also decrease. This is significant because with modern transient stability programs the overall computation time is dependent mainly upon the number of lines and not the number of buses.

The network representing the original power system is naturally very sparse; typically the number of lines is only 1.5–2.0 times the number of buses. When a network reduction is performed and buses are eliminated, additional equivalent lines are introduced into the network. In the worst case, when complete fill-in of the bus admittance matrix occurs; the number of branches will equal $N \times (N - 1)/2$ (N is the number of retained buses). Because N may easily exceed several hundred, it is obvious that elimination of nodes should be applied judiciously; otherwise the equivalent network may have many more branches than the original one.

Sparsity techniques can be applied successfully to the network reduction problem in order to minimize the number of branches introduced into the equivalent network [9]. In sparsity oriented reduction, certain key buses that are special in a topological sense are identified and are retained in the equivalent network. These key buses tend to be buses which either have a high number of connections or buses which connect subareas that have few ties to the rest of the system. Figure 2.12 shows an example network in which two key buses should be retained to enhance sparsity. At present the most effective automatic method for identifying the key buses is to order the bus eliminations using a sparsity oriented scheme and to terminate the bus eliminations when the number of terms in the equivalent admittance matrix starts increasing instead of decreasing. Three alternative bus ordering schemes have been described [10], and the one referred to as Scheme II has been utilized in the DYNRED program. In this scheme, the bus with the minimum number of connections is always eliminated at each step. Figure 2.13 illustrates the application of this method to the WECC system.

The past need for retaining generator buses has been a limiting factor in network reduction even when sparsity techniques are applied. The generator buses tend to be connected radially and scattered throughout the network. Their retention does not enhance sparsity and inhibits the process of deriving a smaller equivalent. For example, if only the generator buses are retained, then the equivalent admittance matrix will be completely full for almost all practical networks. Reduction of the generator buses using coherency has the beneficial effect of reducing the number of buses that have to be retained, and thus enables the possibility of computing an

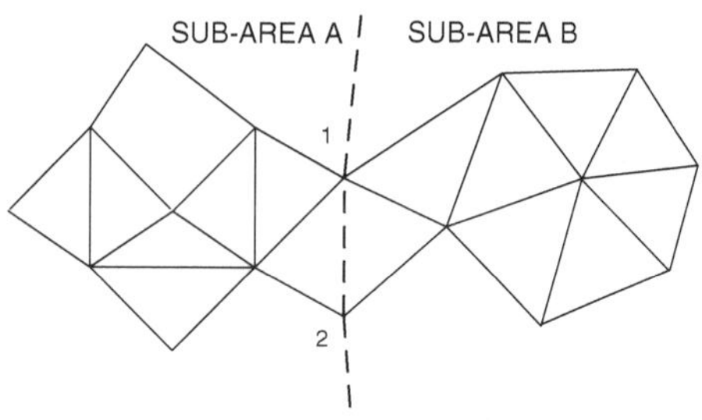

Fig. 2.12 Retention of bus 1 and bus 2 prevents introduction of equivalent lines between subarea A and subarea B

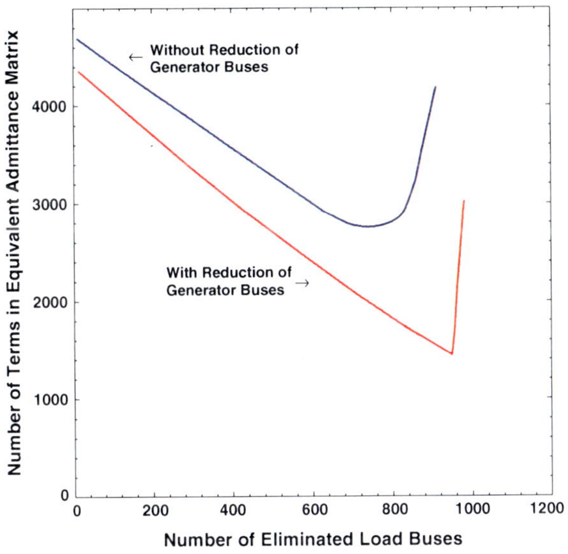

Fig. 2.13 Sparsity-oriented reduction of load buses in WECC system with and without prior reduction of generator buses

equivalent which has many less admittances than would otherwise be possible. To illustrate this, Fig. 2.13 compares the reduction of load buses in the WECC network with and without prior reduction of generator buses using coherency. As shown, a much smaller network is obtained when the generator buses are eliminated first.

The DYNRED program includes the capability for reducing nonlinear loads that consist of constant impedance, constant current (at fixed power factor) and constant power components. The Gaussian elimination procedure is extended to handle the nonlinear constant current and constant power components by approximating them as ideal current sinks. A pair of current sinks is introduced to replace the constant current and constant power components of load at each bus. The base case load flow voltages are used to perform this conversion. The current sinks are reduced using the standard Gaussian elimination formula. Each pair of equivalent current sources on each retained bus is then reconverted to an equivalent constant current component and an equivalent constant power component, respectively.

In order to maintain accuracy with the current sink reduction method, the network is divided into subareas which are roughly coherent. The tie-lines that interconnect the subareas are retained during the reduction. As a result, when a nonlinear load within a subarea is eliminated, the equivalent nonlinear loads which replace this are confined to the same subarea. In order to preserve the sparsity of the equivalent network, it is also necessary that the subareas be loosely connected in a topological sense. At the present time, the subareas are defined by grouping one or more control areas, as illustrated in Fig. 2.14.

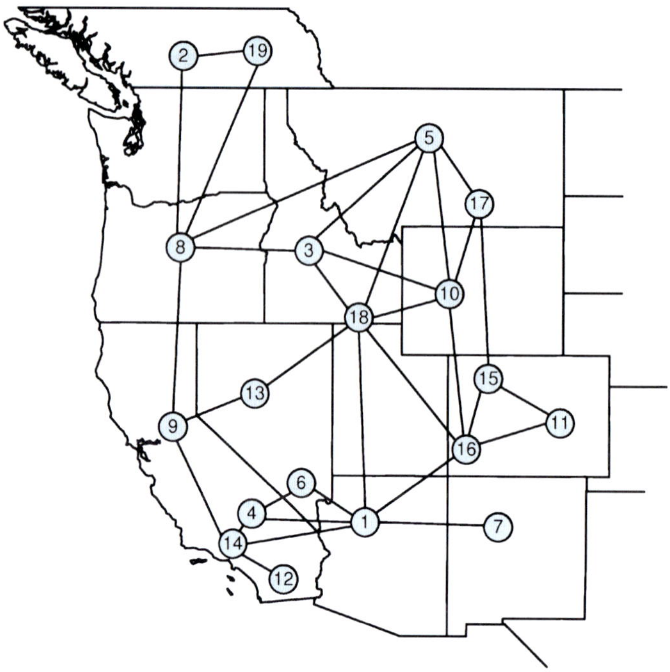

Fig. 2.14 Areas within WECC System

2.5 Dynamic Aggregation of Generating Unit Models

The procedure of coherency-based network and load reduction has the effect of producing an electrically equivalent network model connecting a reduced number of buses. Each of these equivalent buses represents a coherent area of the original power system, and the generating units of each coherent area are now all connected in parallel to the equivalent bus for this area. This is illustrated in Fig. 2.4 for the 39-bus New England system. The order of the differential system representing the dynamic part of the system can be reduced now by grouping the generating units that are in parallel on the same bus and replacing them by an equivalent generating unit. This procedure is called dynamic aggregation. After dynamic aggregation, the system of Fig. 2.4 will appear as the equivalent system of Fig. 2.5. The criteria for an acceptable equivalent model of generating units, from the dynamic viewpoint, is that its electric power output response matches the total electric power output of the units it replaces, and that the voltage response at its terminal bus matches the voltage response of the bus with the individual unit models.

In the simple case where the units to be aggregated are all classical machine models, all with the same model of excitation system and the same model of governor-turbine, it was found empirically that logarithmic averages of the individual parameters of the excitation and governor-turbine models could be calculated and used as the parameters of the equivalent excitation and governor-turbine models [6].

Table 2.1 Generating unit models used from WECC data base

Synchronous machine models
Classical model
Two-axes model with one damping winding and saturation
Excitation system models
Continuously acting DC rotating excitation system
Westinghouse continuously acting brushless excitation system
GE Alterex excitation system
Westinghouse static excitation system
SCR excitation system
Non-continuously acting rheostatic excitation system
Turbine-governor models
Steam governor system
Non reheat turbine
Tandem compound single reheat turbine
Tandem compound double reheat turbine
Cross compound single reheat turbine
Cross compound double reheat turbine
Hydro turbine governor system
Hydro-mechanical governor-turbine
Cross-compound governor-turbine
Power system stabilizer models
Shaft slip input signal
Frequency deviation input signal
Accelerating power input signal

In the general case, the models of the individual unit excitation and governor systems may be different types. The procedure was been implemented with the data base of the Western Electricity Coordinating Council, which includes the machine, governor, turbine, excitation system, and power system stabilizer models listed in Table 2.1. Each generating unit is modeled as a synchronous machine with a turbine, a governor and an excitation system. Some units also have a power system stabilizer (Fig. 2.15).

The aggregation method requires that the units to be aggregated are attached to the same bus and have the same terminal voltage, and assumes that these units are coherent and have the same speed. Thus, every excitation system among a group of coherent units measures the same input voltage signal. Also every governor-turbine system among a group of coherent units measures the same input speed signal.

2.5.1 Aggregation of the Governor-Turbine Systems

The response of each individual governor-turbine system can be calculated for a given perturbation of the speed input signal. The total mechanical power output for

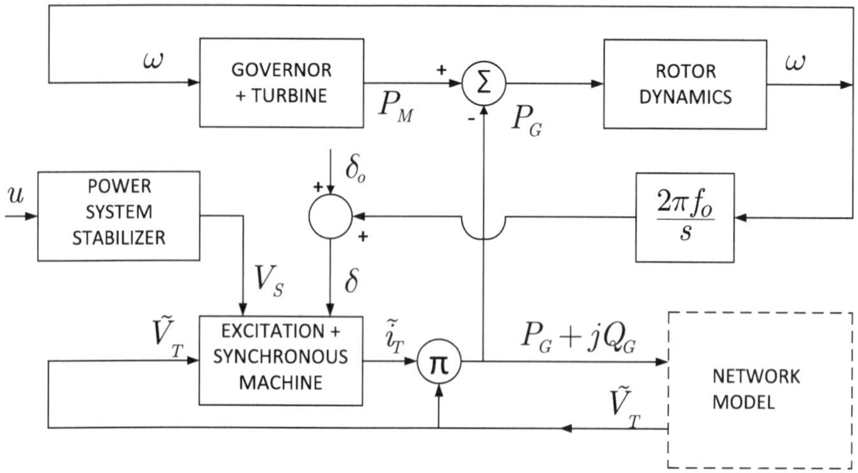

ω p.u. frequency deviation

P_M total mechanical power in p.u.

P_G total active power output in p.u.

Q_G total reactive power output in p.u.

$\tilde{v}_T$ terminal voltage

$\tilde{i}_T$ terminal current

u power system stabilizer input signal

v_s power system stabilizer output

δ angle of machine internal voltage

Fig. 2.15 Generating Unit Model

the coherent group is obtained by summing-up the individual mechanical power output responses.

The problem of the aggregation of the governor-turbine models is to identify the parameters of an equivalent governor-turbine model, with the objective that its output approximates as closely as possible the total mechanical power output, if the same speed signal is applied as input. Two types of speed input signals are considered: periodic inputs of small amplitude and step inputs. For periodic inputs of small amplitude the problem is solved as follows: nonlinearities in the governor-turbine models are ignored; the transfer function of each individual governor-turbine model is calculated for discrete values of the frequency in the range of 0.01–10 Hz, which is of practical importance; and these transfer functions are added. This corresponds to having the individual governor-turbine block-diagrams in parallel as shown on Fig. 2.16. The summation of transfer-functions is called the "aggregated transfer function".

A model for the equivalent governor-turbine is chosen and its parameters are adjusted until its transfer function fits the "aggregated transfer function" within a specified accuracy. The technique for adjusting these parameters is a numerical

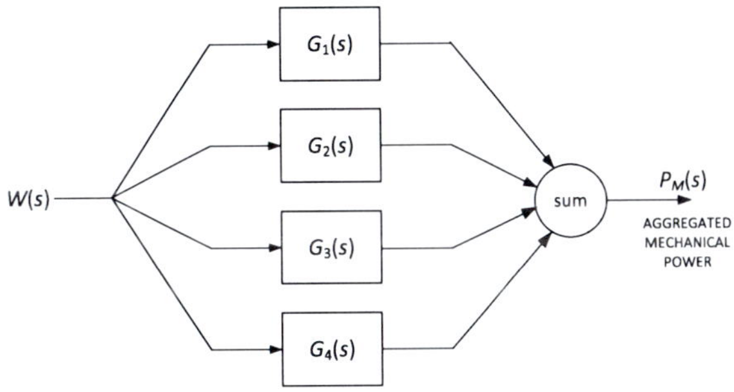

Fig. 2.16 Aggregation of governor turbine systems

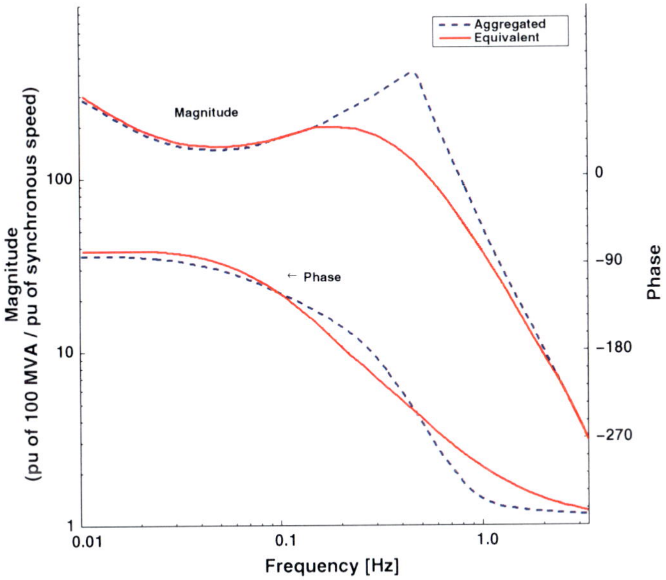

Fig. 2.17 Illustration of bode plots for an aggregated and an equivalent governor turbine transfer function

gradient search. The objective function that is minimized is the sum, over the frequency range, of the squares of the relative error between the equivalent and the aggregated transfer functions. The Bode plot for an equivalent transfer function is shown in Fig. 2.17, together with the Bode plot for the aggregated transfer function.

The nonlinear features of the equivalent model, such as gate limit and gate velocity limit, are evaluated separately by calculating the effect of a step input.

2.5.2 Aggregation of Synchronous Machines

The same principle of transfer-function fitting is applied to the operational admittances, in the direct and in the quadrature axis separately. This is possible since all the synchronous machines of a coherent group exhibit the same terminal voltage.

The mechanical characteristics (rotor dynamics) of the equivalent machine are represented by summing-up the individual inertias and damping factors, respectively.

2.5.3 Aggregation of the Excitation System and Power System Stabilizer Models

The aggregated transfer function to be approximated is a weighted sum of the transfer functions for the individual excitation systems. The weighting factor for an individual excitation system depends upon the parameters of the synchronous machine to which it is connected, and upon the parameters of the equivalent synchronous machine. The weighting factors take into account the fact that the field voltage of larger units has more influence upon the terminal voltage of the coherent group than the field voltage of small units.

Finally, after evaluation of the parameters for the equivalent excitation system, the aggregated power system stabilizer transfer function is calculated and approximated.

2.6 Program Description

The DYNRED program performs the three basic equivalencing steps as outlined below:

- Identification of coherent groups of generators for one or more disturbances. Three programs are used: ACCEL, LINSIM, and GROUP.
- Network Reduction. Two programs are used: GENRED and LODRED.
- Dynamic Aggregation, using the program DYNAGG.

An overall flow diagram illustrating co-ordination of the programs is shown in Fig. 2.18.

The functions of the programs are described respectively:

- ACCEL calculates the generator accelerating powers for a specific fault condition. These are saved on file.
- LINSIM reads the accelerating powers and performs a linear simulation of the disturbance. The approximate swing curves so generated are stored on file.
- GROUP clusters the approximate swing curves for one or several disturbances in parallel and determines the coherent groups of generators. These groups are saved on file.
- GENRED utilizes the coherent groups and performs the coherency-based reduction of generator buses.

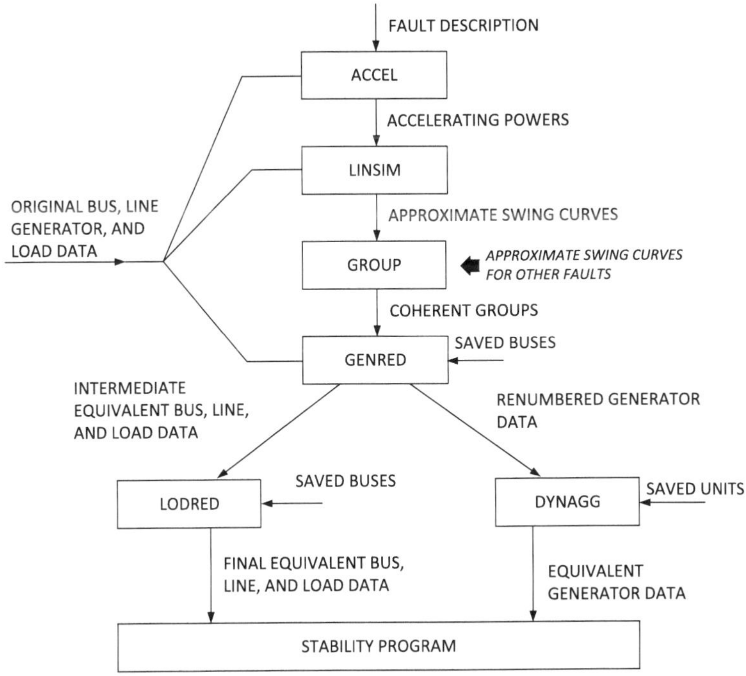

Fig. 2.18 Overview of dynamic equivalencing program

The bus, line, and load data which is output from GENRED describes an intermediate equivalent network in which the generator buses have been reduced but the load buses have not. The generator data is modified only to reflect that the units for a coherent group are now paralleled on a common equivalent bus.

- LODRED performs the reduction of load buses using Gaussian elimination. The bus, line, and load data output from LODRED represents the final equivalent network data.
- DYNAGG reads the modified generator data and combines those units which have the same terminal bus into one or several equivalent units depending upon the types (e.g., steam or hydro) which are present. DYNAGG also requires the bus data produced by either GENRED or LODRED to reconstruct the generating unit initial conditions. The output from GENRED is the final equivalent generating unit data.

The programs are written in Fortran IV and have been executed under the Windows operating system. The programs are available under license from EPRI Grid Operations and Planning Division.

2.7 Conclusions

The coherency-based dynamic equivalents program has been a useful tool to assist utilities in reducing the costs and long computer running times associated with transient stability studies. The program is in a form where it can be implemented by utilities with a modest manpower effort to provide immediate benefits.

The EPRI DYNRED program has been successfully applied to large-scale models of both the Eastern U.S. and Western U.S. interconnected systems. These reduced models are being used by utilities and consultants for system planning and operating studies.

The dynamic equivalents have produced simulation responses which agree very closely with responses obtained by simulating the full system. In the case of the WSCC system, the agreement was obtained over 7.0 s simulation interval.

The computer time for running the dynamic equivalents program is a fraction of the time for a transient stability simulation of the full system. A significant saving in computer time is obtained even if an equivalent is calculated and used to study only a single disturbance condition.

Coherency-based dynamic equivalents can be represented by the same models which are presently used for representing normal power system components. Therefore, they can be used without changes to existing transient stability programs and their implementation does not require a large utility investment.

References

1. R.W. deMello, R. Podmore, K.N. Stanton, Coherency-Based Dynamic Equivalents: Applications in Transient Stability Studies, in *PICA Conference Proceedings* (1975) pp. 23–31
2. R. Podmore and A. Germond, "Dynamic Equivalents for Transient Stability Studies", Systems Control, Inc., Final Report prepared for EPRI Project RP-763, April 1977
3. R. Podmore, Identification of coherent generators for dynamic equivalents. IEEE Trans. Power Apparatus Syst. **PAS–97**(4), 1344–1354 (1978)
4. A. Germond, R. Podmore, Dynamic Aggregation of Generating Unit Models, in *Proceedings of IEEE Winter Power Meeting* (N.Y, New York, January 1977)
5. D. Hackett, Coherency Based Dynamic Equivalent Application Experience - Eastern US Data Bases, in *PICA Conference Proceedings* 1979
6. H.W. Dommel, N. Sato, Fast Transient Stability Solutions. IEEE Trans. Power Apparatus Syst. **PAS–91**, 1643–1650 (1972)
7. B. Stott, O. Alsac, Fast decoupled power flow. IEEE Trans. Power Apparatus Syst. **PAS–93**(3), 859–869 (1974)
8. J.B. Ward, Equivalent circuits for power-flow studies. AIEE Trans. **68**, 373–382 (1949)
9. W.F. Tinney, W.L. Powell, N.M. Peterson, Sparsity-Oriented Network Reduction in *PICA Conference Proceedings* (Minneapolis, 1973), pp. 384–390
10. W.F. Tinney, J.W. Walker, Direct Solutions of Sparse Network Equations by Optimally Ordered Triangular Factorization. in *Proceedings of IEEE* **55** (1967) pp. 1801–1809

Chapter 3
Slow Coherency and Aggregation

Joe H. Chow

Abstract This chapter presents the theory and analysis of slow coherency and aggregation. The main idea is that slow coherency arises from interarea modes, that is, groups of machines swinging together against each other at oscillatory frequencies slower than the local modes of the machines in the same coherent group swinging against each other. We show analytically that this phenomenon can be attributed to the coherent areas being weakly coupled, either because of higher impedance transmission lines, heavily loaded transmission lines, or fewer connections between the coherent areas compared to the connections within a coherent area. These system properties allow the use of singular perturbations to display the time-scale separation of the interarea modes and local modes, resulting in eigenvector-based algorithms to identify coherent machines. In addition, the singular perturbations technique has the capability to provide correction terms for improving reduced-order models in capturing the slow coherent dynamics.

3.1 Introduction

In Chap. 2, Dr. Podmore has provided a discourse on the coherency phenomenon and discussed methods to aggregate the coherent machines to obtain reduced models of large power systems. In this chapter, we will provide the analytical basis for coherency, or to be more specific, slow coherency, because the coherent groups of machines are swinging against each other with respect to the slower interarea modes, as opposed to the higher frequency local modes of machines within each coherent group swinging against each other. This time-scale separation can be exploited using the singular perturbations theory [1].

J. H. Chow (✉)
Rensselaer Polytechnic Institute, Troy, New York, USA
e-mail: chowj@rpi.edu

J. H. Chow (ed.), *Power System Coherency and Model Reduction*,
Power Electronics and Power Systems 94, DOI: 10.1007/978-1-4614-1803-0_3,

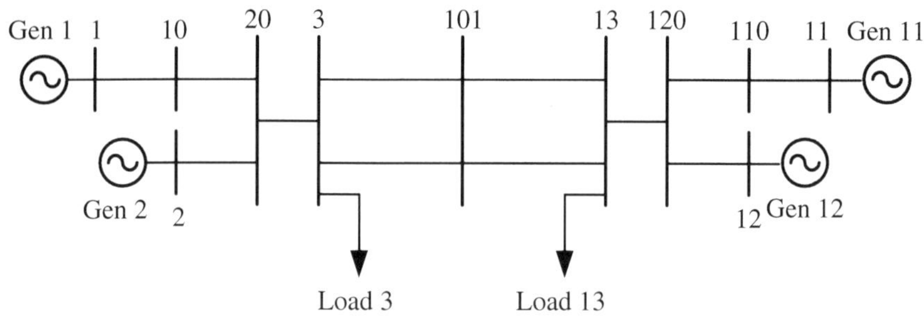

Fig. 3.1 Two-area, four-machine system

The physical origin of slow coherency is perhaps not by accident. Traditional power systems consist of operating regions (such as New York and California). In each region there are large load centers at big cities (such as New York City and Los Angeles), served by large central generating stations, some of which may be close to the loads, but often farther away. The high-voltage transmission systems from 230 to 765 kV have evolved to connect these central generating stations to the loads. As load consumption increases, the individual operating regions have found it more economical to exchange power by building tielines to share baseload and seasonal power resources as well as to rely on each other for reserves. Thus a practical interconnected power system necessarily will have by design strong connections within each operating region and weaker connections between the regions. In the remainder of this chapter, we will use a time-scale analysis to develop the slow-coherency property and algorithms to identify the coherent machines, as well as improvements to reduced-order models.

In this chapter, we will use the popular Klein-Rogers-Kundur two-area, four-machine system [2] shown in Fig. 3.1 to illustrate various aspects of the slow-coherency concepts and algorithms. The data for this system can be found in the appendix of this chapter. If we apply a short duration 3-phase fault at Bus 3, cleared by removing one of the lines from Bus 3 to Bus 101, Generators 1 and 2 will swing coherently against Generators 11 and 12, as shown in the machine speed plot in Fig. 3.2. Note that Fig. 3.2 also shows the faster oscillations between Generators 1 and 2, which are in the same coherent areas.

This chapter is organized in the following sections. Section 3.2 develops the electromechanical model. Section 3.3 establishes the linkage between weakly connected areas in power systems and slow coherency. Two grouping algorithms to identify the coherent groups of machines are developed in Sect. 3.4. Section 3.5 discusses and compares three methods of aggregation, that is, combining generators while retaining the power network structure. In Sect. 3.6, a 48-machine system is used to compare these model reduction methods. Section 3.7 is a brief note on the availability of the "Coherency Toolbox."

The materials of this chapter are based largely on Chow et al. [3, 7]. There is, however, a new extension of the grouping algorithm to identify the load buses for

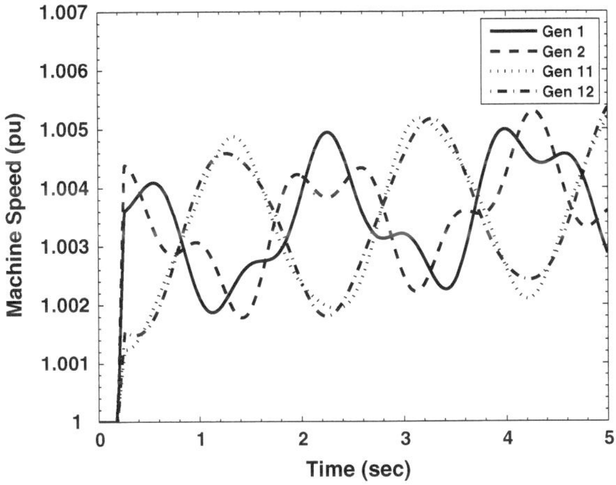

Fig. 3.2 Machine speed response of the two-area, four-machine system

the coherent areas in Sect. 3.4. Also the MATLAB computer code to perform the slow-coherency operations in this chapter is now available.

3.2 Power System Electromechanical Models

To capture the time scales of the electromechanical oscillations in power systems, we will use the classical electromechanical model [8] with loads modeled as constant impedances. Consider an n-machine, N-bus power system. Machine i is modeled as a constant voltage E_i behind a transient reactance x'_{di}. The motion of the machine rotor angle δ_i is modeled as

$$m_i\ddot{\delta}_i = P_{mi} - P_{ei} = P_{mi} - \frac{E_i V_j \sin(\delta_i - \theta_j)}{x'_{di}} = f_i(\delta, V) \tag{3.1}$$

where

$$V_j = \sqrt{V_{j\mathrm{re}}^2 + V_{j\mathrm{im}}^2}, \quad \theta_j = \tan^{-1}\left(\frac{V_{j\mathrm{im}}}{V_{j\mathrm{re}}}\right) \tag{3.2}$$

P_{mi} is the input mechanical power, P_{ei} is the output electrical power, $V_{j\mathrm{re}}$ and $V_{j\mathrm{im}}$ are the real and imaginary parts of the bus voltage phasor at Bus j which is the terminal bus of Machine i, δ is an n-vector of the machine angles, V is the $2N$-vector of the real and imaginary parts of the load bus voltages, $m_i = 2H_i/\Omega$ where H_i is the inertia of the machine most commonly used in power system stability simulation programs, and $\Omega = 2\pi f_0$ is the nominal system frequency in rad/s. Note that damping is neglected in (3.1) so that we can express the electromechanical model in the second-derivative form.

For each load bus j, the active power flow balance at the bus is

$$P_{ej} - \text{Real}\left\{ \sum_{k=1,k\neq j}^{N} (V_{jre} + jV_{jim} - V_{kre} - jV_{kim}) \left(\frac{V_{jre} + jV_{jim}}{R_{Ljk} + jX_{Ljk}} \right)^* \right\} - V_j^2 G_j$$
$$= g_{2j-1} = 0 \tag{3.3}$$

and the reactive power flow balance is

$$Q_{ej} - \text{Imag}\left\{ \sum_{k=1,k\neq j}^{N} (V_{jre} + jV_{jim} - V_{kre} - jV_{kim}) \left(\frac{V_{jre} + jV_{jim}}{R_{Ljk} + jX_{Ljk}} \right)^* \right\}$$
$$- V_j^2 B_j + V_j^2 \frac{B_{Ljk}}{2} = g_{2j} = 0 \tag{3.4}$$

where R_{Ljk}, X_{Ljk}, and B_{Ljk} are the resistance, reactance, and line charging, respectively, of the line connecting buses j and k, P_{ej} and Q_{ej} are generator active and reactive electrical output power, respectively, if bus j is a generator bus, and G_j and B_j are the load conductance and susceptance at bus j. Note that j denotes the imaginary number if it is not used as an index.

Equations (3.1), (3.3), and (3.4) can be combined into a vector form

$$M\ddot{\delta} = f(\delta, V), \quad 0 = g(\delta, V) \tag{3.5}$$

where M is the diagonal machine inertia matrix, f is a vector of acceleration torques, and g is the power flow equation of the transmission network.

The slow coherency is an inherent system property that is valid for both linearized and nonlinear electromechanical models of power systems [3, 9]. Our discussion will be based on linear analysis using eigenvalues and eigenvectors, which will provide much more insight.

We linearize (3.5) about a nominal power flow equilibrium (δ_0, V_0) to obtain the linear model

$$M\Delta\ddot{\delta} = \left.\frac{\partial f(\delta, V)}{\partial \delta}\right|_{\delta_0, V_0} + \left.\frac{\partial f(\delta, V)}{\partial V}\right|_{\delta_0, V_0} = K_1 \Delta\delta + K_2 \Delta V \tag{3.6}$$

$$0 = \left.\frac{\partial g(\delta, V)}{\partial \delta}\right|_{\delta_0, V_0} + \left.\frac{\partial g(\delta, V)}{\partial V}\right|_{\delta_0, V_0} = K_3 \Delta\delta + K_4 \Delta V \tag{3.7}$$

where $\Delta\delta$ is an n-vector of the machine angle deviations from δ_0, and ΔV is a $2N$-vector of the real and imaginary parts of the load bus voltage deviations from V_0. The matrices K_1, K_2, and K_3 consist of the partial derivatives of the power transfer between the machines and their terminal buses. In particular, K_1 is diagonal. The matrix K_4 is the network admittance matrix. This formulation allows for nonconforming loads, which make the 2×2 diagonal blocks of K_4 nonsymmetric.

The analytical expressions of the sensitivity matrices K_i can be derived from (3.1), (3.3), and (3.4). Alternatively, they can be computed numerically by introducing perturbations and evaluating sensitivities [10]. Because K_4 is nonsingular, we can solve (3.6) for

$$\Delta V = -K_4^{-1} K_3 \Delta\delta \tag{3.8}$$

and eliminate ΔV in (3.7) to obtain a linearized electromechanical model reduced to the machine internal nodes

$$M \Delta\ddot{\delta} = K \Delta\delta \tag{3.9}$$

where

$$K = K_1 - K_2 K_4^{-1} K_3 \tag{3.10}$$

The (i, j) entry of K has the form

$$K_{ij} = E_i E_j \left(B_{ij} \cos(\delta_i - \delta_j) - G_{ij} \sin(\delta_i - \delta_j)\right)\Big|_{\delta_0, V_0}, \quad i \neq j \tag{3.11}$$

where E_i is the internal voltage of generator i, and $G_{ij} + jB_{ij}$ is the equivalent admittance between machine i and j. Furthermore, the diagonal entries of K are in the form

$$K_{ii} = -\sum_{j=1, j\neq i}^{n} K_{ij} \tag{3.12}$$

Thus the row sum of K will be equal to zero. In general, $G_{ij} \ll B_{ij}$. The entries K_{ij} of K are known as the synchronizing torque coefficients, as they represent the torque that keeps the machines synchronized and stably connected.

3.3 Two-Time-Scale Power System Models

We define a slow coherent area to be composed of a slow coherent group of machines and the load buses that interconnect these machines. In the next section, we will discuss some algorithms to find the slow coherent areas. Assume that the system (3.6) and (3.7) has r slow coherent areas of machines. For notational convenience, we define

$\Delta\delta_i^\alpha$ = the deviation of the rotor angle of machine i in area α from its equilibrium value

m_i^α – the inertia of machine i in area α.

We also order the machines such that $\Delta\delta_i^\alpha$ from the same coherent areas appears consecutively in $\Delta\delta$.

We attribute the slow-coherency phenomenon to be primarily due to the connections between the machines in the same coherent areas being stiffer than those between different areas. This stiffness can be due to two reasons:

1. The admittances of the external connections B_{ij}^E, that is, connections between the areas are much smaller than the admittances of the internal connections B_{pq}^I, that is, connections within an area. This condition can be represented by the small parameter

$$\varepsilon_1 = \frac{B_{ij}^E}{B_{pq}^I} \tag{3.13}$$

   where E denotes external, I denotes internal, and i, j, p, q are bus indices. This situation also includes heavily loaded high-voltage, long transmission lines between two coherent areas.
2. The number of external connections is much less than the number of internal connections, which can be represented by the small parameter [4]

$$\varepsilon_2 = \frac{\bar{\gamma}^E}{\underline{\gamma}^I} \tag{3.14}$$

   where

$$\bar{\gamma}^E = \max_{\alpha}\{\gamma_\alpha^E\}, \quad \underline{\gamma}^I = \min_{\alpha}\{\gamma_\alpha^I\}, \quad \alpha = 1, \ldots, r \tag{3.15}$$

   γ_α^E = (the number of external connections of area α)/N^α
   γ_α^I = (the number of internal connections of area α)/N^α
   and N^α is the number of buses in area α.

For a large power system, the weak connections between coherent areas would be due to a combination of these two situations. Thus a weak connection parameter indicating the strength of external connections relative to the internal connections can be represented by

$$\varepsilon = \varepsilon_1 \varepsilon_2 \tag{3.16}$$

Using the weak connection parameter ε, we separate the network admittance matrix K_4 into

$$K_4 = K_4^I + \varepsilon K_4^E \tag{3.17}$$

where K_4^I is the matrix of internal connections between the buses in the same coherent areas and K_4^E is the matrix of external connections between different coherent areas.

Including the effect of the weak external connections, the synchronizing torque or connection matrix K of the model (3.10) can be expressed as

$$\begin{aligned} K &= K_1 - K_2(K_4^I + \varepsilon(K_4^I))^{-1} K_3 \\ &= K_1 - K_2(K_4^I(I + \varepsilon(K_4^I)^{-1} K_4^E))^{-1} K_3 \\ &= K^I + \varepsilon K^E \end{aligned} \tag{3.18}$$

where

$$K^I = K_1 - K_2(K_4^I)^{-1}K_3 \tag{3.19}$$

is the block-diagonal matrix of internal connections between the machines in the same coherent areas,

$$K^E = -K_2 K_{4\varepsilon}^E K_3 \tag{3.20}$$

is the matrix of external connections between different coherent areas, and

$$K_{4\varepsilon} = (-(K_4^I)^{-1}K_4^E + \varepsilon((K_4^I)^{-1}K_4^E)^2 + \cdots)(K_4^I)^{-1} \tag{3.21}$$

is an asymptotic expansion in ε. In the separation (3.18), the property that each row of K^I sums to zero is preserved.

We now introduce a transformation to obtain the aggregate variables and the difference variables to reveal the time scales of a power system with the property (3.18). To describe the slow motion, we define for each area an inertia weighted *aggregate variable*

$$y^\alpha = \sum_{i=1}^{n_\alpha} m_i^\alpha \Delta\delta_i^\alpha / m^\alpha, \quad \alpha = 1, 2, \ldots, r \tag{3.22}$$

where m_i^α is the inertia of machine i in area α, n_α is the number of machines in area α, and

$$m^\alpha = \sum_{i=1}^{n_\alpha} m_i^\alpha, \quad \alpha = 1, 2, \ldots, r \tag{3.23}$$

is the aggregate inertia of area α. Denoting by y the r-vector whose αth entry is y^α, the matrix form of (3.22) is

$$y = C\Delta\delta = M_a^{-1}U^T M \Delta\delta \tag{3.24}$$

where

$$U = \text{blockdiag}(u_1, u_2, \ldots, u_r) \tag{3.25}$$

is the grouping matrix with $n_\alpha \times 1$ column vectors

$$u_\alpha = \begin{bmatrix} 1 \ 1 \ \ldots \ 1 \end{bmatrix}^T, \quad \alpha = 1, 2, \ldots, r \tag{3.26}$$

$$M_a = \text{diag}(m^1, m^2, \ldots, m^r) = U^T M U \tag{3.27}$$

is the $r \times r$ diagonal aggregate inertia matrix.

For the fast dynamics, we select in each area a reference machine, say the first machine, and define the motions of the other machines in the same area relative to this reference machine by the *local variables*

$$z_{i-1}^{\alpha} = \Delta\delta_i^{\alpha} - \Delta\delta_1^{\alpha}, \quad i = 2, 3, \ldots, n_\alpha, \quad \alpha = 1, 2, \ldots, r \tag{3.28}$$

Denoting by z^α the $(n_\alpha - 1)$-vector whose ith entry is z_i^α and considering z^α as the αth subvector of the $(n-r)$-vector z, we rewrite (3.28) as

$$z = G\Delta\delta = \text{blockdiag}(G_1, G_2, \ldots, G_r)\Delta\delta \tag{3.29}$$

where G_α is the $(n_\alpha - 1) \times n_\alpha$ matrix

$$G_\alpha = \begin{bmatrix} -1 & 1 & 0 & . & 0 \\ -1 & 0 & 1 & . & 0 \\ . & . & . & . & . \\ -1 & 0 & 0 & . & 1 \end{bmatrix} \tag{3.30}$$

We have thus defined a transformation of the original state $\Delta\delta$ into the aggregate variable y and the local variable z as

$$\begin{bmatrix} y \\ z \end{bmatrix} = \begin{bmatrix} C \\ G \end{bmatrix} \Delta\delta \tag{3.31}$$

The inverse of this transformation is explicitly known

$$\Delta\delta = \begin{bmatrix} U & G^+ \end{bmatrix} \begin{bmatrix} y \\ z \end{bmatrix} \tag{3.32}$$

where

$$G^+ = G^T(GG^T)^{-1} \tag{3.33}$$

is block-diagonal.

Applying the transformation (3.31) to the model (3.9), (3.18), we obtain

$$\begin{aligned} M_a\ddot{y} &= \varepsilon K_a y + \varepsilon K_{ad} z \\ M_d\ddot{z} &= \varepsilon K_{da} y + (K_d + \varepsilon K_{dd})z \end{aligned} \tag{3.34}$$

where

$$\begin{aligned} M_d &= (GM^{-1}G^T)^{-1}, \quad K_a = U^T K^E U \\ K_{da} &= U^T K^E M^{-1} G^T M_d, \quad K_{da} = M_d G M^{-1} K^E U \\ K_d &= M_d G M^{-1} K^I M^{-1} G^T M_d, \quad K_{dd} = M_d G M^{-1} K^E M^{-1} G^T M_d \end{aligned} \tag{3.35}$$

Note that K_a, K_{ad}, and K_{da} are independent of K^I because $K^I U = 0$. System (3.34) is in the *standard singularly perturbed form* [1] showing that y is the slow variable and z is the fast variable. Thus ε is both the weak connection parameter and the singular perturbation parameter, giving rise to slow coherency.

The transformation (3.31) can also be applied to the model (3.6) and (3.7) to obtain

$$\begin{aligned} M_a \ddot{y} &= K_{11} y + K_{12} z + K_{13} \Delta V \\ M_d \ddot{z} &= K_{21} y + K_{22} z + K_{23} \Delta V \\ 0 &= K_{31} y + K_{32} z + (K_4^I + \varepsilon K_4^E) \Delta V \end{aligned} \tag{3.36}$$

where

$$\begin{aligned} &K_{11} = U^T K_1 U, \quad K_{12} = U^T K_1 G^+, \quad K_{13} = U^T K_2, \quad K_{21} = (G^+)^T K_1 U \\ &K_{22} = (G^+)^T K_1 G^+, \quad K_{23} = (G^+)^T K_2, \quad K_{31} = K_3 U, \quad K_{32} = K_3 G^+ \end{aligned} \tag{3.37}$$

Because the load bus elimination process involves the solution of linear equations, it follows that eliminating the ΔV variables would reduce (3.37) to (3.34), that is, the transformation and the load bus elimination commute.

From the singular perturbations theory [1], we can construct an asymptotic expansion in the small parameter ε for the slow subsystem. For our applications, only the zeroth- and the first-order terms are of interest. Observe from (3.34) that the slow variable y is coupled into the fast variable equation through ε. Thus, as a zeroth-order approximation, z can be considered constant and equal to zero. Consequently, (3.37) reduces to

$$\begin{aligned} M_a \ddot{y} &= K_{11} y + K_{13} \Delta V \\ 0 &= K_{31} y + K_4 \Delta V \end{aligned} \tag{3.38}$$

This is the *inertial aggregate model* which is equivalent to linking the internal nodes of the coherent machines by infinite admittances. The model (3.38) is different from the model obtained using the technique in Chap. 2, which aggregates the coherent machines at the terminal buses, not the internal nodes as in (3.38). Thus the inertial aggregate model should be more accurate than the machine terminal bus aggregate model.

If ΔV in (3.38) is eliminated, we obtain the inertial aggregate model as

$$M_a \ddot{y} = \varepsilon K_a y \tag{3.39}$$

For a model more accurate than the inertial aggregate model, we consider z to vary with y. As a first-order approximation, from (3.37) the *quasi-steady-state* of z is obtained as

$$z = -K_{22}^{-1}(K_{21}y + K_{23}\Delta V) \tag{3.40}$$

Eliminating z from (3.37) results in the *slow-coherency (aggregate) model*

$$\begin{aligned} M_a\ddot{y} &= K_{11s}y + K_{13s}\Delta V \\ 0 &= K_{31s}y + K_{4s}\Delta V \end{aligned} \tag{3.41}$$

where

$$\begin{aligned} K_{11s} &= K_{11} - K_{12}K_{22}^{-1}K_{21}, \quad K_{13s} = K_{13} - K_{12}K_{22}^{-1}K_{23} \\ K_{31s} &= K_{31} - K_{32}K_{22}^{-1}K_{21}, \quad K_{4s} = K_4 - K_{32}K_{22}^{-1}K_{23} \end{aligned} \tag{3.42}$$

The major difference between this model and the inertial aggregate model is that in (3.41), the internal nodes of the coherent machines are no longer connected by infinite admittances. Instead, the singular perturbation method introduces impedance correction terms to the connection matrices K_{11}, K_{13}, K_{31}, and K_4. The aggregate models will be discussed further in Sect. 3.5.

To complete the model reduction process, the voltage variables for the load buses that need not be retained can be eliminated from (3.38) or (3.41). The reduced-order model would then consist of the aggregate machines and the critical load buses.

If ΔV in (3.41) is completely eliminated, we obtain the slow-coherency aggregate model as

$$M_a\ddot{y} = \varepsilon K_{as}y \tag{3.43}$$

where

$$\varepsilon K_{as} = K_{11s} - K_{13s}K_{4s}^{-1}K_{31s} \tag{3.44}$$

Consider the two-area system in Fig. 3.1. With the parameters and the loading given in the Appendix, and the ordering of δ as

$$\delta = \begin{bmatrix} \delta_1^1 & \delta_2^1 & \delta_1^2 & \delta_2^2 \end{bmatrix}^T \tag{3.45}$$

representing the rotor angles of Generators 1, 2, 11, and 12, respectively, the inertia matrix in pu on system MVA base is, in a 60 Hz system,

$$M = \text{diag}(117, 117, 91, 143)/(2\pi \times 60) \tag{3.46}$$

and the connection matrix is

$$K = \begin{bmatrix} -9.4574 & 8.0159 & 0.5063 & 0.9351 \\ 8.7238 & -11.3978 & 0.9268 & 1.7472 \\ 0.6739 & 0.9520 & -9.6175 & 7.9917 \\ 1.3644 & 1.9325 & 8.1747 & -11.4716 \end{bmatrix} \tag{3.47}$$

Decomposing K into the internal connection part and the external connection part according to the areas, we obtain

$$K^I = \begin{bmatrix} -8.0159 & 8.0159 & 0 & 0 \\ 8.7238 & -8.7238 & 0 & 0 \\ 0 & 0 & -7.9917 & 7.9917 \\ 0 & 0 & 8.1747 & -8.1747 \end{bmatrix} \tag{3.48}$$

$$\varepsilon K^E = \begin{bmatrix} -1.4414 & 0 & 0.5063 & 0.9351 \\ 0 & -2.6739 & 0.9268 & 1.7472 \\ 0.6739 & 0.9520 & -1.6258 & 0 \\ 1.3644 & 1.9325 & 0 & -3.2969 \end{bmatrix} \tag{3.49}$$

Comparing the off-diagonal entries of εK^E to the smallest value of K^I, we see a ratio of external connections to internal connections ranging from 0.5063/8.0159 to 1.9325/8.0159. The average of the off-diagonal entries of εK^E compared to the average of the off-diagonal entries of K^E is $1.1298/8.2265 = 0.1373$, which can be taken to be ε, from which K^E can be found.

The eigenvalues of the matrix $M^{-1}K$ are

$$\lambda(M^{-1}K) = 0, -14.2787, -60.7554, -62.2531 \tag{3.50}$$

The corresponding eigenvector vectors are

$$v_1 = \begin{bmatrix} 0.5 \\ 0.5 \\ 0.5 \\ 0.5 \end{bmatrix}, \quad v_2 = \begin{bmatrix} 0.4878 \\ 0.4031 \\ -0.5672 \\ -0.5271 \end{bmatrix}, \quad v_3 = \begin{bmatrix} 0.6333 \\ -0.7446 \\ 0.1924 \\ -0.0863 \end{bmatrix}, \quad v_4 = \begin{bmatrix} 0.1102 \\ -0.1494 \\ -0.8098 \\ 0.5566 \end{bmatrix} \tag{3.51}$$

Because (3.9) is in the second-order form, the eigenvalues of the two-area system consist of two poles at the origin corresponding to the system reference mode, an interarea mode of $\sqrt{-14.279} = \pm j3.779$ rad/s of oscillation between the two areas, and the local modes $\sqrt{-60.755} = \pm j7.795$ rad/s and $\sqrt{-62.253} = \pm j7.890$ rad/s. The eigenvectors (3.51) are indications of the mode shapes, that is, the relative motions of the machines. For example, v_1 represents the system mode in which all machines move together in the same direction and proportion, and v_2 represents the 3.779 rad/s interarea mode in which Generators 1 and 2 oscillate against Generators 11 and 12. The local mode oscillations are represented by v_3 and v_4, with v_3 being mostly the oscillation of Generator 1 versus Generator 2, and v_4 being mostly the oscillation of Generator 11 versus Generator 12.

From K^E we compute the initial aggregation model as

$$M_a = \frac{1}{2\pi \times 60} \begin{bmatrix} 234 & 0 \\ 0 & 234 \end{bmatrix}, \quad \varepsilon K_a = \begin{bmatrix} -4.1154 & 4.1154 \\ 4.9227 & -4.9227 \end{bmatrix} \tag{3.52}$$

Table 3.1 Electromechanical modes computed from various models

	Exact model (rad/s)	Inertial aggregation (rad/s)	Error (%)	Slow coherency	Error
Interarea mode	3.779	3.816	0.98	3.799 rad/s	0.53 %
Area 1 local mode	7.795	7.344	−5.79		
Area 2 local mode	7.890	7.393	−6.30		

The eigenvalues of $M_a^{-1} K_a$ are 0 and -14.561, giving rise to an interarea mode frequency of $\sqrt{-14.561} = \pm j3.816$ rad/s. For the fast local dynamics, we obtain

$$M_d = \frac{1}{2\pi \times 60} \begin{bmatrix} 58.500 & 0 \\ 0 & 55.611 \end{bmatrix}, \quad K_d = \begin{bmatrix} -8.3699 & 0 \\ 0 & -8.0628 \end{bmatrix} \tag{3.53}$$

The eigenvalues of $M_d^{-1} K_d$ are -53.939 and -54.660, resulting in the local modes of $\pm j7.3443$ and $\pm j7.3932$ rad/s.

Applying a correction term to the slow subsystem, we obtain

$$\varepsilon K_{as} = K_{11s} - K_{13s} K_{4s}^{-1} K_{31s} = \begin{bmatrix} -4.0698 & 4.0699 \\ 4.8861 & -4.8860 \end{bmatrix} \tag{3.54}$$

so that the approximated interarea mode becomes $\pm j3.799$ rad/s.

A summary of the interarea mode and local mode frequencies from the various models is given in Table 3.1. It is of interest to note that the approximated interarea mode from the inertia aggregation is higher in frequency than that of the exact model because the coherent machines are connected with infinite admittances and the network becomes stiffer (see Sect. 3.5 for additional discussions). On the other hand, the local modes from the initial aggregation are lower in frequencies than those of the exact model because impedances connecting two nodes in a local area through external connections are neglected. That is, the effective stiffness of the connections within an area is reduced, resulting in lower frequencies of oscillations.

3.4 Identification of Slow Coherent Machines

In the illustration of the two-area system example, we remark that the eigenvectors of $M^{-1}K$ show the mode shapes of the electromechanical modes. From the mode shapes, if machines i and j have similar entries in the eigenvector of mode k, we can conclude that these two machines are coherent with respect to that mode. Thus for a group of machines to be slow coherent, their mode shapes with respect to the low frequency interarea modes must be similar. In other words, if V_s is the matrix of the eigenvectors corresponding to the small eigenvalues of $M^{-1}K$ (3.9), then a slow coherent group of machines must have similar row vectors in V_s. For a system with

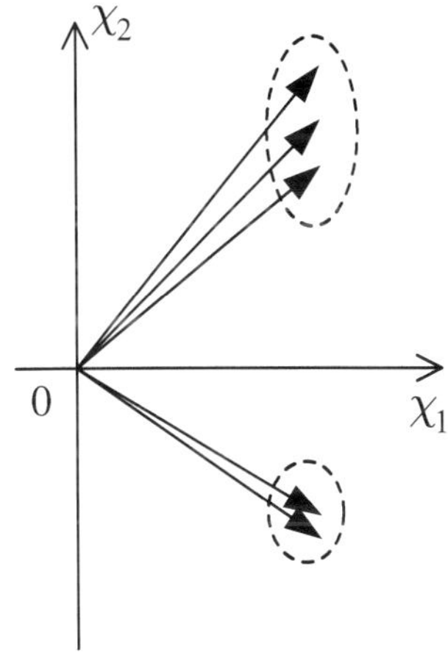

Fig. 3.3 Row vectors of V_s showing 2 coherent groups

r coherent groups, the row vectors of V_s form r clusters in an r-dimensional space. Figure 3.3 illustrates the row vectors for $r = 2$ areas. Thus a practical algorithm to identify the slow coherent groups is to first find the r most linearly independent vectors w^α from V_s and use them as the reference vectors. Then a machine with the row vector w_i will be grouped in the same area with the reference machine whose row vector w^α is closest to w_i.

The representation of the clusters in the r-dimensional space depends on the eigenbasis vectors used in V_s. To avoid this ambiguity we permute the rows of V_s into

$$V_s = \begin{bmatrix} V_{s1} \\ V_{s2} \end{bmatrix} \tag{3.55}$$

such that V_{s1} contains the reference rows. Then the row vectors of V_{s1} are used as unit coordinate vectors in a new coordinate system. This is equivalent to the transformation

$$\begin{bmatrix} V_{s1} \\ V_{s2} \end{bmatrix} V_{s1}^{-1} = \begin{bmatrix} I \\ L \end{bmatrix} \tag{3.56}$$

In the new coordinate system, the coherent clusters are well separated. If there are r coherent areas in the system, then

$$\begin{bmatrix} I \\ L \end{bmatrix} = \begin{bmatrix} I \\ L_g \end{bmatrix} + \begin{bmatrix} 0 \\ O(\varepsilon) \end{bmatrix} \tag{3.57}$$

A function is said to be of $O(\varepsilon)$ if it approaches zero as ε approaches zero. The matrix L_g is similar in structure to a grouping matrix U (3.25) except that the 1 entries may not appear consecutively depending on the machine ordering and it may have zero columns, which correspond to single machine areas. The 1 entries in L_g can be used to group the non-reference machines with the reference machines. The transformed clusters in Fig. 3.3 are shown in Fig. 3.4.

Fig. 3.4 Row vectors of V_s in Fig. 3.3 after transformation

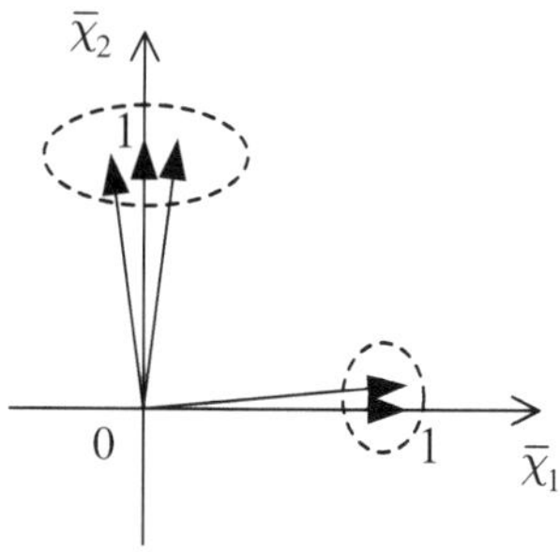

A coherent machine identification algorithm based on (3.57) has been proposed as follows.

Slow-Coherency Grouping Algorithm

1. Choose the number of areas r.
2. Compute a basis matrix V_s of the eigensubspace of the r smallest eigenvalues in magnitude of the model (3.9), including the zero eigenvalue.
3. Apply Gaussian elimination with complete pivoting to V_s and obtain the machines used for the pivots as the reference machines.
4. Order the machines such that the rows of V_{s1} of (3.55) correspond to the reference machines and solve for L from

$$V_1^T L^T = V_2^T \tag{3.58}$$

   using the LU decomposition of V_{s1} already computed in Step 2.
5. Use L to assign the machines to the coherent areas. That is, if the largest positive entry in a row of L is the αth entry, then the machine corresponding to that row is grouped into area α.

We have defined a slow coherent area to be composed of both the coherent machines and the load buses that interconnect those machines. The coherent machine identification algorithm does not directly identify the load buses in a coherent area. However, from (3.8), we can extend the mode shapes of the machine angles in V_s to the voltage angles of the load buses

$$V_\theta = (-K_4^{-1} K_3)_\theta V_s \tag{3.59}$$

The notation $(-K_4^{-1} K_3)_\theta$ is the projection of $-K_4^{-1} K_3$ to the bus voltage sensitivity with respect to the machine angles $\Delta\delta$, because ΔV in (3.6) and (3.7) is in the rectangular coordinates. Thus the Grouping Algorithm can be applied to the rows of the extended slow eigenvectors

$$V_s^e = \begin{bmatrix} V_s \\ V_\theta \end{bmatrix} \tag{3.60}$$

with the proviso that the rows of V_θ would not be used as the reference vector for a coherent group.

For a system with an unknown number of coherent groups, it may be necessary to try several possible values of r. In this case, use a sufficiently large r to compute the eigensubspace of the r smallest eigenvalues. Then Steps 3 to 5 can be repeated for different values of r by extracting the appropriate columns of V_s. The computationally intensive part of the algorithm is Step 2 if the system size is large. Earlier, a sparsity-based Lanczos algorithm has been developed to compute the eigenvectors of the small eigenvalues [11]. More recently, the Arnoldi method [12] has been shown to provide superior computational efficiency, which can be used for the identification of coherent machines in very large power systems.

An advantage of the slow-coherency method is that the coherent groups determined from the slow eigenvectors are disturbance independent, that is, the machines do not switch groups for different disturbances. Thus the slow coherent groups are determined from the base case. In disturbances involving line trips, we expect the boundaries of weak and strong connections to remain the same. If a line connecting two coherent areas is tripped, then the external connecting strength would be weakened further. If a line within a coherent area is tripped, its impact on the overall connection strength would not be significant. However, if a disturbance study requires tripping of multiple lines, then additional slow coherent machine identifications may be required.

We will illustrate the Grouping Algorithm for the two-area system. The slow eigensubspace is given by v_1 and v_2 (3.51) as

$$V_s = \begin{bmatrix} 0.5 & 0.4878 \\ 0.5 & 0.4031 \\ 0.5 & -0.5672 \\ 0.5 & -0.5271 \end{bmatrix} \begin{matrix} \text{Gen 1} \\ \text{Gen 2} \\ \text{Gen 11} \\ \text{Gen 12} \end{matrix} \tag{3.61}$$

The entries in the first column of V_s are all equal. Thus we arbitrarily pick Generator 1 as one of the reference machines and perform a Gaussian elimination on V_s to arrive at

$$V_s^1 = \begin{bmatrix} 0.5 & 0.4878 \\ 0.0 & -0.0846 \\ 0.0 & -1.0550 \\ 0.0 & -1.0149 \end{bmatrix} \begin{matrix} \text{Gen 1} \\ \text{Gen 2} \\ \text{Gen 11} \\ \text{Gen 12} \end{matrix} \tag{3.62}$$

At this point, Generator 11 has the largest magnitude in the second row and is selected as the reference of the second area. Thus

$$V_{s1} = \begin{bmatrix} 0.5 & 0.4878 \\ 0.5 & -0.5672 \end{bmatrix} \begin{matrix} \text{Gen 1} \\ \text{Gen 11} \end{matrix} \tag{3.63}$$

Exchanging the second and third row in V_s to form V_s', we obtain

$$V_s' V_{s1}^{-1} = \begin{bmatrix} 1 & 0 \\ 0 & 1 \\ \underline{0.9198} & 0.0802 \\ 0.0380 & \underline{0.9620} \end{bmatrix} \begin{matrix} \text{Gen 1} \\ \text{Gen 11} \\ \text{Gen 2} \\ \text{Gen 12} \end{matrix} \tag{3.64}$$

The largest positive entries in the rows of Generators 2 and 12 are underlined. Thus we can conclude that Generators 1 and 2 are coherent, and Generators 11 and 12 are coherent, consistent with the observation from the time response shown in Fig. 3.2.

Extending the Grouping Algorithm to the load buses, we compute V_θ using [10] and use it to obtain $V_\theta V_{s1}^{-1}$

$$V_\theta = \begin{bmatrix} 0.5 & 0.4283 \\ 0.5 & 0.3535 \\ 0.5 & 0.2556 \\ 0.5 & 0.3844 \\ 0.5 & -0.5018 \\ 0.5 & -0.4667 \\ 0.5 & -0.3556 \\ 0.5 & 0.3128 \\ 0.5 & -0.0523 \\ 0.5 & -0.4671 \\ 0.5 & -0.4125 \end{bmatrix} \begin{matrix} \text{Bus 1} \\ \text{Bus 2} \\ \text{Bus 3} \\ \text{Bus 10} \\ \text{Bus 11} \\ \text{Bus 12} \\ \text{Bus 13} \\ \text{Bus 20} \\ \text{Bus 101} \\ \text{Bus 110} \\ \text{Bus 120} \end{matrix} \quad V_\theta V_{s1}^{-1} = \begin{bmatrix} \underline{0.9436} & 0.0564 \\ \underline{0.8727} & 0.1273 \\ \underline{0.7800} & 0.2200 \\ \underline{0.9020} & 0.0980 \\ 0.0620 & \underline{0.9380} \\ 0.0953 & \underline{0.9047} \\ 0.2006 & \underline{0.7994} \\ \underline{0.8342} & 0.1658 \\ 0.4880 & 0.5120 \\ 0.0949 & \underline{0.9051} \\ 0.1466 & \underline{0.8534} \end{bmatrix} \tag{3.65}$$

Note that the largest positive entry in each row of $V_\theta V_{s1}^{-1}$ is used to group a bus with the reference machine. If none of the entries for a bus is close to 1, such as Bus 101 whose two entries are roughly equal, that bus should be regarded as a connection bus between the areas. These connection buses are normally not eliminated in the load bus elimination stage, in order to retain network sparsity.

In summary, the 2 coherent areas for the two-area, four-machine systems are given in Table 3.2.

As a second illustration on a larger system, consider the 48-machine NPCC system given in Price [13], which is an expansion of the 16-machine systems with more machines modeled in New York and the neighboring areas. Choosing the nine slowest modes, including the system mode, the Grouping Algorithm yields the 9-area partition shown in Fig. 3.5. Note that the boundaries of the areas are mostly along operating regions, confirming the notion that they form the weak connection boundaries.

Table 3.2 Coherent areas of the two-area system

Area	Generators	Load buses
1	1, 2	1, 2, 3, 10, 20
2	11, 12	11, 12, 13, 110, 120

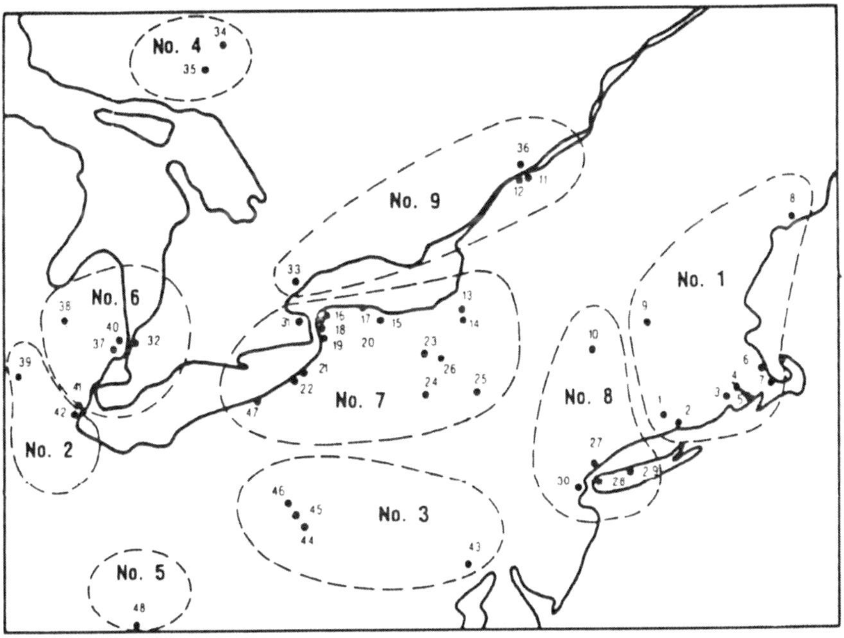

Fig. 3.5 Nine-area partition of the NPCC 48-machine system

3.4.1 A Tolerance-Based Grouping Algorithm

The objective of model reduction is to determine a model that can accurately capture the interarea modes of interest. The parameter r is selected based on the number of areas needed to model the desired interarea modes. However, for a large system, using r coherent areas to model $r - 1$ interarea modes may not yield the desired accuracy on the interarea modes. For large coherent groups, the largest entries in the rows of the L_g may not be close to unity, in which case machines that are not electrically close may still be assigned to the same coherent areas. An example is the 9-area partition of the 48-machine system, where machine 33 is quite far away from the other machines in area 9.

To control the accuracy of coherency, we develop a tolerance-based grouping algorithm to indirectly control the size of the coherent areas. In some sense this is an easier way of obtaining the coherent areas, as one does not need to specify r, the number of areas. Instead, one only needs to specify the interarea modes of interest, and adjust a tolerance value to achieve the desired degree of accuracy.

A tolerance-based grouping algorithm would set a measure of the slow coherency between the machines. Let the columns of V_s be normalized to unity, and define the slow-coherency measure

$$d_{ij} = w_i w_j^T / (|w_i||w_j|) \tag{3.66}$$

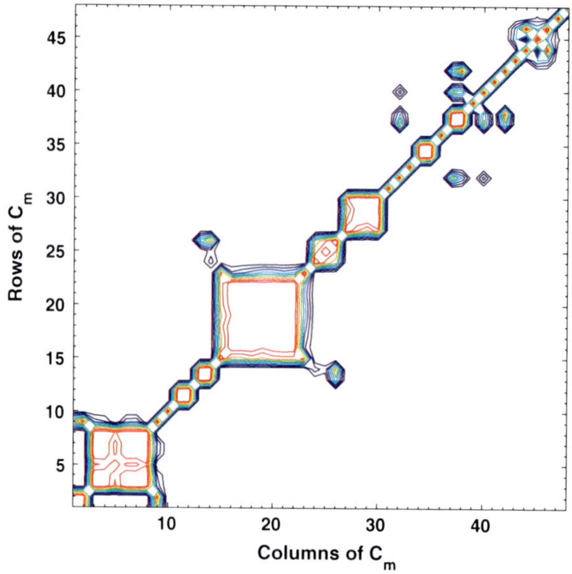

Fig. 3.6 Contour plot of C_m for the NPCC 48-machine system

as the cosine of the angle between w_i and w_j, which are the row of V_s corresponding to machines i and j, respectively. If machines i and j are perfectly coherent with respect to the slow modes, then $w_i = w_j$ and $d_{ij} = 1$. A tolerance, γ, typically in the range of 0.9–0.95, can be selected such that two machines are said to be coherent if $d_{ij} > \gamma$. We define a coherency matrix C_m to be a matrix whose (i, j) entry is given by

$$(C_m)_{ij} = d_{ij} - \gamma. \tag{3.67}$$

A contour plot of C_m of the NPCC system using the eigenvector matrix of the 9 slowest modes and $\gamma = 0.95$ is shown in Fig. 3.6. Note that in Fig. 3.6, the negative values in C_m have been set to 0 to accentuate the coherency. These negative values are not plotted in Fig. 3.6. If the (i, j) grid shows a nonzero value, then machines i and j are coherent. The C_m plot depicts clearly the coherency between machines 1–9 in New England and machines 13–26 in New York.

Based on the coherency measure γ, we formulate a set of coherency rules:

1. Machines i and j are coherent if $(C_m)_{ij} > 0$.
2. If machines i and j are coherent and machines j and k are coherent, then machines i and k are also coherent.
3. A *loose* coherent area J_α is formed by the machines that are coherent under Rules 1 and 2. Let $(C_m)_\alpha$ be a submatrix of C_m corresponding to J_α.
4. If the column sums of $(C_m)_\alpha$ excluding the diagonal entries are all positive, then J_α is a *tight* coherent area.

5. If any of the column sums of $(C_m)_\alpha$ excluding the diagonal entries is negative, then J_α should be decomposed into smaller tight coherent areas.
6. The least coherent machine in J_α corresponds to the columns of $(C_m)_\alpha$ with the smallest sum.
7. The coherency of J_α may be improved by removing the least coherent machine and reassigning it to a different area.
8. Given two partitions I_1 and I_2 of J_α, I_1 is tighter than I_2 if the sum of the off-diagonal entries of $(C_m)_\alpha$ corresponding to I_1 is larger than that of I_2.

These coherency rules can be used to construct the following algorithm.

Tolerance-Based Coherency Grouping Algorithm

1. Find the loose coherent areas using Rules 1–3.
2. For each loose coherent area J_α,
   a. Use Rule 4 to determine whether it is also a tight coherent area, which requires no further decomposition.
   b. If the area is not tight, decompose the area into tight coherent areas. Start the decomposition by identifying the least coherent machine using Rule 6 and reassigning it using Rule 8. Continue until the loose coherent area has been decomposed into tight coherent areas and no improvement is possible under Rule 8.

When the tolerance-based coherency grouping algorithm is applied to the NPCC system with L containing the eigenvectors of the 9 slowest modes and $\gamma = 0.95$, 17 slow-coherent areas are found. The machine groups are listed in Table 3.3 and shown in Fig. 3.7. For a comparison, the slow-coherency grouping algorithm with L containing the eigenvectors of the 17 slowest modes is used to find 17 areas for the NPCC system. The resulting areas are also listed in Table 3.3. Note that the first machine listed in each group is the reference machine for that particular group. These two 17-area partitions have a lot of similarities, but also some differences.

To assess the significance between the two different 17-area partitions, and to show the improvement of the 17-area partition over the 9-area partition, we perform the aggregation based on these partitions and compute the slow eigenvalues of the linearized reduced-order models. For a comprehensive illustration, we use both the inertial aggregation and slow-coherency aggregation. In the inertial aggregations, nonlinear reduced networks are first constructed using the algorithms described in the next section. Eigenvalues are then computed from the linearized models. For the slow-coherency aggregation, eigenvalues are directly computed from (3.41). The frequencies of the 8 slowest modes (excluding the system mode) are shown in Tables 3.4 and 3.5.

From Tables 3.4 and 3.5, we conclude that the 17-area partition from the slow-coherency grouping algorithm does not necessarily provide a better eigenvalue

Table 3.3 Seventeen-area partitions using tolerance-based slow coherency and slow coherency

Area	9 Slowest modes, $\gamma = 0.95$	Slow coherency, 17 areas
1	3 4 5 6 7 8	6 3 7 9
2	1 2 9	1 2
3	10	8
4	11 12	11 12
5	13 14 24 25 26	13 10 14 24 25 26
6	15 16 17 18 19 20 21 22 23	16 17 18 19 20 21 22
7	27 28 29 30	29 27 28 30
8	31	15 23
9	32 37 38 40 42	32 31 33 37 38
10	33	40
11	34 35	34 35
12	36	36
13	39	39 42
14	41	41
15	43 44 45 46	44 43 45 46 47
16	47	5 4
17	48	48

approximation than the 9-area partition. However, the 17-area tolerance-based partition shows a substantial improvement over both the 9-area and 17-area partitions from the slow-coherency grouping algorithm.

3.5 Generator and Network Aggregation

Thus far we have discussed slow coherency using linearized models. However, it is also important to develop nonlinear reduced models based on slow coherency so that they can be used for dynamic security assessment, such as transient stability simulations of faults and line trips. Although linearized models are derived for the inertial and slow-coherency aggregations, aggregates with conventional network and machine models can be reconstructed from the linearized reduced models. For completeness, we will discuss these two aggregation techniques, as well as the Podmore aggregation method in Chap. 2, and show the improvements between the techniques.

The Podmore aggregation for two machines A and B is shown in Fig. 3.8 where a and b are the generator terminal buses. In this technique, it is assumed that coherency occurs at the generator terminal buses a and b (Fig. 3.8a) . As a result these buses are tied together to a common bus q with infinite admittances (Fig. 3.8b). The voltage at bus q can be set either to an average of the voltages at buses a and b or a weighted average with respect to the active and reactive power generation. To preserve the power flow, ideal transformers with no leakage reactance and complex turns ratios $\alpha_a \arg(\phi_a)$ and $\alpha_b \arg(\phi_b)$ link the buses a and b to bus q, respectively. The phases

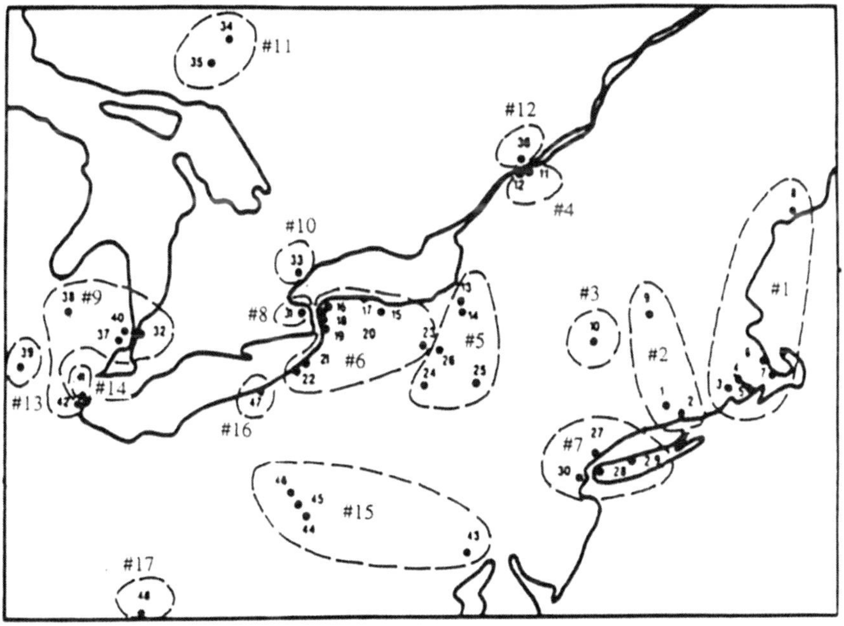

Fig. 3.7 Seventeen-area partition of the NPCC 48-machine system

Table 3.4 Interarea modes from inertial aggregation

Full model	Grouping algorithm 9 areas		Grouping algorithm 17 areas		Tolerance-based 17 areas	
Frequency (Hz)	Frequency (Hz)	Error (%)	Frequency (Hz)	Error (%)	Frequency (Hz)	Error (%)
0.2697	0.2958	9.7	0.2958	9.7	0.2747	1.8
0.3815	0.4049	6.1	0.4277	12.1	0.3932	3.0
0.4873	0.5378	10.4	0.5352	9.8	0.5034	3.3
0.5328	0.5847	9.7	0.5822	9.3	0.5472	2.7
0.7069	0.8315	17.6	0.7582	7.3	0.7123	0.8
0.7405	0.8514	15.0	0.7995	8.0	0.7515	1.5
0.8040	0.8918	10.9	0.8397	4.4	0.8365	4.0
0.8411	0.9678	15.1	0.8592	2.1	0.8562	1.8

ϕ_a and ϕ_b can be represented separately as phase shifters. The machines A and B are then aggregated into a single equivalent machine with an inertia m_{eq} and a transient reactance $(x'_d)_{\text{eq}}$ equal to

$$m_{\text{eq}} = m_A + m_B, \quad (x'_d)_{\text{eq}} = (1/(x'_d)_A + 1/(x'_d)_B)^{-1} \tag{3.68}$$

Table 3.5 Interarea modes of slow-coherency aggregation

Full model	Slow coherency 9 areas		Slow coherency 17 areas		Tight coherency 17 areas	
Frequency (Hz)	Frequency (Hz)	Error (%)	Frequency (Hz)	Error (%)	Frequency (Hz)	Error (%)
0.2697	0.2772	2.8	0.2786	3.3	0.2709	0.4
0.3815	0.4087	7.1	0.4348	14.0	0.3862	1.2
0.4873	0.5554	14.0	0.5580	14.5	0.4904	0.6
0.5328	0.5745	7.8	0.5999	12.6	0.5297	0.6
0.7069	0.8532	20.7	0.7629	7.9	0.7119	0.7
0.7405	0.8545	15.4	0.7987	7.9	0.7463	0.8
0.8040	0.8827	9.8	0.8369	4.1	0.8304	3.3
0.8411	0.9708	15.4	0.8606	2.3	0.8485	0.9

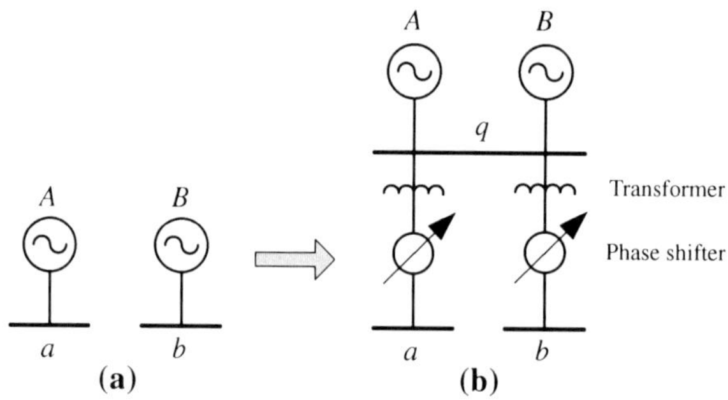

Fig. 3.8 Podmore aggregation: (a) two coherent generators, (b) generators connected to a common bus

where m_A and m_B are the inertias and $(x'_d)_A$ and $(x'_d)_B$ the transient reactances of machines A and B, respectively.

In the initial aggregate model (3.38), the dynamics inside a coherent area are assumed to be infinitely fast and thus neglected. This is akin to connecting the states of (3.38), which are the internal nodes of the machines, with infinite impedances. To illustrate, this inertial aggregation for two machines A and B is shown in Fig. 3.9 where a and b are the generator terminal buses. In the inertial aggregation technique, the machine internal node voltages E'_A and E'_B, are computed (Fig. 3.9a). These two generator internal nodes are tied together to a common bus p with infinite admittances (Fig. 3.9b). The voltage at bus p, E'_{eq}, can be set either to an average of E'_A and E'_B or a weighted average with respect to the active and reactive power generation. To preserve the power flow, ideal transformers with complex turns ratios $\alpha_a \arg(\phi_a)$ and $\alpha_b \arg(\phi_b)$ and zero impedances link the buses a and b to bus p.

The linking of the internal nodes creates an equivalent generator with multiple terminal buses, which is not a conventional power system network representation. Taking $(x'_d)_{\text{eq}}$ from (3.68) as an equivalent transient reactance, the network is extended beyond node p by two buses with reactances of $-(x'_d)_{\text{eq}}$ and $(x'_d)_{\text{eq}}$, as shown in Fig. 3.9c. The node r serves as the internal node of the equivalent machine,

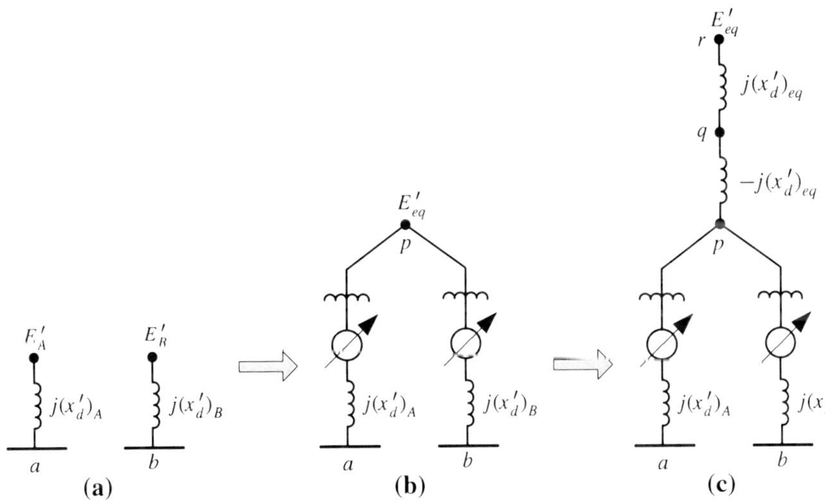

Fig. 3.9 Inertial aggregation: (a) two coherent generators, (b) generator internal node connected, (c) aggregate generator transient reactance created

and the node q the generator terminal bus. The node p can be eliminated if desired. Finally, the inertia of the equivalent machine, m_{eq}, is computed from (3.68).

The construction of the slow-coherency model (3.41) is more involved. There are two important observations. First recall that (3.41) requires a singular perturbation correction involving only the network within the coherent area, which we refer to as the "per coherent area" aggregation. This property does not hold if additional corrections are made to (3.41). The second observation is that we need to derive a nonlinear power system model from the linearized model (3.41).

To perform the network aggregation, the coherent areas of generators and load buses must first be specified, using the generator and load bus grouping algorithms from Sect. 3.3. Then the boundary buses are identified as the load buses with direct connections to the neighboring areas, and all the other load buses are the internal buses. In the per coherent area aggregation procedure, the model of each coherent area is linearized and aggregated, and the non-essential internal load buses are eliminated. The linearization of the coherent areas is similar in concept to the modal dynamic equivalencing technique [14, 15] of linearizing the external networks. Then the reduced models of the coherent areas are reconnected and the non-essential boundary buses are eliminated. Finally, the resulting linear model is used to reconstruct a network model with transmission lines and load parameters. We shall first discuss these operations and summarize the procedure in an algorithm.

For notational convenience, we assume that the coherent machines appear consecutively in δ and the load buses in each coherent area also appear consecutively in V. The nonlinear model of coherent area α, $\alpha = 1, 2, \ldots, r$, is given by

$$
\begin{aligned}
M^\alpha \ddot{\delta}^\alpha &= f^\alpha(\delta^\alpha, V^\alpha) \\
I^\alpha &= g^\alpha(\delta^\alpha, V^\alpha)
\end{aligned}
\tag{3.69}
$$

where δ^α is an n_α-vector of the machine angles, V^α is a $2N_\alpha$-vector of the real and imaginary part of load bus voltages, M^α is the inertia matrix, f^α is a vector of acceleration torques, and g^α is the loadflow equation of the power network in area α. The current injection I^α is nonzero at a boundary bus and represents the connections to the neighboring areas. These injections need not be computed as the current injections at the boundary buses will cancel as the coherent areas are reconnected to form the complete system. We linearize (3.69) about a nominal power flow equival$(\delta_0^\alpha, V_0^\alpha)$ to obtain the model

$$
\begin{aligned}
M^\alpha \ddot{\delta}^\alpha &= \frac{\partial f^\alpha(\delta^\alpha, V^\alpha)}{\partial \delta^\alpha}\Delta\delta^\alpha + \frac{\partial f^\alpha(\delta^\alpha, V^\alpha)}{\partial V^\alpha}\Delta V^\alpha = K_1^\alpha \Delta\delta^\alpha + K_2^\alpha \Delta V^\alpha \\
\Delta I^\alpha &= \frac{\partial g^\alpha(\delta^\alpha, V^\alpha)}{\partial \delta^\alpha}\Delta\delta^\alpha + \frac{\partial g^\alpha(\delta^\alpha, V^\alpha)}{\partial V^\alpha}\Delta V^\alpha = K_3^\alpha \Delta\delta^\alpha + K_4^\alpha \Delta V^\alpha
\end{aligned}
\tag{3.70}
$$

where $\Delta\delta^\alpha$ is an n_α-vector of the machine angle deviations from δ_0^α, and ΔV^α is a $2N_\alpha$-vector of the real and imaginary parts of the load bus voltage deviations from V_0^α. The matrices K_i^α are similar to the matrices K_i in (3.5). As in (3.70), ΔI^α also need not be computed.

Applying a partition of the transformation (3.31) as

$$
\begin{bmatrix} y_\alpha \\ z^\alpha \end{bmatrix} = \begin{bmatrix} C_\alpha \\ G_\alpha \end{bmatrix} \Delta\delta^\alpha \tag{3.71}
$$

whose inverse is

$$
\Delta\delta^\alpha = \begin{bmatrix} u_\alpha & G_\alpha^+ \end{bmatrix} \begin{bmatrix} y_\alpha \\ z^\alpha \end{bmatrix} \tag{3.72}
$$

where C_α and G_α^+ are the αth diagonal blocks of C and G^+, respectively, to (3.70), we obtain the standard singularly perturbed form of area α as

$$
\begin{aligned}
m_a^\alpha \ddot{y}_\alpha &= K_{11}^\alpha y_\alpha + K_{12}^\alpha z^\alpha + K_{13}^\alpha V^\alpha \\
M_d^\alpha \ddot{z}_\alpha &= K_{21}^\alpha y_\alpha + K_{22}^\alpha z^\alpha + K_{23}^\alpha V^\alpha \\
\Delta I^\alpha &= K_{31}^\alpha y_\alpha + K_{32}^\alpha z^\alpha + K_4^\alpha V^\alpha
\end{aligned}
\tag{3.73}
$$

where M_d^α is the αth diagonal block of M_d, and

$$
\begin{aligned}
&K_{11}^\alpha = u_\alpha^T K_1^\alpha u_\alpha, \quad K_{12}^\alpha = u_\alpha^T K_1^\alpha G_\alpha^+, \quad K_{13}^\alpha = u_\alpha^T K_2^\alpha \\
&K_{21}^\alpha = (G_\alpha^+)^T K_1^\alpha u_\alpha, \quad K_{22}^\alpha = (G_\alpha^+)^T K_1^\alpha G_\alpha^+, \quad K_{23}^\alpha = (G_\alpha^+)^T K_2^\alpha \\
&K_{31}^\alpha = K_3^\alpha u_\alpha, \quad K_{32}^\alpha = K_3^\alpha G_\alpha^+
\end{aligned}
\tag{3.74}
$$

From (3.73), the slow-coherency aggregate model can be obtained as

$$\begin{aligned} m_a^\alpha \ddot{y}_\alpha &= K_{11s}^\alpha y_\alpha + K_{13s}^\alpha \Delta V^\alpha \\ \Delta I^\alpha &= K_{31s}^\alpha y_\alpha + K_{4s}^\alpha \Delta V^\alpha \end{aligned} \tag{3.75}$$

where

$$\begin{aligned} K_{11s}^\alpha &= K_{11}^\alpha - K_{12}^\alpha (K_{22}^\alpha)^{-1} K_{21}^\alpha \\ K_{13s}^\alpha &= K_{13}^\alpha - K_{12}^\alpha (K_{22}^\alpha)^{-1} K_{23}^\alpha \\ K_{31s}^\alpha &= K_{31}^\alpha - K_{32}^\alpha (K_{22}^\alpha)^{-1} K_{21}^\alpha \\ K_{4s}^\alpha &= K_4^\alpha - K_{32}^\alpha (K_{22}^\alpha)^{-1} K_{23}^\alpha \end{aligned} \tag{3.76}$$

To complete the model reduction process, the voltage variables for the internal load buses that need not be retained can be eliminated from (3.69). The reduced order model of the αth area would then consist of the aggregate machine, the retained load buses and the boundary buses

$$\begin{aligned} m_a^\alpha \ddot{y}_\alpha &= \bar{K}_{11s}^\alpha y_\alpha + \bar{K}_{13s}^\alpha \Delta \bar{V}_\alpha \\ \Delta \bar{I}_\alpha &= \bar{K}_{31s}^\alpha y_\alpha + \bar{K}_{4s}^\alpha \Delta \bar{V}_\alpha \end{aligned} \tag{3.77}$$

After the aggregate models (3.77) for all the coherent areas have been obtained, they are reconnected to form the aggregate system model

$$\begin{aligned} M_a \ddot{y} &= \bar{K}_{11s} y + \bar{K}_{13s} \Delta \bar{V} \\ 0 &= \bar{K}_{31s} y + \bar{K}_{4s} \Delta \bar{V} \end{aligned} \tag{3.78}$$

where $\bar{K}_{ijs}$ are diagonal block matrices of the corresponding matrices $\bar{K}_{ijs}^\alpha$, and the current injections I^α's are now modeled by the connections between the boundary buses as the off-diagonal blocks of $\bar{K}_{4s}$. It is shown in Date [16], Date and Chow [6] that (3.78) and (3.41) are the same dynamic model. This invariance property is due to the fact that from a singular perturbation analysis, the first-order correction terms to the slow subsystem involve only the connections from the generator internal nodes to the terminal buses. For higher order correction terms, the per area aggregation concept is no longer applicable since the impedance corrections will depend on parameters in the other areas.

The slow-coherency aggregation for two machines A and B is illustrated in Fig. 3.10. In slow-coherency aggregation, the machine internal node voltages E'_A and E'_B are computed to obtain the linearized model (3.78) (Fig. 3.10a). In the construction of (3.78), only the fast variable z is eliminated, while all the bus voltage variables are retained. This allows the reconstruction from the connection matrices K_{11s}, K_{13s}, K_{31s}, and K_{4s} of a power network consisting of impedances and phase shifters (Fig. 3.10b). Although branch parameters can be reconstructed from the connection matrices, the recovered network, in general, would not have a balanced

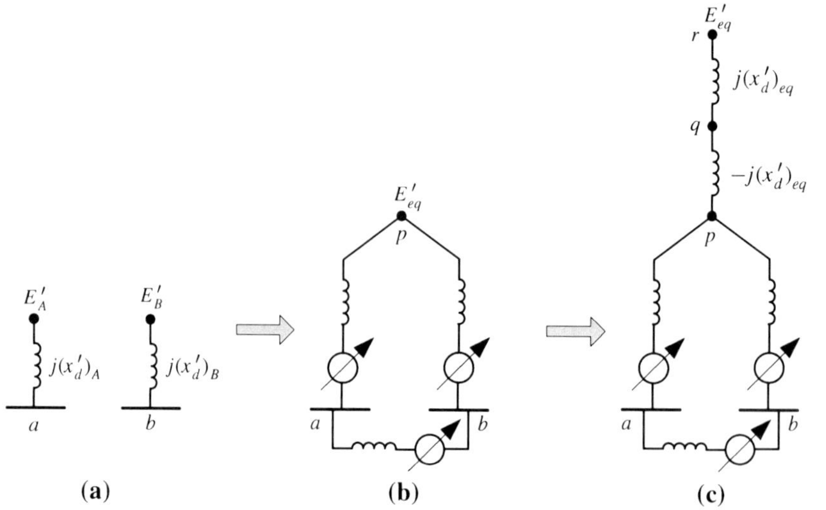

Fig. 3.10 Slow-coherency aggregation: (a) two coherent generators, (b) generator internal node connected with finite admittance between the generator buses, (c) aggregate generator transient reactance created

load flow. For tightly connected areas, the load flow mismatch would be small and loads can be added to the generation terminal buses to balance the load flow. The elimination of the fast variable z results in K_{4s} being a dense matrix. Thus in the reconstruction, all the generator terminal buses in the same area will be interconnected. This interconnection is dependent only on the parameters within an coherent area, and represents the improvement to the inertial aggregate. For practical reasons, such as implementing voltage regulator control, it is desirable to have the machine connected to only one terminal bus. Thus for each aggregate machine, we modify (3.78) by inserting two buses which are connected with reactances of x'_d and $-x'_d$ between the machine internal node and the remaining network (Fig. 3.10c). The reactance x'_d is an aggregate machine transient reactance, which can be computed as the MVA-weighted average of the individual machine transient reactances.

Finally, all the non-essential buses including the boundary buses can be eliminated from the model to form

$$\begin{aligned} M_a \ddot{y} &= \bar{K}_{11} y + \bar{K}_{12} \Delta \bar{V} \\ 0 &= \bar{K}_{21} y + \bar{K}_{22} \Delta \bar{V} \end{aligned} \tag{3.79}$$

As the last step, the linear aggregate model (3.79) has to be converted into a physical power system model whose data can be used directly by conventional stability analysis and simulation programs. The parameters of interest are, for each transmission line, the line resistance, reactance, transformer ratio and phase shifter angle, and for each load bus, the active and reactive parts of the constant impedance, current and power type loads. The aggregate network needs to retain the dynamics represented by the aggregate sensitivity matrices $\bar{K}$'s, as well as preserve the network loadflow on the portions of the system retained in full detail.

If the nonlinear network representation of (3.79) is in the form

$$M_a \ddot{y} = f_a(y, \bar{V}) \tag{3.80}$$

$$0 = g_a(y, \bar{V}) \tag{3.81}$$

where f_a and g_a are the same types of nonlinear functions as those in (3.5) but with aggregate network parameters, then the aggregate sensitivity matrices $\bar{K}$'s in (3.79) must satisfy

$$\bar{K}_{ij} = \hat{K}_{ij}, \quad i, j = 1, 2 \tag{3.82}$$

where

$$\hat{K}_{11} = \left.\frac{\partial f_a(y, \bar{V})}{\partial y}\right|_{y_0, \bar{V}_0}, \quad \hat{K}_{12} = \left.\frac{\partial f_a(y, \bar{V})}{\partial V}\right|_{y_0, \bar{V}_0}$$

$$\hat{K}_{21} = \left.\frac{\partial g_a(y, \bar{V})}{\partial y}\right|_{y_0, \bar{V}_0}, \quad \hat{K}_{22} = \left.\frac{\partial g_a(y, \bar{V})}{\partial V}\right|_{y_0, \bar{V}_0} \tag{3.83}$$

With the introduction of the aggregate machine terminal buses, the matrix equalities in (3.83) will be exactly satisfied except for

$$\bar{K}_{22} = \hat{K}_{22} \tag{3.84}$$

Thus the aggregate network reconstruction process is to solve for the aggregate network parameters to satisfy (3.84), subject to the constraint of achieving a balanced loadflow. The presence of an adequate number of network parameters will be essential for an exact solution. In addition, there are constraints on the various entries of $\hat{K}_{22}$ which are functions of these physical parameters. As a result some network parameters do not provide any additional degrees of freedom for the solution. Thus we use, for a line, only 3 out of 4 parameters, namely, the line resistance, reactance and phase shifter angle, and for a load, 4 out of 6 load parameters, namely, the constant impedance and constant current active and reactive types of loads.

In general, the equality (3.84) will not be satisfied exactly. As a result, we propose a two-step least-squares optimization scheme. We denote the line and load parameters by the vectors γ and β, respectively, and the set of physically realizable parameter values of γ and β by $\mathcal{G}$ and $\mathcal{B}$, respectively.

In the first step, the parameters for each line are solved independently from an optimization problem. For the ith line, the problem is to solve for

$$\min_{\gamma \in \mathcal{G}} \|\hat{K}^i(\gamma) - \bar{K}^i\| \tag{3.85}$$

where $\hat{K}^i(\gamma)$ and $\bar{K}^i$ are the portions of the matrices $\hat{K}_{22}$ and $\bar{K}_{22}$ due to the ith line.

In the second step, the parameters of each load are solved independently from a similar optimization problem. For the jth load, the problem is to solve for

$$\min_{\beta \in \mathscr{B}} \|\hat{K}^j(\beta) - \bar{K}^j\|, \quad \text{subject to} \quad S^j(\beta) = 0 \tag{3.86}$$

where $\hat{K}^j(\beta)$ and $\bar{K}^j$ are the 2×2 diagonal blocks of $\hat{K}_{22}$ and $\bar{K}_{22}$ due to the jth load, and $S^j(\beta)$ is the function representing the current balance of the jth load.

The optimization scheme would yield an aggregate power system model with physical machine, line, and load parameters. In addition to being a close approximate of the slow dynamics of the original system, the aggregate model also preserves the power flow as well as approximates the network flow sensitivities of the original system.

The *slow-coherency aggregation algorithm* is now summarized as follows.

Slow-Coherency Aggregation Algorithm

Step 1: For coherent area α, $\alpha = 1, 2, \ldots, r$,

1. Linearize the coherent area model (3.69) to obtain the model (3.70).
2. Apply the transformation (3.71) to (3.70) and eliminate the fast variables to obtain (3.77).
3. Eliminate the non-essential internal buses of (3.77) to form (3.77).

Step 2.

1. Reconnect the aggregate models of the coherent areas to form (3.78).
2. Introduce aggregate generator terminal buses as needed.
3. Eliminate all the non-essential buses to form (3.79).

Step 3.

1. For each line, solve the optimization (3.85) to obtain the line parameters.
2. For each load, solve the optimization (3.86) to obtain the load parameters.

We remark that an inertial aggregate model can be obtained from the same algorithm except that in Step 1.2, the fast variables are assumed to be identically zero. The network aggregation algorithms will be illustrated in the next section.

To summarize, the slow-coherency aggregation method represents an improvement over the generator terminal bus aggregation technique by providing impedance corrections to the aggregate models and a more accurate loadflow sensitivity. The additional computation effort required by the slow-coherency aggregate is modest. An immediate consequence of the impedance corrections is the improved low frequency approximation by a slow-coherency aggregate network. The improved loadflow sensitivity is important in using the aggregate model for nonlinear simulations.

We note that simulations using the slow-coherency aggregate model are equivalent to the nonlinear simulation technique (A3) in Winkelman et al. [17], which used a specialized simulation program because the approach eliminated all the load buses. Given the simulation experience in Newell et al. [18], Troullinos et al. [19] using the DYNRED programs in Podmore and Germond [20], the slow-coherency aggregation technique can potentially achieve the same level of approximations with smaller reduced models.

3.6 Simulation Studies

To provide a comparison of the nonlinear aggregate models obtained from the three techniques, we apply the techniques to the 17-area tight coherency partition of the NPCC system. The disturbance considered is a 6-cycle short-circuit fault at the Medway bus (Bus 7 in the data set) located in Area 1, which is cleared by removing the line from Medway to the Sherman Road bus (Bus 6 in the data set). The study region in which local machine dynamics are of interest is designated to be Areas 1 and 2. As a result, all the machines in these two areas are retained individually and no aggregation is required. The aggregation algorithms are applied to the other coherent areas to combine areas with more than one coherent machine into single equivalent machine areas. As a result, all the aggregate models contain 24 machines, of which 7 are equivalent machines.

The Medway disturbance is simulated on the full model and the three aggregate models. The responses for machines 1 and 4 are shown in Figs. 3.11 and 3.12, respectively. Note that the Podmore aggregate model provides good approximations up to 4 seconds, after which the error in the slow interarea mode frequency approximation becomes evident. The inertial aggregate model shows an improvement over the Podmore aggregate model. The slow-coherency aggregate shows a small improvement over the inertial aggregate model. From the eigenvalue analysis results in Tables 3.4 and 3.5, more improvement would have been expected. However, the addition of loads to balance the power flow in the slow-coherency algorithm introduces an approximation that reduces the improvement.

For a practical power system, a reduced model may consist of a mixture of individual machines, inertial aggregates, and slow-coherency aggregates. In general, no aggregation is performed within the study region such that the machines in the study region are kept separately. Furthermore, areas immediately adjacent to the study region, the so-called buffer regions, are also kept in detail. Only the areas remote to the study region are aggregated. In Chow et al. [21], a study is reported in which various aggregates are used for the external areas to achieve an optimal aggregation.

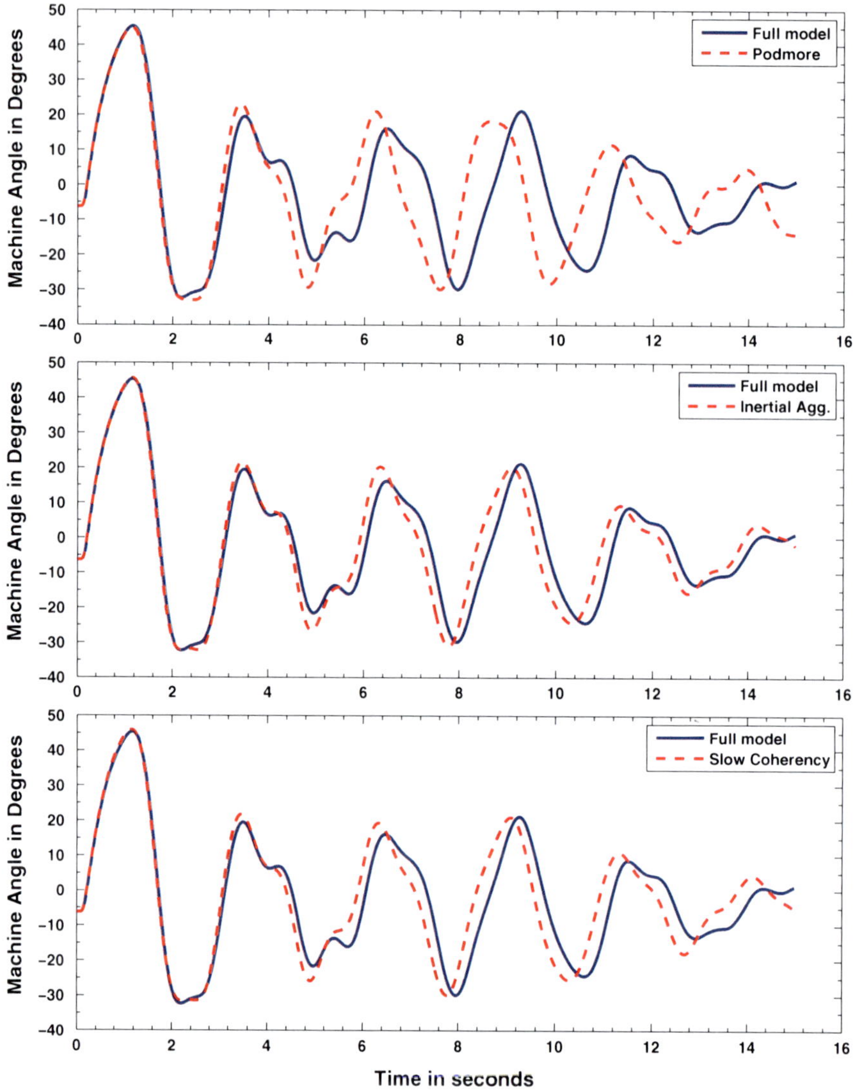

Fig. 3.11 Time response of machine 1 for the Medway disturbance

3.7 Coherency Toolbox

Users of the Power System Toolbox [10] can try out the various coherency and aggregation algorithms discussed in this chapter using the "Coherency Toolbox." The Coherency Toolbox contains functions to perform the two grouping algorithms presented in Sect. 3.4. It also provides functions to perform the Podmore aggregation, the initial aggregation, and the slow-coherency aggregation. A load bus

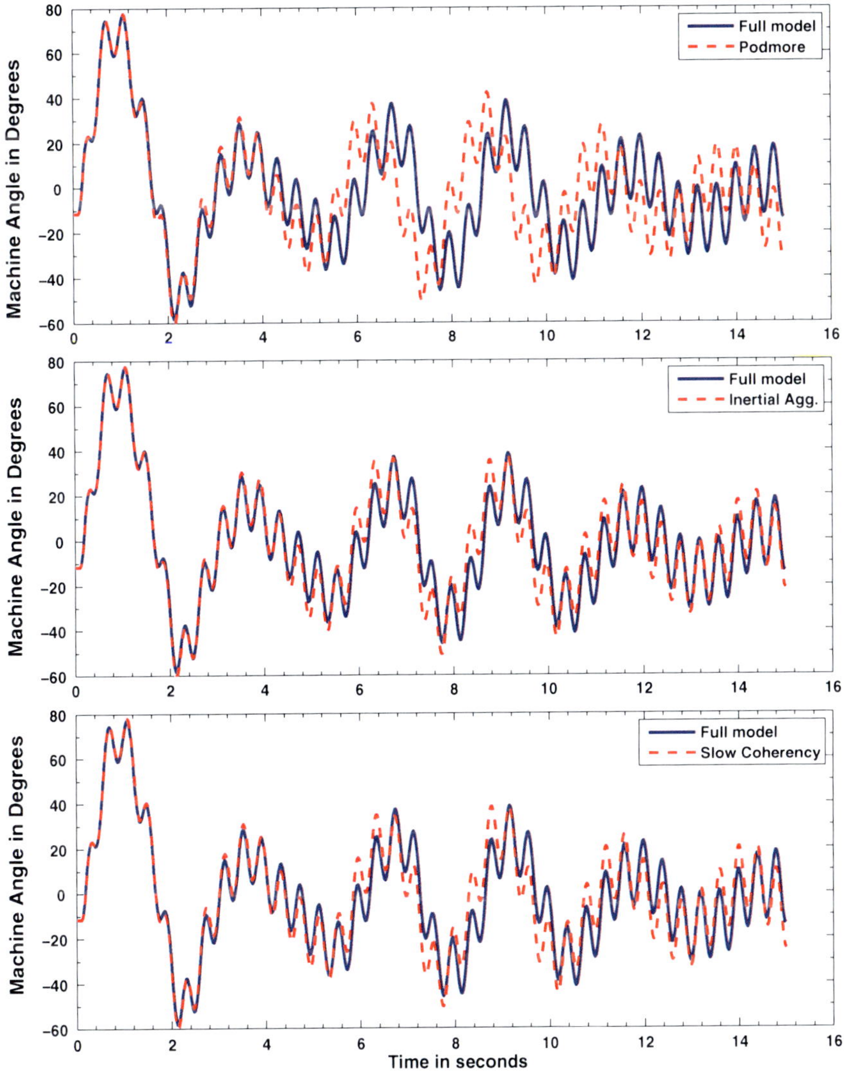

Fig. 3.12 Time response of machine 4 for the Medway disturbance

reduction function is also available [22]. Three system data are included: the Klein-Rogers-Kundur 2-area system, and the NPCC 16-machine and 48-machine systems.[1] The software can be obtained via the author's website www.ecse.rpi.edu/~chowj. A manual is available to provide some guidance in using the software.

[1] The 16-machine system has been used in several coherency papers such as Chow et al. [3].

3.8 Conclusions

In this chapter, we have provided an analytical foundation for the slow-coherency phenomenon in power systems. The main idea is that coherency is due to the slow interarea oscillations arising from weak connections. The singular perturbations method is used to develop this slow-coherency result, from which grouping algorithms based on eigenvalues and eigenvectors can be developed. An extension to group load buses is introduced. The method has the advantage of having the coherent groups of machines not dependent on the disturbance.

The singular perturbations method can also be used to generate more accurate aggregate models, with correction terms required only from within a coherent area. For a user of such techniques, there will be a number of tradeoffs. If the area to be aggregated is small, then the inertial aggregation technique would provide adequate accuracy. If the area to be aggregated is large, a user may consider using the slow-coherency aggregation with more corrections.

In addition to obtaining reduced-order models, other applications of slow coherency includes interconnected power system islanding as a means for power system emergency control [23]. The singular perturbations method for slow coherency can be extended to other network systems and models of Markov chains [24].

3.9 Appendix

The Appendix contains the data for the 2-area, 4-machine Klein-Rogers-Kundur system [2], on 100 MVA base, with modifications of the generator parameters and power flow data.

1. Bus parameters with solved power flow solutions:

Bus	$\vert\tilde{V}\vert$, pu	$\angle\tilde{V}$, deg	P_{gen}, pu	Q_{gen}, pu	P_{load}, pu	Q_{load}, pu
1	1.0500	0.1117	7.000	1.492	0	0
2	1.0500	−9.306	7.000	2.810	0	0
3	0.9917	−23.49	0	0	12.76	0.5
10	1.0323	−6.089	0	0	0	0
11	1.0500	0.000	9.247	2.344	0	0
12	1.0500	−16.60	5.000	3.225	0	0
13	0.9813	−29.27	0	0	14.76	0.5
20	1.0114	−15.63	0	0	0	0
101	1.0242	−26.59	0	0	0	0
110	1.0234	−8.263	0	0	0	0
120	1.0019	−21.15	0	0	0	0

2. Line parameters:

From bus	To bus	Circuit	R, pu	X, pu	B, pu
1	10	1	0.0	0.0167	0.00
2	20	1	0.0	0.0167	0.00
3	20	1	0.001	0.0100	0.0175
3	101	1	0.011	0.110	0.1925
3	101	2	0.011	0.110	0.1925
10	20	1	0.0025	0.025	0.0437
11	110	1	0.0	0.0167	0.0
12	120	1	0.0	0.0167	0.0
13	101	1	0.011	0.11	0.1925
13	101	2	0.011	0.11	0.1925
13	120	1	0.001	0.01	0.0175
110	120	1	0.0025	0.025	0.0437

3. Generator parameters:

Generator	Bus	Rated MVA	x'_d, pu	H, pu
1	1	900	0.25	6.5
2	2	900	0.25	6.5
11	11	1100	0.25	6.5
12	12	700	0.25	6.5

References

1. P.V. Kokotović, H. Khalil, J. O'Reilly, *Singular Perturbation Methods in Control: Analysis and Design* (Academic Press, 1986)
2. M. Klein, G.J. Rogers, P. Kundur, A fundamental study of inter-area oscillations in power systems. IEEE Trans. Power Syst. **6**, 914–921 (1991)
3. J.H. Chow, G. Peponides, P.V. Kokotović, B. Avramović, J.R. Winkelman, *Time-Scale Modeling of Dynamic Networks with Applications to Power Systems* (Springer, New York, 1982)
4. J.H. Chow, P.V. Kokotović, Time scale modeling of sparse dynamic networks. IEEE Trans. Autom. Control, **AC-30** 714–722 (1985)
5. J.H. Chow, R.A. Date, H. Othman, W.W. Price, Coherency aggregation of large power systems. in *Application of Eigenanalysis and Frequency Domain Methods for System Dynamic Performance* (IEEE Power Engineering Society, 1990)
6. R.A. Date, J.H. Chow, Aggregation properties of linearized two-times scale power networks. IEEE Trans. Circuits Syst. **38**, 720–730 (1991)
7. J.H. Chow, New algorithms for slow coherency aggregation of large power systems, in eds. by J.H. Chow, R.J. Thomas, P.V. Kokotović. *Systems and Control Theory for Power Systems, IMA Volumes in Mathematics and its Applications, vol 64* (Springer, 1994)
8. P.W. Sauer, M.A. Pai, *Power System Dynamics and Stability* (Stipes Publishing, 2007)
9. G. Peponides, P.V. Kokotović, J.H. Chow, Singular perturbations and time scales in nonlinear models of power systems. IEEE Trans. Circuits Syst. **CAS-29** 758–767 (1982)
10. J.H. Chow, K.W. Cheung, A toolbox for power system dynamics and control engineering education. IEEE Trans. Power Syst. **7**, 1559–1564 (1992)
11. J.H. Chow, J. Cullum, R.A. Willoughby, A sparsity-based technique for identifying slow-coherent areas in large power systems. IEEE Trans. Power Apparatus Syst. **PAS-103** 463–473 (1983)
12. L. Wang, A. Semlyen, Applications of sparse eigenvalue techniques to the small signal stabiity analysis of large power systems. IEEE Trans. Power Syst. **5**, 635–642 (1990)

13. W.W. Price, A Study of the Accuracy and Computation Savings of Modal Analysis Dynamic Equivalents. Report to NPCC-10 Working, Group (1974)
14. W.W. Price, E.M. Gulachenski, P. Kundur, F.J. Lange, G.C. Loehr, B.A. Roth, R.F. Silva, Testing of the modal dynamic equivalents technique. IEEE Trans. Power Apparatus Syst. **PAS-97** 1366–1372 (1978)
15. W.W. Price, B.A. Roth, Large-scale implementation of modal dynamic equivalents. IEEE Trans. Power Apparatus Syst. **PAS-100** 3811–3817 (1981)
16. R.A. Date, *Dynamic Model Reduction of Large Scale Systems: Applications to Power Systems*, MS Thesis, Rensselaer Polytechnic Institute, Troy, New York, May, 1989
17. J.R. Winkelman, J.H. Chow, B.C. Bowler, B. Avramović, P.V. Kokotović, An analysis of inter-area dynamics of multi-machine systems. IEEE Trans. Power Apparatus Syst. **PAS-100** 754–763 (1981)
18. R.J. Newell, M.D. Risan, L. Allen, K.S. Rao, D.L. Stuehm, Utility experience with coherency-based dynamic equivalents of very large systems. IEEE Trans. Power Apparatus Syst. **PAS-104** 3056–3063 (1985)
19. G. Troullinos, J. Dorsey, H. Wong, J. Myers, Reducing the order of very large power system models. IEEE Trans. Power Syst. **3**, 127–133 (1988)
20. R. Podmore, A. Germond, *Dynamic Equivalents for Transient Stability Studies*, Systems Control, Inc., Final Report prepared for EPRI Project RP-763, April 1977
21. J.H. Chow, R.J. Galarza, P. Accari, W.W. Price, Inertial and slow coherency aggregation algorithms for power system dynamic model reduction. IEEE Trans. Power Syst. **10**, 680–685 (1995)
22. J.B. Ward, Equivalent circuits for power-flow studies. AIEE Trans. **68**, 373–382 (1949)
23. H. You, V. Vittal, X. Wang, Slow coherency-based islanding. IEEE Trans. Power Syst. **19**, 483–491 (2004)
24. P.V. Kokotović, Subsystems, time-scales and multimodeling. Automatica **17**, 789–795 (1981)

Chapter 4
Excitation System Aggregation

Joe H. Chow and Ricardo J. Galarza

Abstract Constructing a dynamic equivalent for a power system involves several steps: the partition of the system into coherent areas, the aggregation of coherent generator buses, and the aggregation of the coherent generators and their control devices. These steps have been discussed in previous chapters, including a method in Chap. 2 to aggregate the exciter models using frequency response. In this chapter, we investigate a trajectory sensitivity method to tune the aggregate exciter parameters of the equivalenced model. The optimal parameters of the aggregated exciter yielding the least error are used to evaluate the sensitivity technique against the aggregation method in the DYNRED program and a weighted MVA-based method. A three-machine system with one coherent area satisfying the theoretical coherency conditions is used to investigate the impact of the variations of the individual generator, network, and exciter parameters on the aggregate exciter model parameters. The technique is then applied to the exciter aggregation of a larger NPCC 48-machine system.

4.1 Introduction

As illuminated in the previous chapters, the modern approach of dynamic equivalencing involves the following steps:

1. identification of the groups of coherent generators,
2. aggregation of the generator buses,

J. H. Chow (✉) · R. J. Galarza
Rensselaer Polytechnic Institute, Troy, New York, USA
e-mail: chowj@rpi.edu

R. J. Galarza
PSM Consulting, Inc., Guilderland, New York, USA
e-mail: rgalarza@psm-consulting.com

J. H. Chow (ed.), *Power System Coherency and Model Reduction*,
Power Electronics and Power Systems 94, DOI: 10.1007/978-1-4614-1803-0_4,

3. aggregation of the generator models and their associated control devices,
4. reduction of the load buses.

Steps 1, 2, and 4 have been discussed quite extensively in the previous chapters. In most practical applications of dynamic equivalencing (see Chap. 7), the equivalenced generators in the external coherent areas are modeled with classical or electromechanical models using an weighted MVA method to obtain an equivalent inertia and transient reactance. In some applications, such as a simulation of the power system lasting 10–15 s, it may be important to maintain some equivalent impact in the reduced models of the generator control devices, such as excitation systems and governors. Chapter 2 describes a method of using the frequency response of transfer functions to tune the parameters of the aggregate exciter, governor, and power system stabilizer models.

In this chapter, we develop a sensitivity method to determine optimal parameters for the aggregate exciter model [1]. The results are compared to the DYNRED results. In addition, a second goal of the investigation is to develop simple rules for computing aggregate exciter parameters that closely approximate the optimal parameters.

The remainder of the chapter is organized as follows. Section 4.2 describes the input–output model of a coherent area and the application of the sensitivity technique to identify the reduced model parameters. In Sect. 4.3, the concept of theoretical coherency is presented for a coherent area with detailed machine models. In Sect. 4.4 we present the exciter aggregation results of a three-machine system to test the sensitivity method. Section 4.5 contains the results of the exciter aggregation of the Northeast Power Coordinating Council (NPCC) 48-machine system.

4.2 Coherent Area Input–Output Model and Sensitivity Method

In aggregating the control devices, we assume that the groups of coherent machines are already known. The coherent groups that are not in the study area will constitute the external system. Based on the slow coherency concept that the coherent machines are strongly connected among themselves and weakly connected between the different coherent groups, we can decouple the detailed dynamic model aggregation of the coherent groups. That is, to a first approximation, the aggregation of a coherent group can be done independently of the other coherent groups.

To develop the reduced model of a coherent group, we define a slow-coherent area to consist of:

1. the coherent generators and their control devices,
2. the generator terminal buses,
3. the load buses strongly connected to the generator buses, and
4. all the transmission lines connecting the generator and load buses in the area.

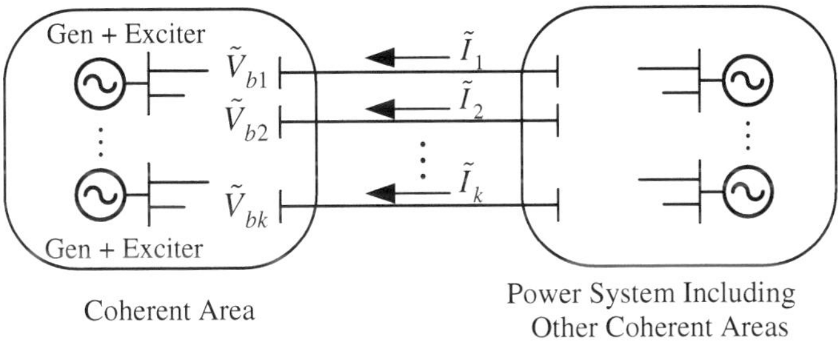

Fig. 4.1 Coherent area input–output model

Figure 4.1 shows a coherent area connected to the rest of the system. The buses in the coherent area with lines connected to the study area and the other coherent areas are called boundary buses. Their voltages are denoted by $\tilde{V}_{bj}, j = 1, 2, \ldots, k$. Within the coherent area, all the buses are interconnected, that is, there are no isolated buses or groups of buses.

A nonlinear input–output model for coherent area i can be expressed as

$$\dot{x}_i = f_i(x_i, V_i, I_i) \tag{4.1}$$

$$0 = g_i(x_i, V_i, I_i) \tag{4.2}$$

where the state vector x_i contains all the generator and exciter state variables in the area, the current injection vector I_i

$$I_i = [I_1\ I_2\ \cdots\ I_k]^T \tag{4.3}$$

is the input, and the boundary bus voltage vector V_b

$$V_b = [V_{b1}\ V_{b2}\ \cdots\ V_{bk}]^T \tag{4.4}$$

which is a subvector of the bus voltage vector V_i, is the model output. The nonlinear vector function f_i represents the dynamics of the generators and exciters present in the area i. Note that each vector V_{bi} or I_i consists of a real part and an imaginary part, the same notation used in Sect. 3.2. The network equations are represented by the vector of nonlinear functions g_i. Disturbances applied in the study area will affect the current injections I_i, which will perturb the coherent area from its equilibrium condition. The resulting boundary bus voltages will then affect the power flows between the coherent area and the rest of the system, which will lead to further variations in I_i.

The choice of I_i and V_{bi} as the input vector and the output vector, respectively, is based on modeling and simulation procedures commonly found in power system simulation programs [2]. At each integration step, the generator currents are computed and injected into the power network. Then the new voltage solution of the power network is obtained using these generator currents and injections due to the loads. Thus using the current injections I_i as inputs to the coherent area, the voltage phasors of all the buses within the coherent area can be computed. The boundary bus voltage phasors can be used as output variables to tune the parameters of the equivalent excitation system model.

The exciter aggregation problem is shown in Fig. 4.2 and is formulated as follows. First aggregate the detailed models of the generators of coherent area i into a single generator using a MVA-weighted-average algorithm (Chaps. 2 and 3). Then apply an aggregate exciter to the aggregate generator so that the reduced nonlinear model of the coherent area is modeled by

$$\dot{x}_{ri} = f_{ri}(x_{ri}, V_{ri}, I_i, \alpha_i) \tag{4.5}$$

$$0 = g_{ri}(x_{ri}, V_{ri}, I_i) \tag{4.6}$$

where the state vector x_{ri} has a smaller dimension than that of x_i, and the bus voltage vector V_{ri} could have a dimension different from that of V_i. However, the dimensions of the current injection vector and the boundary bus voltage vector remain unchanged. The exciter parameters are contained in the vector α_i. The objective of the exciter aggregation problem is to tune α_i so that the boundary bus voltages V_{rbi} of the reduced model (4.5), (4.6) match well V_{bi} with the full model (4.1), (4.2) for a selected set of disturbances in the study area.

The tuning of the exciter parameters of the reduced model to match the boundary bus voltages of the full model is a nonlinear optimization problem. Here we will utilize the trajectory sensitivity technique [3] to find the optimal exciter parameters. The procedure for applying the trajectory sensitivity technique is depicted in Fig. 4.3. A more detailed description and application of the technique can be found in [4, 5]. The trajectory sensitivity method has been applied to machine parameter estimation [6], exciter model parameter estimation [7, 8], reduced transformer models [9], and hybrid system simulation [10].

The parameter tuning procedure requires the following steps.

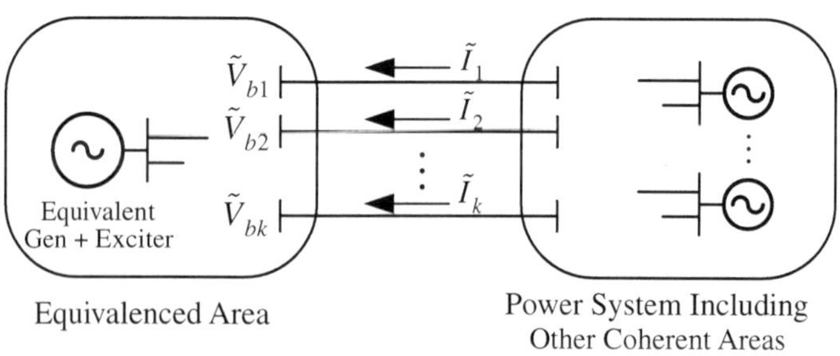

Fig. 4.2 Coherent area input–output model

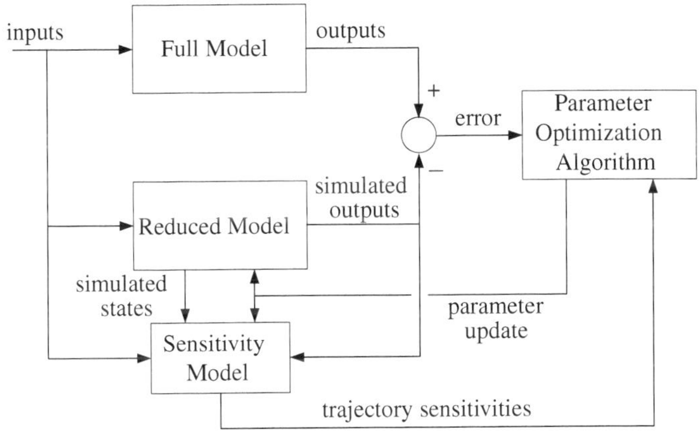

Fig. 4.3 Sensitivity-based reduced-order model parameter identification

Parameter tuning algorithm

1. Simulate the time response of the full model subject to a particular disturbance in the study area
2. Store the current injections I_i and the boundary bus voltages V_{bi}
3. Aggregate the coherent generators into an equivalent generator using a weighted MVA-weighted average of the generator parameters in the coherent area
4. Add to the equivalent generator an exciter model of a type similar to the exciters in the coherent area. Set the nominal exciter model data to be the same as the most dominant unit in the area. Typically this is the largest unit or the one with the most responsive exciter
5. Simulate the reduced model using I_i as the input to obtain V_{rbi}
6. Select a subset of the exciter model parameters α_i and use the trajectory sensitivity method to tune them to match V_{rbi} to V_{bi}

The algorithm for performing this procedure was coded using functions in the Power System Toolbox [2]. To use the program, a user has to provide the full model, the disturbances, and the coherent areas. The simulation of the equivalenced model (4.5) and (4.6) in Fig. 4.3 is accomplished by lifting the equivalenced model data from the complete data set and simulating its response to I_i modeled as time-varying constant current injections at the boundary buses. The procedure is tested using a small system as described in Sect. 4.4.

4.3 Theoretical Coherency

In general, a nonlinear model of a coherent area cannot be reduced exactly to an input-output model consisting of one equivalent generator. However, under certain ideal conditions, an exact reduced model can be achieved. The theoretical coherency concept in [11] denotes such a special case. If the coherent area satisfies the theoretical coherency conditions, a disturbance external to the area will not excite any relative motions between the machines in the coherent area. To an observer outside the coherent area, the motions of these machines are seen as if they were originated from one single machine. Thus, it presents an ideal case in which generators can be aggregated exactly in an input–output sense. This is an important idea because if we know that a coherent area satisfies approximately the theoretical coherency conditions, then the reduced model can provide a good approximation of the coherent area. In addition, the parameters of the equivalent model would be similar to those of the individual machines. Thus coherent areas close to satisfying theoretical coherency conditions provide good test cases for the sensitivity method of parameter tuning.

To show the theoretical coherency conditions, consider a coherent area consisting of two identical machines as shown in Fig. 4.4. This coherent area is connected to the rest of the power system through the boundary Bus 3, and the flow exchange with the external system is denoted by the current injection I_1. We let $Y_{13} = Y_{23}$ and the loading of the two generators be identical, that is, $V_1 = V_2$ and $I_{g1} = I_{g2}$. The admittance Y_{12} of the line connecting the generator terminal buses can be arbitrary.

The generators in Fig. 4.4 have the same MVA rating and generator and exciter models. The dynamic models are represented as

$$\dot{x}_j = f\left(x_j, V_j\right), \ I_{gj} = h\left(x_j, V_j\right), \quad j = 1, 2 \tag{4.7}$$

where x_j is the vector of the state variables for machine j, and the network equation is

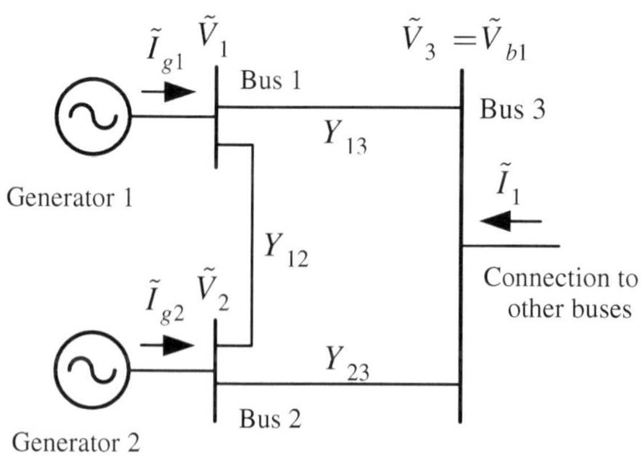

Fig. 4.4 An area illustrating theoretical coherency

$$\begin{bmatrix} Y_{11} & Y_{12} & Y_{13} \\ Y_{21} & Y_{22} & Y_{23} \\ Y_{31} & Y_{32} & Y_{33} \end{bmatrix} \begin{bmatrix} \tilde{V}_1 \\ \tilde{V}_2 \\ \tilde{V}_3 \end{bmatrix} = \begin{bmatrix} \tilde{I}_{g1} \\ \tilde{I}_{g2} \\ \tilde{I}_1 \end{bmatrix} \tag{4.8}$$

where $Y_{11} = Y_{22}$. In (4.8), we use the complex phasor form for the voltages and currents as the admittance terms Y_{ij} are complex. For any deviation of the injected current $\tilde{I}_1$ from its equilibrium value, the dynamic response of the system will satisfy

$$x_1(t) = x_2(t) = x(t), \ \tilde{V}_1(t) = \tilde{V}_2(t) = \tilde{V}(t), \ \tilde{I}_{g1}(t) = \tilde{I}_{g2}(t) \tag{4.9}$$

Thus the first and second network equations in (4.8) can be expressed as

$$(Y_{11} + Y_{12})\,\tilde{V} + Y_{13}\tilde{V}_3 = \tilde{I}_{g1} \tag{4.10}$$

Then (4.8) reduces to

$$\begin{bmatrix} 2\,(Y_{11} + Y_{12}) & 2Y_{13} \\ 2Y_{13} & Y_{33} \end{bmatrix} \begin{bmatrix} \tilde{V} \\ \tilde{V}_3 \end{bmatrix} = \begin{bmatrix} \tilde{I}_g \\ \tilde{I}_1 \end{bmatrix} \tag{4.11}$$

where $\tilde{I}_g = 2\tilde{I}_{g1}$. The machine dynamics are now governed by

$$\dot{x} = f\,(x, V)\,, \quad I_g = 2h\,(x, V) \tag{4.12}$$

Equation (4.12) represents a machine with the same model and parameters as Generators 1 and 2, except that it has double the MVA rating. The model (4.11), (4.12) forms an exact equivalent of the coherent area for any variation of the injected current I_1 caused by a disturbance. The theoretical coherency concept can be readily extended to coherent areas with more than 2 machines.

Summarizing, a set of sufficient conditions for theoretical coherency consists of:

1. identical machine and control device models, parameters, and power output levels,
2. identical admittances connecting the generator buses to each boundary bus.

There are no restrictions on the admittances between the generator buses and between the boundary buses. These conditions, however, are in general not satisfied in the coherent areas of a real power system.[1] As a result, it is necessary to study practical aggregation techniques for generators and control devices, to account for the variations of model and network parameters from the theoretical coherency conditions.

---

[1] A practical means of check theoretical coherency is that if any large disturbance external to a coherent area does not significantly excite the local modes within the coherent area, then the theoretical coherency conditions are approximately satisfied by the coherent area. Figure 3.1 illustrates this situation for the coherent areas in the 2-area, 4-machine system.

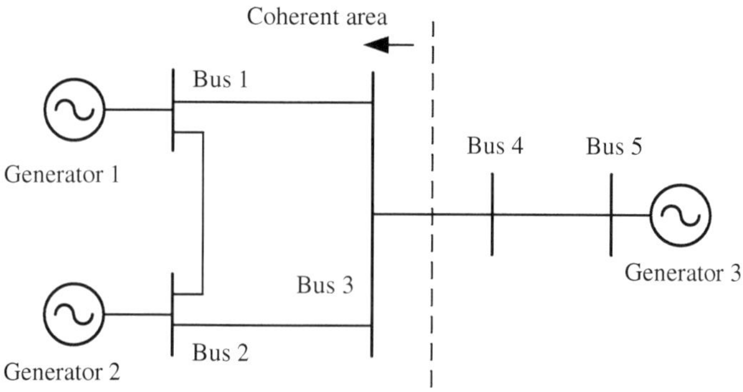

Fig. 4.5 A three-machine test system

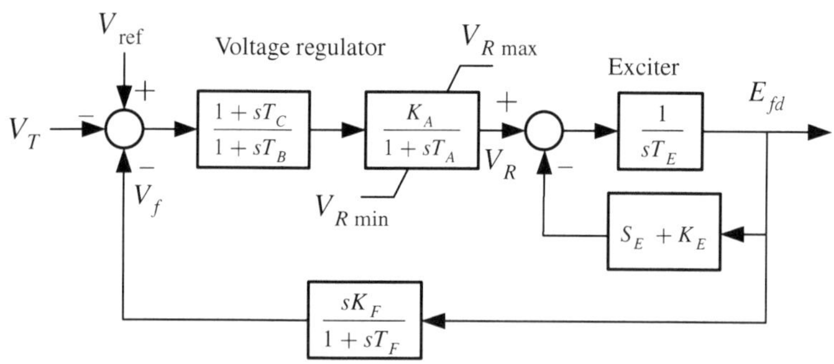

Fig. 4.6 Exciter model DC1A block diagram

4.4 A Three-Machine Test System

In this section, we investigate a small power system having a coherent area whose nominal models and parameters satisfy the theoretical coherency conditions. Then we perturb some of the parameters so that not all the theoretical coherency conditions are satisfied. For the perturbed model, we use the trajectory sensitivity method to find the optimal aggregate exciter parameters for the reduced model to produce a good approximation of the boundary bus voltage. The objective is to develop fundamental insights on exciter aggregation, which can then be applied to larger systems.

Figure 4.5 shows a three-machine system whose data can be found in the Appendix. Generators 1 and 2 and Buses 1, 2, and 3 form the coherent area, with Bus 3 being the boundary bus. For the nominal parameters, the coherent area satisfies the theoretical coherency conditions and Generators 1 and 2 can be aggregated exactly. Generator 3 is considered as the study area. The exciter models used for Machines 1 and 2 are Type DC1A [12], shown in Fig. 4.6, which is a common exciter model type in the stability data of the US power systems.

In a coherent area not satisfying the theoretical coherency conditions, the accuracy of the reduced model of the coherent area will depend strongly on many factors, including:

1. the machine sizes and power output levels,
2. the admittances of the lines connecting the machine terminal buses to the boundary buses,
3. the exciter models and parameters,
4. the type of disturbances.

Based on the above list, we introduce the following perturbations to the nominal parameters to study the impact of the variations on the tuning of aggregate exciter parameters for the reduced model:

1. Change the MVA bases of Machines 1 and 2 such that $MVA_2 = 2 \times MVA_1$. The power output levels are also changed accordingly.
2. Change the terminal voltage V_1 at Bus 1 from 1.05 to 1.03 pu.
3. Change the impedance and charging of the line connecting Buses 1 and 3 (Line 1–3) to 1/3 of its nominal values.
4. Change the exciter voltage regulator gain, K_A.

These perturbed parameters represent substantial violations of the theoretical coherency conditions.

For each perturbation, the coherent area is reduced to a single-machine system. The performance of the reduced-order model is evaluated using two different disturbances outside the coherent area:

1. a three-phase, three-cycle short-circuit fault on Bus 5 without line removal (denoted as *SC*),
2. a 0.5 pu increase in the reactive power loading on Bus 4 (denoted as Q).

The *SC* disturbance tests the transient performance of the aggregate exciter. In the Q disturbance the post-fault operating condition is different from the pre-fault operating condition. Thus the Q disturbance tests the steady-state regulation of the aggregate exciter. A summary of all the cases is given in Table 4.1.

For the given disturbances, the trajectory sensitivity technique was used to tune the exciter parameters to match the boundary bus voltage response of the reduced model to that of the full model. Our first conclusion was that the dominant parameter is the aggregate regulator gain K_A, which will be denoted by K_{Ar}. Once an optimal K_{Ar} was found, tuning the other exciter parameters resulted in only small improvement. Thus we will only present the results of tuning K_{Ar} in this chapter, with the other exciter parameters set to their nominal values.

Table 4.2 summarizes the results of parameter tuning for both the *SC* and Q disturbances. For comparison purposes, we also include the aggregate regulator gains K_{Ar} obtained using the DYNRED program.[2]

The main conclusions are listed as follows:

[2] DYNRED computes K_{Ar} using a weighted least-squares frequency response technique.

Table 4.1 Summary of cases

Case	Description
A	Theoretical coherency
A_1	$K_{A1} = 299$, $K_{A2} = 100$
B	$MVA_2 = 2 \times MVA_1$
B_1	B and A_1
C	$V_1 = 1.03$ pu
C_1	C and A_1
D	Line 1–3: 1/3 of nominal values
D_1	D and A_1
E	C and D
E_1	C, D, and A_1

Table 4.2 Tuning aggregate exciter voltage regulator gain

Case	Individual K_A		Optimal K_{Ar}		DYNRED
	K_{A1}	K_{A2}	SC	Q	Aggregate K_{Ar}
A	299	299	299	299	299
A_1	299	100	170	182	208
B	299	299	299	299	299
B_1	299	100	164	175	175
C	299	299	299	299	299
C_1	299	100	200	200	208
D	299	299	299	299	299
D_1	299	100	200	200	208
E	299	299	277	283	299
E_1	299	100	180	180	208

1. Case A is the theoretical coherency case. The reduced model response matches exactly the full model response when the nominal exciter parameters are used for the aggregate exciter model.
2. In Case A_1, the theoretical coherency conditions are no longer satisfied. The algorithm yields different optimal values of K_{Ar} for the two disturbances. This shows that the optimal values are disturbance dependent. The optimal K_{Ar}'s are close to the average of the individual K_A's.
3. In Cases B, C, and D when $K_{A1} = K_{A2}$, the optimal aggregate K_{Ar} achieved by the algorithm has the same value.
4. In Case B_1 when the individual exciter K_A's are different, the optimal aggregate K_{Ar}'s obtained can be approximated by using a MVA-base weighted average

$$K_{Ar} = (1 \times K_{A1} + 2 \times K_{A2})/3 = 166$$

This result shows the effect of machine ratings on the aggregate K_{Ar}.

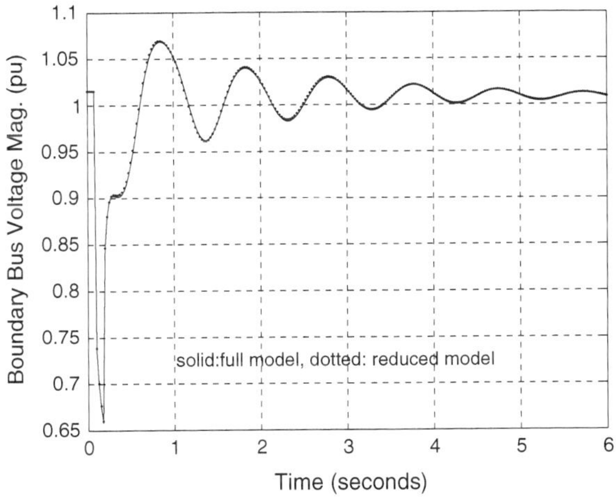

Fig. 4.7 Case E_1, full versus reduced model—optimal K_{Ar}

5. The optimal aggregate K_{Ar}'s achieved in Cases C_1 and D_1 are only slightly higher than those obtained in Case A_1, indicating that the terminal voltage and line impedance variations do not significantly impact the aggregate K_{Ar}.
6. Cases E and E_1 are combinations of Cases C and D. However, the optimal aggregate K_{Ar}'s have lower values than those achieved in Cases C and D. This illustrates that the optimal aggregate K_{Ar} depends nonlinearly on the deviation from the theoretical coherency conditions.
7. The aggregate K_{Ar}'s obtained from DYNRED are usually within 10% of the optimal values obtained from the trajectory sensitivity method. Thus these values can be used instead of the optimal values.

We now illustrate the approximation achieved by the reduced models for Case E_1 using nonlinear time simulation of the *SC* disturbance. We show the results for the reduced model with three different sets of exciter model parameters:

1. optimal K_{Ar} with the other exciter parameters at their nominal values,
2. all aggregate exciter parameters from DYNRED,
3. K_{Ar} from DYNRED with the other exciter parameters at their nominal values.

In the DYNRED model, the time constants of the exciter parameters are obtained from a least-squares frequency-response method and are in general different from the nominal values.

The comparisons of the time responses of the full and reduced models are shown in Figs. 4.7, 4.8, and 4.9. The reduced model with the optimal K_{Ar} is almost an exact match of the full model, and the reduced model with only K_{Ar} from DYNRED also shows a good match. The reduced model with all exciter parameters from DYNRED is somewhat less accurate.

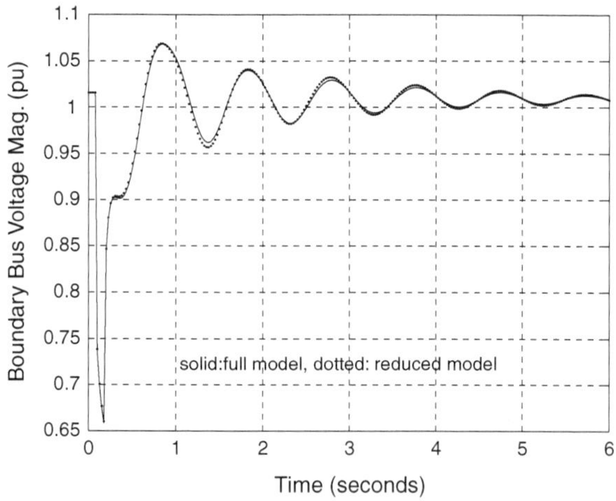

Fig. 4.8 Case E_1, full versus reduced model—K_{Ar} from DYNRED

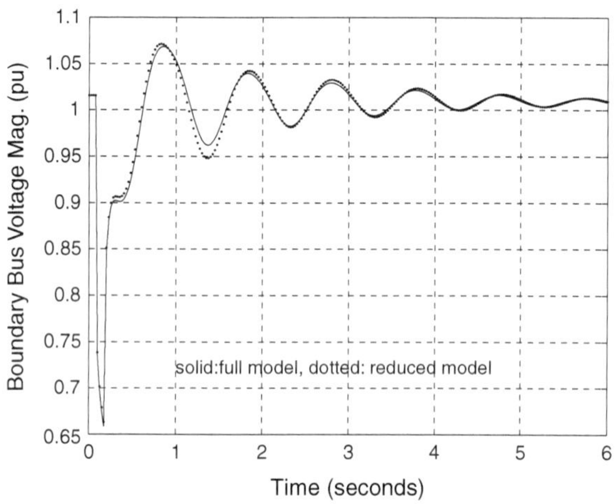

Fig. 4.9 Case E_1, full versus reduced model—DYNRED parameters

Only the type DC1A exciter is investigated here. The other two common exciter types, AC (alternator supplied rectifier excitation system) and ST (static excitation system), contain nonlinearities arising from the rectifier circuits, limits, and saturations. These exciters would exhibit substantial nonlinear transients if they are equipped on machines close to a disturbance. However, exciter aggregation should only be performed for coherent machines away from the disturbance. As a result, these nonlinearities may not be significant, and K_A may still be the most important parameter. Further work is required to investigate the aggregation of the AC and ST exciter types.

Table 4.3 Coherent machine groups and models

Area	Machine number	Dynamic models		
		DM	CM	EXC
1	3,4,5,6,7,8	6	0	6
2	1,2,9	3	0	3
3	10	1	0	1
4	11,12	2	0	2
5	13,14,24,25,26	2	3	2
6	15,16,17,18,19,20,21,22,23	7	2	7
7	27,28,29,30	3	1	3
8	31	0	1	0
9	32,37,38,40,42	1	4	0
10	33	0	1	0
11	34,35	0	2	0
12	36	0	1	0
13	39	1	0	0
14	41	0	1	0
15	43,44,45,46	0	4	0
16	47	0	1	0
17	48	0	1	0

4.5 NPCC 48-Machine System

The exciter aggregation results from the three-machine system are applied to a larger NPCC 48-machine system used in Chap. 3 [13].

Following the dynamic equivalencing procedure, the 48-machine system was partitioned into 17 coherent areas using the tolerance-based coherency algorithm in Chap. 3 with the 9 slowest modes and a tolerance of 0.95. The machines in each area are listed in Table 3.3. Here we expanded the Table (4.3) to show the dynamic models used: CM: classical machine model; DM: detailed machine model; and EXC: IEEE DC1A exciter model. For example, Area 5 has 5 machines, of which two are represented by detailed models and the other three by classical models. All the machines with detailed models are equipped with exciters. A map with the geographical location of the machines and coherent areas is shown in Fig. 3.7.

Areas 1 and 2, which correspond to the New England power system, are taken to be the study area. The external system consists of all the other areas. In the study area, all 9 machines are kept in full detail. In the external system, both machines in Area 4 has exciter models. Areas 5, 6, and 7 consist of both detailed machine models with exciters as well as machines with classical models. DYNRED will aggregate the classical models separately from the detailed machine models. We follow this idea and subdivide Areas 5, 6, and 7 into two areas. The subareas with detailed machine and exciter models are denoted as 5*e*, 6*e*, and 7*e*, respectively. To focus on the exciter aggregation, we only aggregate the external areas containing exciters. In addition, we test the aggregation by constructing dynamic equivalence with only one aggregated area. We also test a reduced model with all four areas (Areas 4, 5e, 6e, and 7e) aggregated. The resulting reduced 31-machine system (denoted as the R31 model) consists of

1. the study area: 9 machines (Areas 1 and 2) in full detail
2. the external system:
   - 4 aggregate machines with exciter models,
   - 1 machine in Area 10 with an exciter model,
   - 17 machines with classical models.

The network is aggregated using the inertial aggregation method (see Sect. 3.5).

Although we can use the trajectory sensitivity method to tune the aggregate exciter parameters individually for Areas 4, 5e, 6e, and 7e, we make use of the MVA weighted average results from the three-machine system. Let K_{Aj} be the regulator gain of the exciter for machine j, and n_g be the number of generators with exciter models in the coherent area. The exciter parameter K_{Ar} for the aggregated machine is computed as follows

$$K_{Ar} = \frac{\sum_{j=1}^{n_g} \text{MVA}_j \times H_j \times K_{Aj}}{\sum_{j=1}^{n_g} \text{MVA}_j \times H_j} \tag{4.13}$$

where MVA_j and H_j are the machine j MVA base and inertia constant, respectively. The machine inertia is included in (4.13) because in the system data, the MVA base of the machines in the external system has all been set to 100. The ratio of MVA bases between the machines can be recovered approximately by multiplying the MVA base of 100 with H_j.

For each of the equivalenced systems, we evaluate three different means of obtaining the aggregated exciter parameters:

1. K_{Ar} from (4.13), and the other aggregate exciter parameters from the exciter of the most dominant machine in the area.
2. K_{Ar} from DYNRED, and the other aggregate exciter parameters from the exciter of the most dominant machine in the area.
3. All aggregated exciter parameters from DYNRED.

These models are evaluated by applying a 3-phase, 6-cycle short-circuit fault on Bus 7 (Medway) in the study area. The fault is cleared by removing the line from Bus 7 to Bus 6 (Medway to Sherman Road). As a measure of the accuracy of the reduced models, we compute the error function $J_a(k)$

$$J_a(k) = \frac{1}{T}\int_0^T \mid \delta_{rk} - \delta_k \mid \, \mathrm{d}t \tag{4.14}$$

for machine k in a time period from 0 to T, where δ_{rk} and δ_k are the rotor angles of machine k in the reduced and full model, respectively.

Table 4.4 Error functions for machine 1—different aggregated exciter parameters

Area aggregated	Error $J_a(1)$		
	K_{Ar} from (4.13)	K_{Ar} from DYNRED	DYNRED
4	0.0217	0.0217	1.291
5*e*	0.2702	0.2700	0.2718
6*e*	0.1128	0.1128	0.1182
7*e*	0.0321	0.0319	0.0324
R31	2.012	2.013	3.046

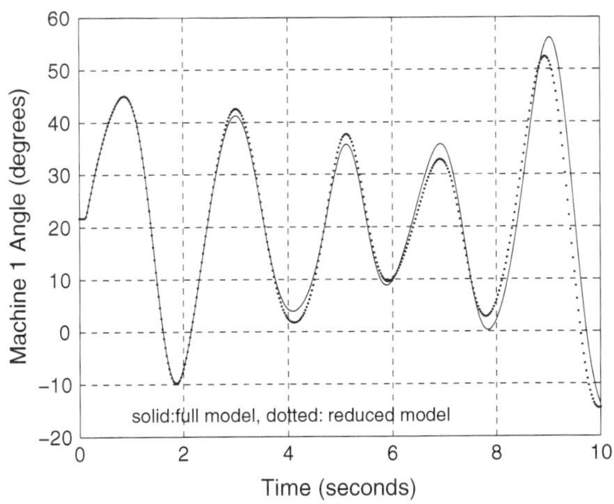

Fig. 4.10 Time response of machine 1, full versus reduced 31-machine system (MVA weighted)

Table 4.4 shows the error functions of Machine 1 for the disturbance simulation. All three methods yield comparable results when only Areas 5*e*, 6*e*, and 7*e* are aggregated one at a time. However, for Area 4, even though the two machines have identical generator and exciter models and parameters, DYNRED yields a set of aggregate exciter time constants that are different from those of the individual exciter time constants. This area is located next to the study area, and hence has a large influence on the study area. Thus the all-DYNRED-parameter model shows a much larger error $J_a(1)$ than the other two schemes. The last row in Table 4.4 shows the performance of the reduced 31-machine system R31. The error from the all-DYNRED model is again higher because of Area 4.

Figures 4.10 and 4.11 are plots of the Machine 1 response of the full model versus the R31 model using the MVA-weighted K_{Ar} and the all-DYNRED exciter parameters, respectively. The MVA-weighted K_{Ar} model shows a better approximation compared to the DYNRED model. The simulation for the reduced model using K_{Ar} from DYNRED is very similar to that of MVA-weighted K_{Ar} and hence is not shown here. We also found that tuning the aggregate exciter parameter K_{Ar} using the trajectory sensitivity method did not improve the performance of the reduced model very much. Hence the value of K_{Ar} used here is close to the optimal.

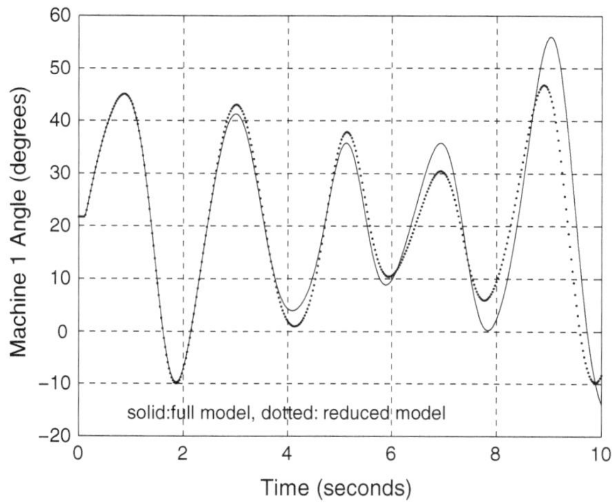

Fig. 4.11 Time response of machine 1, full versus reduced 31-machine system (DYNRED parameters)

A more detailed analysis of the aggregation results of the 48-machine systems can be found in [5].

4.6 Conclusions

In this chapter, we have reported on an investigation of the aggregation of exciters for use in constructing power system dynamic equivalents. Starting from a small three-machine system under theoretical coherency conditions, several perturbations have been applied to investigate the effect of these variations on the aggregate exciter parameters. For the type DC1A exciter model, we have concluded that the most important parameter is the regulator gain K_A. Furthermore, we show that an MVA-base weighted K_{Ar} is a close approximate of the optimal K_{Ar}. A similar value of K_{Ar} is computed by DYNRED. Thus we recommend using either the DYNRED or the MVA-base weighted K_{Ar}, with the time constants of the aggregate exciter set to those from the exciter of the most dominant unit in the coherent area.

The sensitivity method used in this chapter serves as a tool for studying the impact of the different exciter parameters and as a benchmark for evaluating other methods. The results here show that the much simpler weighted MVA base method provides close to optimal K_{Ar}. Thus it may not be always necessary to use the sensitivity method to obtain the optimal aggregate K_{Ar} if computation time is of concern.

Only the type DC1A exciter is investigated in this chapter. Further work is required to investigate the optimal aggregation of other types of exciters as well as turbine-governor models.

Appendix: Three-machine system data

The power flow data of the 3-machine system in Fig. 4.5 are given in Tables 4.5 and 4.6, on 100 MVA base.

Table 4.5 Three-machine system bus data

Bus #	Voltage		Generation		Load	
	Magnitude pu	Phase degree	Active pu	Reactive pu	Active pu	Reactive pu
1	1.0500	22.51	0.0	0.0	4.0	1.0073
2	1.0500	22.51	0.0	0.0	4.0	1.0073
3	1.0055	10.98	6.0	1.2	0	0
4	1.0186	5.39	0.0	0.0	0	0
5	1.0300	0.00	2.0	0.4	0.2125	0.7873

Table 4.6 Three-machine system line data

Line		R, pu	X, pu	B, pu	Tap
From bus	To bus				
1	2	0.0000	0.0330	0.0	
1	3	0.0056	0.0536	0.2	1.014
2	3	0.0056	0.0536	0.2	1.014
3	4	0.0056	0.0536	0.2	
4	5	0.0056	0.0536	0.2	

The parameters of the generators are: base = 450 MVA, $x_d = 1.87$ pu, $x_q = 1.77$ pu, $x_l = 0.21$ pu, $x'_d = 0.365$ pu, $x'_q = 0.61$ pu, $x''_d = 0.28$ pu, $x''_q = 0.28$ pu, $T'_{do} = 7.3$ s, $T'_{qo} = 0.73$ s, $T''_{do} = 0.032$ s, $T''_{qo} = 0.056$ s, $H = 4.74$ pu, $S(1.0) = 0.072$, and $S(1.2) = 0.282$.

The parameters of the IEEE type DC1A exciter model are: $K_A = 299$, $T_A = 0.02$ s, $K_E = 1$, $T_E = 0.785$ s, $E_1 = 2.87$ pu, $S_E(E_1) = 0.67$, $E_2 = 3.82$ pu, $S_E(E_2) = 0.91$, $K_F = 0.015$, $T_F = 1$ s, $T_B = 0$ s, $T_C = 0$ s, $V_{\mathrm{Rmax}} = 7.3$ pu, and $V_{\mathrm{Rmin}} = -7.3$ pu.

References

1. R.J. Galarza, J.H. Chow, W.W. Price, A.W. Hargrave, P.M. Hirsch, Aggregation of exciter models for constructing power system dynamic equivalents. IEEE Trans. Power Syst. **13**(3), 782–788 (1998)
2. J.H. Chow, K.W. Cheung, A toolbox for power system dynamics and control engineering education and research. IEEE Trans. Power Syst. **7**, 1559–1564 (1992)
3. P.M. Frank, *An Introduction to Sensitivity Theory* (Academic Press, New York, 1978)

4. J.H. Chow, S.H. Javid, J.J. Sanchez-Gasca, C.E.J. Bowler, J.S. Edmonds, Torsional model identification for Turbine-generators. IEEE Trans. Energy Convers. **EC–1**, 83–90 (1986)
5. R.J. Galarza, Power System Dynamic Equivalencing: Advanced Analysis and Improvements, Ph.D. Thesis, Rensselaer Polytechnic Institute, Aug 1996.
6. J.H. Chow, M.T. Glinkowski, R.J. Murphy, T.W. Cease, N. Kosaka, Generator and exciter parameter estimation of Fort Patrick Henry Hydro Unit 1. IEEE Trans. Energy Convers. **14**, 923–929 (1999)
7. S.M. Benchluch, J.H. Chow, A trajectory sensitivity method for the identification of nonlinear excitation system model. IEEE Trans. Energy Convers. **8**, 150–164 (1993)
8. J.W. Feltes, S. Orero, B. Fardanesh, E. Uzunovic, S. Zelingher, N. Abi-Samra, Deriving model parameters from field test measurements. IEEE Comput. Appl. Power **15**(4), 30–36 (2002)
9. R.J. Galarza, J.H. Chow, R.C. Degeneff, Transformer model reduction using time and frequency domain sensitivity techniques. IEEE Trans. Power Delivery **10**, 1052–1059 (1995)
10. I. Hiskens, M.A. Pai, Trajectory sensitivity analysis of hybrid systems. IEEE Trans. Circuits Syst. I Fundam. Theory Appl. **47**(2), 204–220 (2000)
11. U. DiCaprio, Conditions for theoretical coherency in multi-machine power systems. Automatica **17**, 687–701 (1981)
12. D.C. Lee, IEEE recommended practice for excitation system models for power system stability studies. IEEE Standards Std 421.5, March 19, 1992.
13. W.W. Price, G.E. Boukarim, J.H. Chow, R.J. Galarza, A.W. Hargrave, B.J. Hurysz, R. Tapia, Improved Dynamic Equivalencing Software, Report No. RP2447-02, Electric Power Research Institute, 1996.

Chapter 5
A Hybrid Dynamic Equivalent Using ANN-Based Boundary Matching Technique

Vijay Vittal and Feng Ma

Abstract In this chapter, a hybrid dynamic equivalent consisting of both a coherency-based conventional equivalent and an artificial neural network (ANN)-based equivalent is developed and analyzed. The ANN-based equivalent complements the coherency-based equivalent at all the boundary buses of the retained area. It is designed to compensate for the discrepancy between the full system model and the reduced equivalent developed using any commercial software package, such as the dynamic reduction program (DYNRED), by providing appropriate power injections at all the boundary buses. These injections are provided by the ANN-based equivalent which is trained using the outputs from a trajectory sensitivity simulation of the system responses to a candidate set of disturbances. The proposed approach is tested on a system representing a portion of the Western Electricity Coordinating Council (WECC) system. The case study shows that the hybrid dynamic equivalent can enhance the accuracy of the coherency-based dynamic equivalent without significantly increasing the computational effort.

5.1 Introduction

With the evolution of heavily interconnected power systems, it is computationally burdensome to represent the entire system in detail to conduct numerous stability studies [1]. This is especially true for real-time power system transient stability assessment as there is a strict limitation on the size of the system that can be simulated. Therefore, it is necessary to build a reduced order model which can preserve

V. Vittal (✉)
Arizona State University, Arizona, USA
e-mail: vijay.vittal@asu.edu

F. Ma
ISO New England Inc., Holyoke, USA
e-mail: fma@iso-ne.com

J. H. Chow (ed.), *Power System Coherency and Model Reduction*, Power Electronics and Power Systems 94, DOI: 10.1007/978-1-4614-1803-0_5,

the dynamic properties of the specific subsystem that is of interest. To account for this challenge, power system dynamic equivalents have received renewed attentions recently and have been widely applied for the purpose of reducing the computational effort of dynamic security assessment.

Dynamic equivalents are commonly developed using a coherency-based approach [2–8] in which a retained area and an external area are demarcated first. Then the coherency patterns of the generators in the external area are evaluated. A commonly used method is to introduce faults on the retained area boundary and to group the generators with similar dynamic responses in the external area. The other methods, such as weak-link method [2] and slow coherency-based method [3], have also been proposed and implemented. As a result of generator coherency identification, the coherent generators in the external area are aggregated. Network reduction is then performed at the interface between the retained area and the external area to suitably interconnect the equivalent generators.

When building a dynamic equivalent, the retained area boundary definition can significantly affect the equivalencing accuracy of the reduced system model. As more components are included in the retained area, more attributes related to the dynamic characteristics of the retained area can be preserved. In conventional dynamic equivalent applications, however, the retained area and external area are arbitrarily determined without thoroughly examining the system dynamic behaviors. An improperly defined retained area boundary can have a detrimental impact on the effectiveness of the equivalenced model in preserving dynamic characteristics of the original unreduced system. Under realistic situations, generator coherency information obtained under one particular operating condition might not be applicable to another system operating condition. When operating condition changes, the errors resulting from generator aggregation cannot be completely eliminated. This is especially true for the classical generator aggregation method [4] because the detailed representation of the generators, exciters, and governors are neglected in forming the equivalent generator models. Furthermore, the accuracy of the equivalent model of the external area might be even more questionable as detailed information about the network and component models within the external system is often inaccessible to a single entity in the deregulated electric utility environment.

Recently, artificial neural network (ANN) technique has been successfully applied to the subject of dynamic equivalents due to its superior capability of capturing arbitrary input-and-output mappings from training samples. The ANN-based equivalent, represented by a set of neural networks, is a "black box" in nature. It only needs the measurements at the retained area boundary buses instead of the detailed model information in the external area. In [9–11], different types of neural networks, such as bottleneck network and recurrent network, have been studied. In an attempt to effectively capture the dynamic characteristics of the external area, these efforts demonstrate the need for a complicated neural network structure to achieve reasonable equivalencing accuracy.

In order to overcome the disadvantage of a complex neural network structure, a hybrid dynamic equivalent model that combines the advantages of both the coherency-based and ANN-based equivalent methods has been proposed in this chapter. The data for neural network training are generated using a trajectory sensitivity-based approach [12, 13]. In the proposed hybrid model, the ANN-based equivalent works in conjunction with the coherency-based equivalent formed using the dynamic reduction package (DYNRED) [14] at the retained area boundary buses. By providing the desirable supplementary compensation in terms of power injections at the boundary buses, the ANN-based equivalent compensates for the discrepancy between the full system and the conventionally reduced system responses. The test on a portion of the Western Electricity Coordinating Council (WECC) system shows that the hybrid dynamic equivalent method can improve the accuracy of the coherency-based equivalent for both trained and untrained cases. The proposed approach also shows great potential to improve the equivalent models for online dynamic security assessment (DSA) using measurements collected by the synchronized phasor measurements units (PMUs) at the boundary buses.

5.2 Coherency-Based Dynamic Equivalent

As shown in Fig. 5.1, the study area, buffer area, and external area are defined prior to building a dynamic equivalent:

- *Study area*: the core subsystem that is of direct interest and must be retained in detail. For the reduced system to effectively represent the full system behavior, the

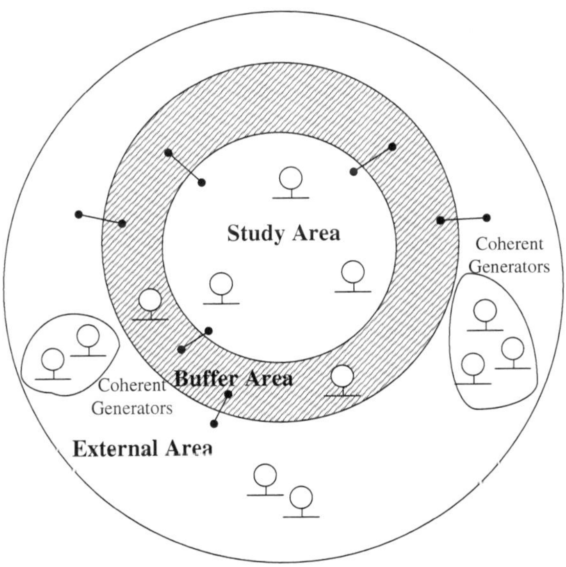

Fig. 5.1 Definitions of study area, buffer area, and external area

power flow and dynamic responses within this area are supposed to be the same as those in the full system representation.

- *Buffer Area*: the area that is geographically or electrically close to the retained area. It has significant impact on study area due to its tight electrical coupling with the study area. For better equivalencing accuracy, this area along with its component models is retained in detail.
- *External area*: this is the portion of the system for which detailed information on the system responses is not required and where the models can be replaced by proper equivalent models.

In practice, all the models in the study area and buffer area are retained; therefore, both areas together are denoted as the retained area. After dynamic equivalencing process, the reduced system consisting of both the retained area and equivalent model of the external area can be formulated. In the literature, the methods based on generator coherency have been extensively studied. The basic idea is to aggregate the generators in the external area that present similar dynamic characteristics. The extent of the similarity is measured by generator coherency. For example, if two generators present similar rotor angle responses following a system disturbance, they are considered tightly coherent. Otherwise, they are considered weakly coherent. To evaluate generator coherency, the most intuitive approach is to compare the generator responses following certain system disturbances. Based on this idea, the authors in [5] proposed a linear simulation method for coherency evaluation. For example, two generators are considered tightly coherent when the maximum deviation of their rotor angle responses subjected to a specific disturbance is smaller than a predefined threshold value. Although classical generator models are used for coherency evaluation, the simulation-based method is still time-consuming when different system disturbances need to be investigated on a large-scale inter-connected power system. The weak-link method was first introduced in [2]. Unlike the linear simulation method, it measures the coupling of generators directly based on the system state matrix. A group of generators are considered coherent if the coupling coefficients among them are high. In [3], the slow coherency technique based on the singular perturbation theory was implemented to separate slow and fast dynamics in large power systems and to identify the coherent generators from the perspective of the slow dynamic process. As a simplified realization of the slow coherency method, a coherency index is developed based on the similarity of the mode shapes associated with a set of specific slow oscillation modes within the system [4].

After identifying the coherent generators that belong to a group, the parameters of an equivalent generator model are aggregated from individual coherent generators. The frequency-domain method [6] and the structure preserving method [7] have been proposed in the literature. When forming the equivalenced generators with detailed representation, however, these methods might lead to problematic parameters in certain cases. Therefore, the classical aggregation method [4] is widely implemented in practice. By this method, the equivalenced generator is formed using a classical representation. Its inertia is the sum of individual generator inertias, and its transient reactance is the parallel combination of individual transient reactances.

The aggregated damping coefficient is computed based on the user-defined ratio of aggregated damping to aggregated inertia.

As the final step in building an equivalent system, the electrical network in the external area is reduced. During this process, the buses in the external area are eliminated using the Gaussian elimination. In the meantime, detailed load models are aggregated and replaced by appropriate equivalent load models.

5.3 Hybrid Dynamic Equivalent

In the coherency-based dynamic equivalent, the coherent generators in the external area are aggregated and replaced by appropriate classical equivalent generator models. Because of the errors accumulated in the generator aggregation process, a large retained area is commonly defined to achieve reasonable equivalencing accuracy. However, this improvement might become insignificant when the retained area itself is already large enough [15, 16]. This limitation becomes more significant as the detailed information about the network and component models within the external system is often inaccessible to a single entity in the deregulated electric utility environment. To address this challenge, a novel hybrid dynamic equivalent, comprised of both a coherency-based equivalent and an ANN-based equivalent, is developed. The evolution of the proposed hybrid dynamic equivalent is shown in Fig. 5.2.

Using the voltages at the boundary buses as inputs, the ANN-based equivalent dynamically adjusts its power injections to minimize the voltage response mismatches at all the boundary buses between the unreduced system and the conventionally reduced system. With a properly trained neural network, the hybrid reduced system is expected to more accurately represent the dynamic characteristics of the retained area. It is to be noted that in the proposed hybrid equivalent, the coherency-based (conventional) equivalent still plays a dominant role, and the ANN-based equivalent is merely used to capture additional dynamic characteristics that have not been captured in the conventionally reduced system. The hybrid equivalent connection configuration allows great flexibility for practical implementation. For example, the hybrid reduced system can convert back to the conventionally reduced system with the ANN-based equivalent disconnected.

5.3.1 Training Data for ANN-Based Equivalent

As shown in Fig. 5.2, the ANN-based equivalent complements the conventional equivalent developed using DYNRED. At each time instant, the training samples containing the connection bus voltages together with the power injections from the ANN-based equivalent need to be obtained. In the literature, the trajectory sensitivity method [12, 13] provides an effective approach to estimate system variable deviations with respect to single or multiple system parameter changes. Therefore this method

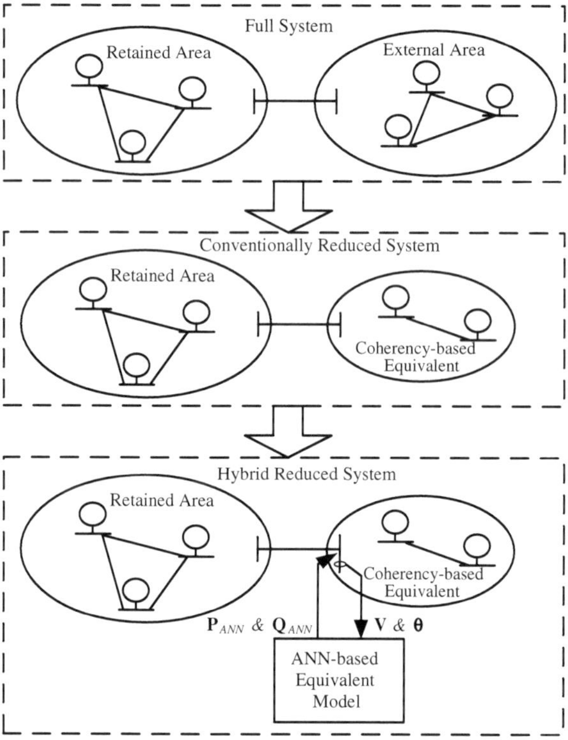

Fig. 5.2 Evolution of the hybrid dynamic equivalent

is utilized to determine the gradient for computing the desirable power injections of the ANN-based equivalent that matches the boundary voltage responses in the reduced systems to the unreduced system.

In general, a power system can be represented by a set of differential algebraic equations (DAEs) that is given by:

$$\begin{aligned} \dot{\mathbf{x}} &= \mathbf{f}(\mathbf{x}, \mathbf{y}, \beta) \\ \mathbf{0} &= \mathbf{g}(\mathbf{x}, \mathbf{y}, \beta) \end{aligned} \tag{5.1}$$

where $\mathbf{x}$ is the state variable vector, $\mathbf{y}$ is the algebraic variable vector, β is a system parameter subject to change, and β_0 is the initial value of β. The vector of differential equations $\mathbf{f}$ represents the dynamics of system components, and the vector of algebraic equations $\mathbf{g}$ represents the power flow balance equation at each bus in the system.

To numerically solve the DAEs in (5.1) in a time-domain simulation, an implicit method like the trapezoidal rule is applied. At the time instant n, the following equations need to be solved to determine $\mathbf{x}_n$ and $\mathbf{y}_n$:

$$\begin{aligned}\mathbf{F}(\mathbf{x}_n,\mathbf{y}_n) &= \mathbf{x}_n - \mathbf{x}_{n-1} - \tfrac{1}{2}\Delta t\left[\mathbf{f}(\mathbf{x}_n,\mathbf{y}_n,\beta) + \mathbf{f}(\mathbf{x}_{n-1},\mathbf{y}_{n-1},\beta)\right] = \mathbf{0}\\ \mathbf{G}(\mathbf{x}_n,\mathbf{y}_n) &= \mathbf{g}(\mathbf{x}_n,\mathbf{y}_n,\beta) = \mathbf{0}\end{aligned} \tag{5.2}$$

where Δt is the integration time step.

To solve the set of nonlinear equations in (5.2), the Newton-Raphson (NR) iterative algorithm is implemented. At the kth iteration, the following equation can be formed:

$$\begin{bmatrix}\mathbf{F}_{\mathbf{x}}^k & \mathbf{F}_{\mathbf{y}}^k\\ \mathbf{G}_{\mathbf{x}}^k & \mathbf{G}_{\mathbf{y}}^k\end{bmatrix}\begin{bmatrix}\Delta\mathbf{x}_n^k\\ \Delta\mathbf{y}_n^k\end{bmatrix} = \begin{bmatrix}-\mathbf{F}(\mathbf{x}_n^k,\mathbf{y}_n^k)\\ -\mathbf{G}(\mathbf{x}_n^k,\mathbf{y}_n^k)\end{bmatrix} \tag{5.3}$$

where $\mathbf{F}_{\mathbf{x}}^k = \partial\mathbf{F}/\partial\mathbf{x} = \mathbf{I} - 0.5\Delta t\,\partial\mathbf{f}/\partial\mathbf{x}$, $\mathbf{F}_{\mathbf{y}}^k = \partial\mathbf{F}/\partial\mathbf{y} = -0.5\Delta t\,\partial\mathbf{f}/\partial\mathbf{y}$, $\mathbf{G}_{\mathbf{x}}^k = \partial\mathbf{G}/\partial\mathbf{x} = \partial\mathbf{g}/\partial\mathbf{x}$, and $\mathbf{G}_{\mathbf{y}}^k = \partial\mathbf{G}/\partial\mathbf{y} = \partial\mathbf{g}/\partial\mathbf{y}$ are the partial derivative matrices formulated at $\mathbf{x} = \mathbf{x}_n^k$ and $\mathbf{y} = \mathbf{y}_n^k$. $\mathbf{I}$ is an identity matrix.

Then the system variables are updated to $\mathbf{x}_n^{k+1} = \mathbf{x}_n^k + \Delta\mathbf{x}_n^k$ and $\mathbf{y}_n^{k+1} = \mathbf{y}_n^k + \Delta\mathbf{y}_n^k$, respectively. The entire iterative procedure terminates when the increment is smaller than a specified tolerance.

For β that is sufficiently close to β_0, the perturbed trajectory solution can be determined by taking the derivatives of the system variables with respect to β. Accordingly, a new set of linear DAEs can be formed:

$$\begin{aligned}\dot{\mathbf{x}}_\beta &= \mathbf{f}_{\mathbf{x}}\mathbf{x}_\beta + \mathbf{f}_{\mathbf{y}}\mathbf{y}_\beta + \mathbf{f}_\beta\\ \mathbf{0} &= \mathbf{g}_{\mathbf{x}}\mathbf{x}_\beta + \mathbf{g}_{\mathbf{y}}\mathbf{y}_\beta + \mathbf{g}_\beta\end{aligned} \tag{5.4}$$

where $\mathbf{x}_\beta = \partial\mathbf{x}/\partial\beta$ and $\mathbf{y}_\beta = \partial\mathbf{y}/\partial\beta$ are the trajectory sensitivities of $\mathbf{x}$ and $\mathbf{y}$ at $\beta = \beta_0$, respectively. $\mathbf{f}_{\mathbf{x}} = \partial\mathbf{f}/\partial\mathbf{x}$, $\mathbf{f}_{\mathbf{y}} = \partial\mathbf{f}/\partial\mathbf{y}$, $\mathbf{f}_\beta = \partial\mathbf{f}/\partial\beta$, $\mathbf{g}_{\mathbf{x}} = \partial\mathbf{g}/\partial\mathbf{x}$, $\mathbf{g}_{\mathbf{y}} = \partial\mathbf{g}/\partial\mathbf{y}$, and $\mathbf{g}_\beta = \partial\mathbf{g}/\partial\beta$.

Similar to (5.3), the trapezoidal rule is also applied to solve (5.4). At the time instant n, the following linear equation holds:

$$\begin{aligned}\mathbf{x}_{\beta,n} - \mathbf{x}_{\beta,n-1} - \frac{\Delta t}{2}\begin{bmatrix}\mathbf{f}_{\mathbf{x},n}\mathbf{x}_{\beta,n} + \mathbf{f}_{\mathbf{y},n}\mathbf{y}_{\beta,n} + \mathbf{f}_{\beta,n}\\ +\mathbf{f}_{\mathbf{x},n-1}\mathbf{x}_{\beta,n-1} + \mathbf{f}_{\mathbf{y},n-1}\mathbf{y}_{\beta,n-1} + \mathbf{f}_{\beta,n-1}\end{bmatrix} &= \mathbf{0}\\ \mathbf{g}_{\mathbf{x},n}\mathbf{x}_{\beta,n} + \mathbf{g}_{\mathbf{y},n}\mathbf{y}_{\beta,n} + \mathbf{g}_{\beta,n} &= \mathbf{0}\end{aligned} \tag{5.5}$$

where $\mathbf{f}_{\mathbf{x},n}$, $\mathbf{f}_{\mathbf{y},n}$, $\mathbf{g}_{\mathbf{x},n}$, and $\mathbf{g}_{\mathbf{y},n}$ are the partial derivative matrices formed at the last iteration of (5.3). To determine the unknowns, namely $\mathbf{x}_{\beta,n}$ and $\mathbf{y}_{\beta,n}$, (5.5) can be re-formatted as:

$$\begin{aligned}&\begin{bmatrix}\mathbf{I} - \frac{\Delta t}{2}\mathbf{f}_{\mathbf{x},n} & \frac{\Delta t}{2}\mathbf{f}_{\mathbf{y},n}\\ \mathbf{g}_{\mathbf{x},n} & \mathbf{g}_{\mathbf{y},n}\end{bmatrix}\begin{bmatrix}\mathbf{x}_{\beta,n}\\ \mathbf{y}_{\beta,n}\end{bmatrix}\\ &= \begin{bmatrix}\mathbf{x}_{\beta,n-1} + \frac{\Delta t}{2}\mathbf{f}_{\beta,n} + \frac{\Delta t}{2}\left(\mathbf{f}_{\mathbf{x},n-1}\mathbf{x}_{\beta,n-1} + \mathbf{f}_{\mathbf{y},n-1}\mathbf{y}_{\beta,n-1} + \mathbf{f}_{\beta,n-1}\right)\\ -\mathbf{g}_{\beta,n}\end{bmatrix}\end{aligned} \tag{5.6}$$

It can be seen that the additional computational effort for calculating the trajectory sensitivity is manageable as (5.6) is a linear equation, and all the entries on the left-hand side have already been computed when calculating the initial trajectory with $\beta = \beta_0$ at the time instant n in (5.2). Based on the specific changing pattern of β, the remaining unknown coefficients can be easily determined. Without loss of generality, it is assumed that at the time instant n the active power injected from ANN-based equivalent to Bus j, namely $P_{ANN\,j,n}$, is the changing system parameter. It can be shown that:

- $P_{ANN\,j,n}$ is the parameter variable defined specifically for the time instant n, and it does not have any impact on the system dynamics at previous time instants. Therefore at the time instant n-1, $\partial \mathbf{x}/\partial P_{ANN\,j,n} = \mathbf{0}$, $\partial \mathbf{y}/\partial P_{ANN\,j,n} = \mathbf{0}$, and $\partial \mathbf{f}/\partial P_{ANN\,j,n} = \mathbf{0}$. Accordingly, $\mathbf{x}_{\beta,n-1} = \mathbf{0}$, $\mathbf{y}_{\beta,n-1} = \mathbf{0}$, and $\mathbf{f}_{\beta,n-1} = \mathbf{0}$.
- $\mathbf{g}_{\beta,n}$ is the derivatives of $\mathbf{g}$ with respect to $P_{ANN\,j,n}$ at the time instant n. At Bus j, the active power balance equation, namely g_{Pj}, is:

$$g_{Pj} = P_{j,n} + P_{ANN\,j,n} - \sum_{k=1}^{N} V_{j,n} V_{k,n} \left(G_{jk} \cos\theta_{jk,n} + B_{jk} \sin\theta_{jk,n} \right) = 0 \quad (5.7)$$

  where $P_{j,n}$ is the existing active power injection at Bus j, $V_{j,n}\angle\theta_{j,n}$ and $V_{k,n}\angle\theta_{k,n}$ are the voltages at Bus j and Bus k, respectively. $G_{jk} + \mathrm{j}B_{jk}$ are (j,k)th element of the system admittance matrix. N is the total number of buses in the system. It can be proven that $\mathbf{g}_{\beta,n}$ is a column vector with 1 at the entry corresponding to g_{pj} and zeros elsewhere.
- $\mathbf{f}_{\beta,n}$ denotes the derivatives of $\mathbf{f}$ with respect to $P_{ANN\,j,n}$ at the time instant n. Because $P_{ANN\,j,n}$ is independent of any dynamic states that are represented in $\mathbf{f}$ and is only modeled as an additional power injection at Bus j, it can be proven that $\mathbf{f}_{\beta,n} = \mathbf{0}$.

Let B denote the set comprising the boundary buses, then (5.6) can be directly solved for $\partial V_{i,n}/\partial P_{ANN\,j,n}$ and $\partial\theta_{i,n}/\partial P_{ANN\,j,n}$ ($i \in B$ and $j \in B$) at the time instant n. The sensitivities of the boundary bus voltages with respect to the reactive power injections of the ANN-based equivalent, namely $\partial V_{i,n}/\partial Q_{ANN\,j,n}$ and $\partial\theta_{i,n}/\partial Q_{ANN\,j,n}$ can be obtained in a similar manner. The only difference is that $\mathbf{g}_{\beta,n}$ in (5.6) now becomes a column vector with 1 at the entry corresponding to the reactive power balance equation of Bus j and zeros elsewhere.

An example of the iterative procedure to compute the required power compensations from the ANN-based equivalent at the time instant n is illustrated in Fig. 5.3. $\mathbf{V}_{ri,n-1}$ and $\mathbf{V}_{fi,n-1}$ are the voltage at Bus i at the previous time instant $n-1$. At the beginning of the procedure, it is assumed that $\mathbf{V}_{ri,n-1} = \mathbf{V}_{fi,n-1}$.

During the process shown in Fig. 5.3, the power injections are updated iteratively to modify the voltage responses at the boundary buses in the reduced system. At the last iteration, the voltage matches the response obtained in the full system representation.

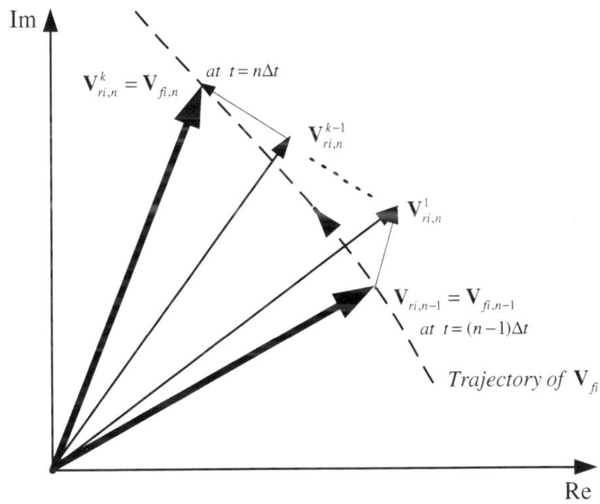

Fig. 5.3 Example of the training data formulation iterations

For a specific system disturbance, the general procedure to build the training samples includes the following steps:

- Step 1: $n = n + 1$, the power injections of the ANN-based equivalent are initialized to the same values obtained at the previous time instant $n - 1$.
- Step 2: Calculate the voltage response in the reduced system with the power injections of the ANN-based equivalent fixed by solving (5.2) for the present time instant. Determine the voltage mismatch at the boundary buses, namely $\Delta \mathbf{V}_n = [\mathbf{V}_{f,n} - \mathbf{V}_{r,n}]$ and $\Delta \boldsymbol{\theta}_n = [\boldsymbol{\theta}_{f,n} - \boldsymbol{\theta}_{r,n}]$. The subscripts f and r denote the full system and reduced system, respectively. If the mismatch is smaller than a specified tolerance, then go to Step 4; otherwise go to Step 3.
- Step 3: Calculate the power injection increments of the ANN-based equivalent by solving the following equation

$$\begin{bmatrix} \Delta \mathbf{P}_{ANN,n} \\ \Delta \mathbf{Q}_{ANN,n} \end{bmatrix} = \begin{bmatrix} \partial \mathbf{V}_{r,n} / \partial \mathbf{P}_{ANN,n} & \partial \mathbf{V}_{r,n} / \partial \mathbf{Q}_{ANN,n} \\ \partial \boldsymbol{\theta}_{r,n} / \partial \mathbf{P}_{ANN,n} & \partial \boldsymbol{\theta}_{r,n} / \partial \mathbf{Q}_{ANN,n} \end{bmatrix}^{-1} \begin{bmatrix} \Delta \mathbf{V}_n \\ \Delta \boldsymbol{\theta}_n \end{bmatrix} \tag{5.8}$$

  where the coefficient matrix elements on the right-hand side, namely $\partial V_{ri,n}/\partial P_{ANNj,n}$, $\partial \theta_{ri,n}/\partial P_{ANNj,n}$, $\partial V_{ri,n}/\partial Q_{ANNj,n}$, and $\partial \theta_{ri,n}/\partial Q_{ANNj,n}$ ($i \in B$ and $j \in B$), are the trajectory sensitivities computed in (5.6). Then, the power injections of the ANN-based equivalent are updated to $\mathbf{P}_{ANN,n} = \Delta \mathbf{P}_{ANN,n} + \mathbf{P}_{ANN,n}$ and $\mathbf{Q}_{ANN,n} = \Delta \mathbf{Q}_{ANN,n} + \mathbf{Q}_{ANN,n}$. Go to Step 2.
- Step 4: if $n < n_f$ (n_f is the time instant until which the responses are compared), then go to Step 1. Otherwise terminate the procedure.

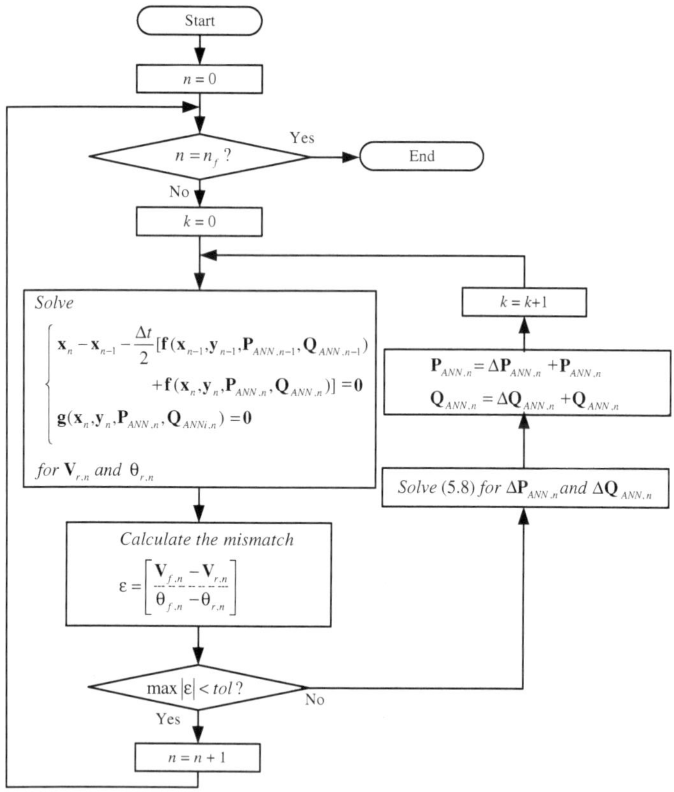

Fig. 5.4 Flowchart for building an ANN training set

To minimize the number of modifications to the existing transient simulation software packages, the flowchart shown in Fig. 5.4 is proposed. Prior to the proposed routine, the voltage responses at the boundary buses in the full system representation are simulated and stored for the same system disturbance of interest. The voltage responses are then used as the reference in determining the proper power injections of the ANN-based equivalent.

For a given disturbance, the following training set can be formed:

$$\text{Training Set} = \left(\begin{bmatrix} \mathbf{V}_{r,n} \\ \boldsymbol{\theta}_{r,n} \end{bmatrix}, \begin{bmatrix} \mathbf{P}_{ANN,n} \\ \mathbf{Q}_{ANN,n} \end{bmatrix} \right), n = 0, \ldots, n_f \tag{5.9}$$

It can be seen that the proposed algorithm is a type of boundary matching technique specifically for the dynamic simulation. Throughout simulation, the trajectory sensitivity can be calculated without extensively increasing the computational burden. Besides, the trajectory sensitivity with respect to a specific system parameter is independent of other parameters. The computational efficiency in Step 3 can be further enhanced using either the parallel or the cluster computation techniques. Another advantage of the proposed method is that the real-time data collected by PMUs installed at the boundary buses can be directly used as the reference in forming the training set. It allows the ANN-based equivalent to capture the real discrepancy between the planning models that are used for building the conventionally reduced system and the exact system models under realistic operating conditions, leading to improved models for on-line DSA.

5.3.2 Formulation of the ANN-Based Equivalent

As a dynamic system, the present outputs of the ANN-based equivalent depend not only on the present inputs but also on the inputs and outputs at previous instants. To accommodate this characteristic, a fully recurrent neural network (FRNN) [17] is utilized. A typical structure of the proposed ANN-based equivalent is shown in Fig. 5.5. Unlike a typical multilayer perceptron network (MLP), a feedback loop is added between the output and input sides in the FRNN. Both bus voltages and power injection predictions are fed into the network through the tapped-delay-line-memories with l and m delayed units, respectively. At the time instant n, the ANN-based equivalent can be represented by a nonlinear equation vector defined as:

$$\mathbf{S}_{ANN,n} = \mathbf{h}(\mathbf{V}_{ANN,n}, \mathbf{V}_{ANN,n-1}, \ldots, \mathbf{V}_{ANN,n-l}, \mathbf{S}_{ANN,n-1}, \ldots, \mathbf{S}_{ANN,n-m}) \quad (5.10)$$

where $\mathbf{V}_{ANN,n-l}$ is the voltage vector at the time instant $n-$l, and $\mathbf{S}_{ANN,n-m} = [\mathbf{P}_{ANN,n-m}{}^{\mathrm{T}}\ \ \mathbf{Q}_{ANN,n-m}{}^{\mathrm{T}}]^{\mathrm{T}}$ is the power injection vector at the time instant $n-m$. The vector $\mathbf{h}$ is composed of nonlinear mapping functions determined by the embedded MLP.

For the FRNN to achieve a better performance of capturing the input-and-output mapping stored in the training set and to limit the number of neuron parameters to be optimized, an embedded MLP with three layers of neurons in Fig. 5.6 is formed.

In Fig. 5.6, $\mathbf{x}_1 = [\mathbf{V}_{ANN,n}{}^{\mathrm{T}}, \mathbf{V}_{ANN,n-1}{}^{\mathrm{T}}, \ldots, V_{ANN,n-l}{}^{\mathrm{T}}, \mathbf{S}_{ANN,n-1}{}^{\mathrm{T}}, \ldots, S_{ANN,n-m}{}^{\mathrm{T}}]^{\mathrm{T}}$ is the input vector; $\mathbf{x}_2$ and $\mathbf{x}_3$ are the internal state vectors; $\mathbf{W}(1)$, $\mathbf{W}(2)$, and $\mathbf{W}(3)$ are the weighting matrices; $\mathbf{b}(1)$, $\mathbf{b}(2)$, and $\mathbf{b}(3)$ are the bias vectors; and ϕ_1, ϕ_2, and ϕ_3 are the activation functions. The subscript 1, 2, and 3 denote layers 1, 2, and 3, respectively. Assuming $n_1 \times 1$, $n_2 \times 1$, $n_3 \times 1$, and $n_4 \times 1$ are the dimensions of $\mathbf{x}_1$, $\mathbf{x}_2$, $\mathbf{x}_3$, and $\mathbf{S}_{ANN,n}$, respectively, for layer 1 it can be proven that:

$$x_{2k} = \phi_1\left(\sum_{j=1}^{n_1} w_{kj}(1)x_{1j} + b_k(1)\right), k = 1, \ldots, n_2 \quad (5.11)$$

where x_{1j} is the jth element of $\mathbf{x}_1$, $w_{kj}(1)$ is the (k, j)th element of $\mathbf{W}(1)$, $b_k(1)$ is the kth element of $\mathbf{b}(1)$, and x_{2k} is the kth element of $\mathbf{x}_2$. Similarly, the following equations hold for layers 2 and 3:

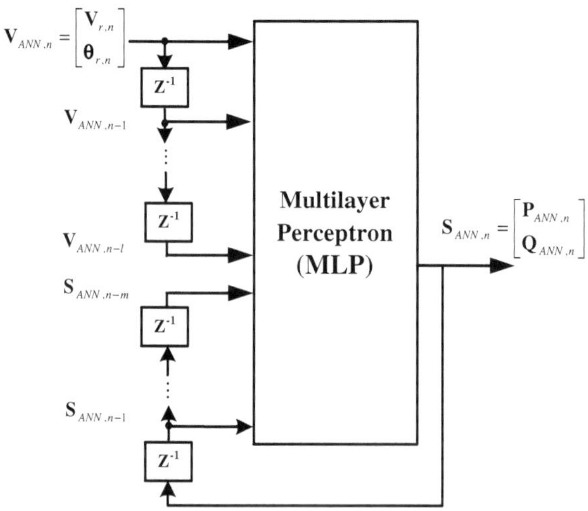

Fig. 5.5 ANN-based equivalent using fully recurrent neural network (RFNN)

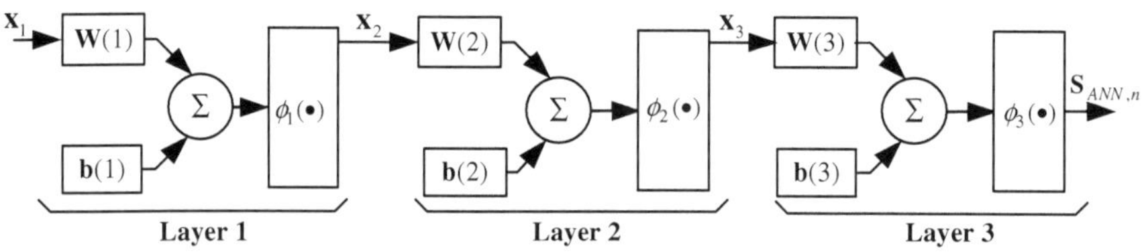

Fig. 5.6 Structure of the embedded MLP in FRNN

$$x_{3k} = \phi_2\left(\sum_{j=1}^{n_2} w_{kj}(2)x_{2j} + b_k(2)\right), k = 1, \ldots, n_3 \tag{5.12}$$

$$S_{ANN,nk} = \phi_3\left(\sum_{j=1}^{n_3} w_{kj}(3)x_{3j} + b_k(3)\right), k = 1, \ldots, n_4 \tag{5.13}$$

Then the characteristic equation vector **h** in (5.10) can be determined uniquely as the combination of (5.11)–(5.13). It is noted that at the time instant n, the inputs $[\mathbf{V}_{ANN,n-1}{}^{\mathrm{T}}, \ldots, \mathbf{V}_{ANN,n-l}{}^{\mathrm{T}}]^{\mathrm{T}}$ and $[\mathbf{S}_{ANN,n-1}{}^{\mathrm{T}}, \ldots, S_{ANN,n-m}{}^{\mathrm{T}}]^{\mathrm{T}}$ have already been calculated and can be treated as constants. Then (5.10) can be simplified to a nonlinear mapping from $\mathbf{V}_{ANN,n}$ to $\mathbf{S}_{ANN,n}$.

For the training purpose, the FRNN in Fig. 5.5 is equivalenced to a static feed-forward network after disconnecting the feedback loop. The equivalent neural network is shown in Fig. 5.7.

In the equivalent network, the exact inputs and outputs at any time instants are available in the training set. Therefore, more efficient back-propagation training

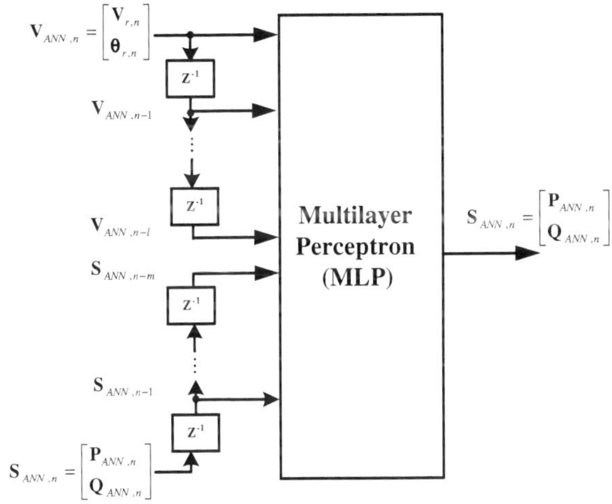

Fig. 5.7 Equivalent neural network for training purpose

methods can be implemented. For example, by forming the approximated Hessian matrix of the prediction error with respect to the weights and biases, the Levenberg-Marquardt (LM) algorithm can achieve a second-order convergence speed in searching for the optimum weights and biases. It provides an efficient solution for the problems involving the network with up to a few hundred weights [18].

5.3.3 Integration of ANN-Based Dynamic Equivalent

When considering the ANN-based equivalent, the entire reduced system becomes a hybrid system. It includes both the continuous system (described by a set of DAEs) and the discrete system (ANN-based equivalent). The ANN-based equivalent can be modeled as a set of adjustable $P - Q$ sources using the terminal voltages as inputs. To model the hybrid reduced system, the impacts of the ANN-based equivalent on the power balance equations $\mathbf{g}$ and the partial derivative matrix $\mathbf{g}_\mathrm{y}$ need to be considered. At the time instant n, the new power balance equations at Bus $i \in B$ are given by:

$$
\begin{aligned}
g_{Pi} &= P_{i,n} + P_{ANNi,n} - \sum_{j=1}^{N} V_{i,n}V_{j,n}(G_{ij}\cos\theta_{ij,n} + B_{ij}\sin\theta_{ij,n}) \\
g_{Qi} &= Q_{i,n} + Q_{ANNi,n} - \sum_{j=1}^{N} V_{i,n}V_{j,n}(G_{ij}\sin\theta_{ij,n} - B_{ij}\cos\theta_{ij,n})
\end{aligned}
\tag{5.14}
$$

Corresponding to the power injections defined as $P_{ANNi,n}$ and $Q_{ANNi,n}$, additional partial derivatives, namely $\partial P_{ANNi,n}/\partial V_{j,n}$, $\partial P_{ANNi,n}/\partial \theta_{j,n}$, $\partial Q_{ANNi,n}/\partial V_{j,n}$, and $\partial Q_{ANNi,n}/\partial \theta_{j,n} (j \in B)$, need to be formulated. For instance, $\partial P_{ANNi,n}/\partial V_{j,n}$ is given by:

$$\frac{\partial P_{ANNi,n}}{\partial V_{j,n}} = \left(\frac{\partial P_{ANNi,n}}{\partial \mathbf{x}_3}\right)\left(\frac{\partial \mathbf{x}_3}{\partial \mathbf{x}_2}\right)\left(\frac{\partial \mathbf{x}_2}{\partial \mathbf{x}_1}\right)\left(\frac{\partial \mathbf{x}_1}{\partial V_{j,n}}\right) \tag{5.15}$$

Assuming ϕ_1 and ϕ_2 are tan-sigmoid activation functions and ϕ_3 is a liner activation function, the elements of each derivative matrix are given by:

$$\partial P_{ANNi,n}/\partial x_{3m} = w_{km}(3) \tag{5.16}$$

where k is the index of $P_{ANNi,n}$ in $\mathbf{S}_{ANN,n}$.

$$\partial x_{3m}/\partial x_{2p} = \left[1 - \left(\sum_{j=1}^{n_2} w_{kj}(2) x_{2j} + b_k(2)\right)^2\right] w_{mp}(2) \tag{5.17}$$

where $m = 1,\ldots, n_3$, and $p = 1,\ldots, n_2$.

$$\partial x_{2m}/\partial x_{1p} = \left[1 - \left(\sum_{j=1}^{n_1} w_{kj}(1) x_{2j} + b_k(1)\right)^2\right] w_{mp}(1) \tag{5.18}$$

where $m = 1,\ldots, n_2$, and $p = 1,\ldots, n_1$.

$$\partial \mathbf{x}_1/\partial V_{j,n} = [0, \ldots, 1, \ldots, 0]^T \tag{5.19}$$

where the element of $\mathbf{x}_1$ at the entry associated with $V_{j,n}$ is 1 and zeros elsewhere.

The same procedure as in (5.15)–(5.19) are followed to model the impacts of $Q_{ANNi,n}$ but are omitted for the sake of brevity. The computed derivative elements are appended to the corresponding entries of the original derivative matrix $\mathbf{g}_y$ formulated in the dynamic simulation software package. The detailed flowchart of the proposed integration method is depicted in Fig. 5.8.

During the iterative procedure in Fig. 5.8, both the power balance equations and system Jacobian matrix are updated to model the power injections of the ANN-based equivalent. It is to be noted that in a realistic setting, the number of boundary buses connected with the ANN-based equivalent are substantially smaller than the total number of buses in the entire reduced system. In addition, the appended derivatives in (5.16)–(5.19) can also be readily obtained due to the superior derivative characteristics of the activation functions. Therefore, additional computational effort to simulate the hybrid reduced system can be limited.

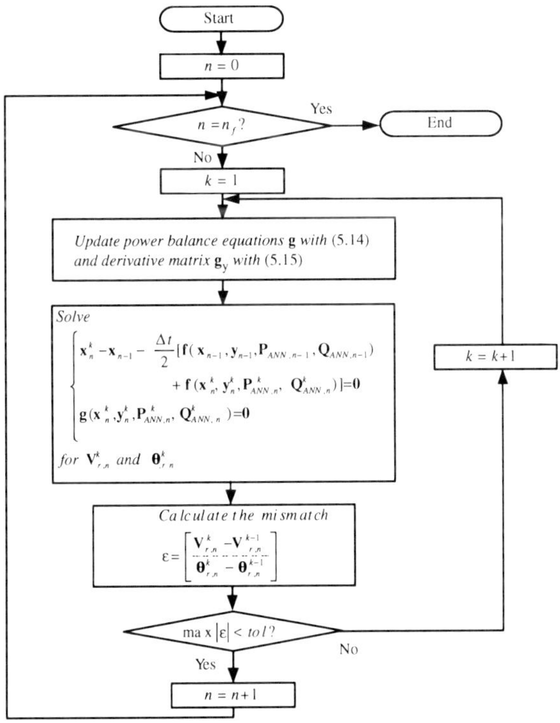

Fig. 5.8 Flowchart of hybrid reduced system simulation

5.4 Case Study

The system representing a portion of the WECC system is tested to validate the proposed approach. A schematic diagram of the test system is presented in Fig. 5.9. Within the entire test system, the detailed generator unit models with exciters, governors, and power system stabilizers (PSS) are used.

In Fig. 5.9, the network at 230 kV and above within the retained area is detailed. The remaining portion is defined as the external area. After loading the power flow and dynamic data into DYNRED, all the coherent generators are identified using the weak-link method [2]. In the external area, a total of 115 groups of coherent generators are identified with 24 of them having only one generator in each group. In each group, the coherent generators are aggregated and replaced by a classical equivalent generator model, followed by the network reduction and load aggregation. In the end, a conventionally reduced system, consisting of both the study area and the equivalenced external area, is formulated. The summary of the full system and the conventionally reduced system is given in Table 5.1.

It is observed from Table 5.1 that a significant reduction in the system size can be achieved after the dynamic equivalencing procedure. In the conventionally reduced

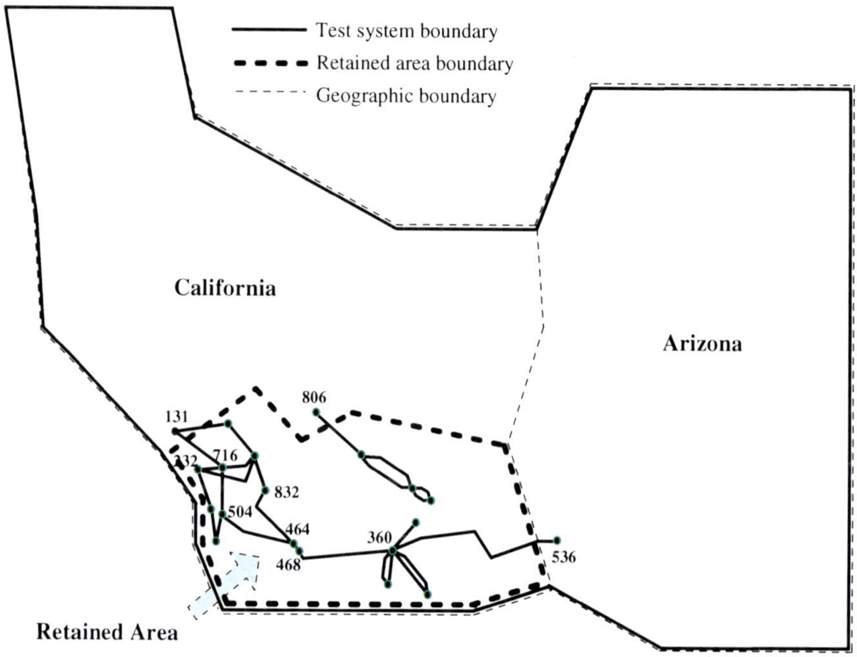

Fig. 5.9 Schematic diagram of the test system

Table 5.1 Summary of full system and the conventionally reduced system

	Full system	Conventionally reduced system	% of full system (%)
Buses	2240	806	36.0
Branches	3015	1573	52.2
Loads	913	481	52.7
Generators	342	145	42.4

system, 57.6% of the generators in the full system have been eliminated. Similar reduction in the number of buses, branches, and loads can also be found.

To validate the equivalencing accuracy of the conventionally reduced system, a three-phase fault is applied on Bus 360 at 0.5 s, and it is cleared after 0.1 s by tripping the 500 kV line from Bus 360 to Bus 468. For the studied case, the root mean square error (RMSE) [8] is computed to numerically measure the mismatch between the dynamic responses obtained in the full system and the reduced systems. For Generator i, the RMSE_i is defined as:

$$RMSE_i = \sqrt{\frac{1}{T}\int_0^T \left(\delta_i^{full}(t) - \delta_i^{reduced}(t)\right)^2 dt} \quad \text{(in degree)} \tag{5.20}$$

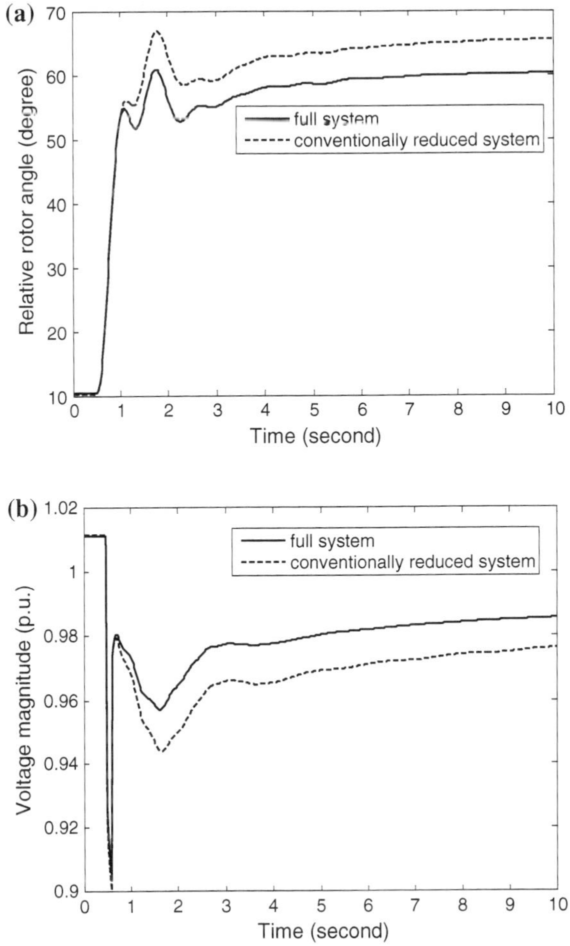

Fig. 5.10 System response comparison for Case OP1_2. **a** Relative rotor angle responses of Generator 15 **b** Voltage responses at Bus 131

where $\delta_i^{full}(t)$ and $\delta_i^{reduced}(t)$ are the relative rotor angle of Generator i obtained in the full system and reduced system, respectively. T is the entire simulation duration.

In simulation, the relative rotor angle of Generator 15 in the retained area that has the largest RMSE and the voltage magnitude at one of the boundary buses (Bus 131) are plotted in Fig. 5.10.

It is seen from Fig. 5.10 that the conventionally reduced system formed in DYNRED cannot fully capture the dynamic characteristics of the retained area, and the discrepancy becomes significantly larger as the simulation evolves. To account for the observed mismatch between the full system and the conventionally reduced system, the ANN-based equivalent is formed. Specifically, all the buses (*in the case*

Table 5.2 Tested contingencies in training set

Contingency	3-Phase faulted bus	Clearing time (ms)	Tripped line	
			From bus	To bus
CON_1	360	60	360	468
CON_2	360	100	360	468
CON_3	832	100	464	832
CON_4	832	160	464	832
CON_5	716	100	232	716
CON_6	716	160	232	716

considered there are only three buses) on the original retained area boundary, namely 131, 806, and 536, are connected to the ANN-based equivalent. The retained boundary is formed based on the original balancing area definition. More details about how to formulate a right-sized retained area boundary for improved equivalencing accuracy can be found in [15, 16]. In most applications, a fixed coherency-based dynamic equivalent model is utilized for different operating conditions. Therefore, an effective hybrid dynamic equivalent requires the ANN-based equivalent to be capable of providing essential compensations when the system condition changes. In forming the training set, six typical contingencies summarized in Table 5.2 are evaluated under both the base system condition (OP1) and the revised system condition (OP2). In the full system under the revised operating condition, the loads and generations in the retained area are increased by 10 % and the remaining imbalance is compensated by the generators in the external area based on their own capabilities. The same dispatching pattern is applied to the conventionally reduced system under OP2, in which the equivalent model obtained under the base operating condition OP1 is used. For all training cases, the simulation duration is 10 s.

For each contingency in Table 5.2, two training cases are formed for both operating conditions OP1 and OP2. For example, Case OP1_1 denotes the case in which contingency CON_1 is applied under the operating condition OP1. For the cases considered, the trajectory sensitivity-based approach is applied to compute the desired power injections of the ANN-based equivalent. Using the fixed integration step of 0.02 s, a total of 6012 training samples are created for the training set. In Case OP1_2, for example, the required power injections of the ANN-based equivalent at each connection bus are presented in Fig. 5.11.

For the same case, the number of iterations required in the trajectory-sensitivity approach for the required power injections to converge is shown in Fig. 5.12.

It is seen from Fig. 5.12 that because the trajectory-sensitivity-approach provides accurate estimation of the voltage deviations at the connection buses, one iteration is often needed to obtain the power injections that can match the boundary bus voltage responses. At the worst scenarios, such as the fault occurrence and clearing instants, only two iterations are needed. The FRNN network in Fig. 5.13 is formed to model the ANN-based equivalent.

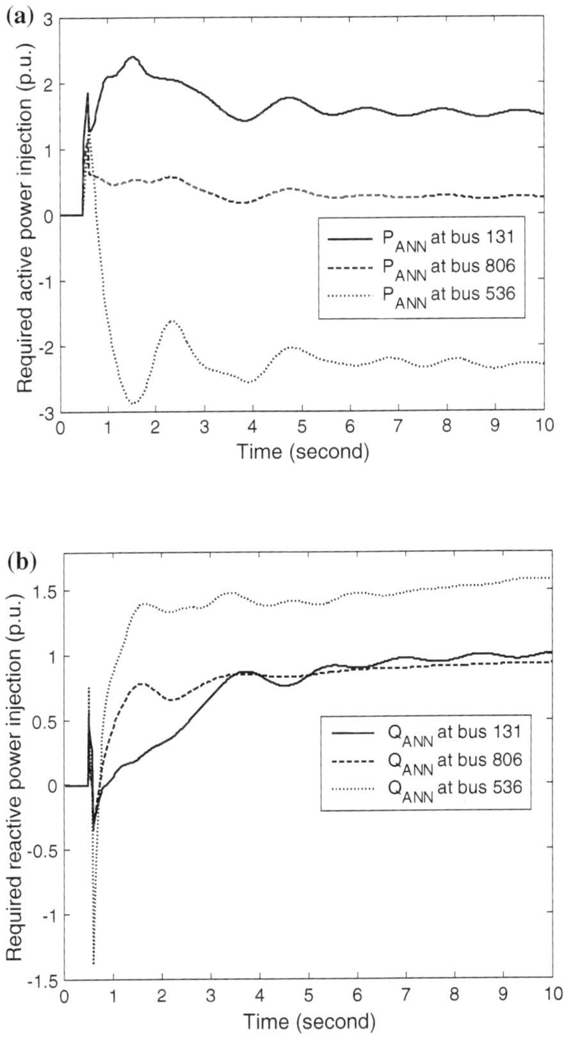

Fig. 5.11 Required power injections of ANN-based equivalent for Case OP1_2. **a** Required active power injections at Buses 131, 806, and 536 **b** Required reactive power injections at Buses 131, 806, and 536

In Fig. 5.13, $n_1 = 84$, $n_2 = 8$, $n_3 = 8$, $n_4 = 6$, $l = 6$, and $m = 7$. To avoid the potential saturation issues with the tan-sigmoid activation functions at layers 1 and 2, the following pre-processing functions are defined,

$$V'_{ANNi,n} = V_{ANNi,n}/3, \quad \theta'_{ANNi,n} = \theta_{ANNi,n}/3 \tag{5.21}$$

$$P'_{ANNi,n} = P_{ANNi,n}/6, \quad Q'_{ANNi,n} = Q_{ANNi,n}/6 \tag{5.22}$$

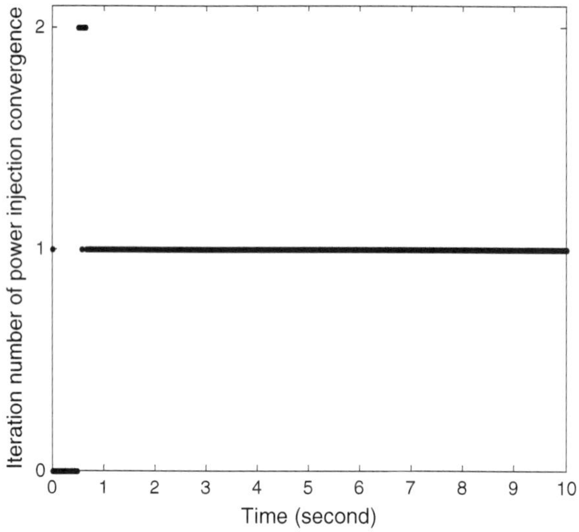

Fig. 5.12 Iteration number of power injection convergence for Case OP1_2

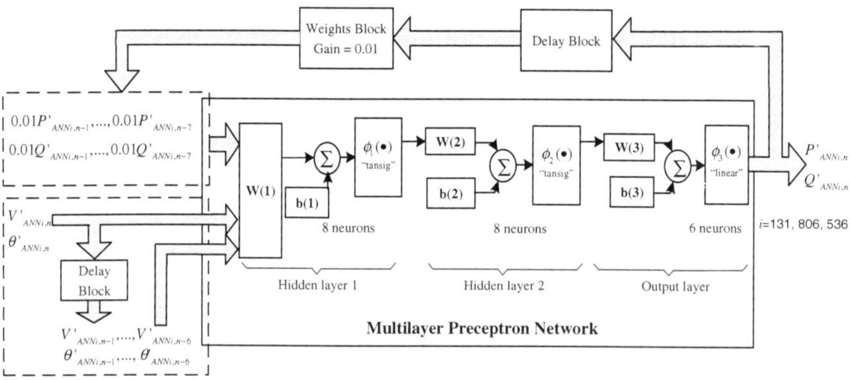

Fig. 5.13 Diagram of the ANN-based equivalent

where i is the boundary bus number, namely 131, 806, and 536 (see Fig. 5.9).

Using the above pre-processing functions, the inputs and outputs are mapped to the unsaturated section of the tan-sigmoid activation function. In the meantime, in order to improve the numerical stability of the FRNN, a gain of 0.01 is applied to the feedbacks from the output side. After forming the pre-processed training data and opening the feedback loop, the LM routine in MATLAB [18] is deployed to optimize the unknown parameters in the neural network to simulate the input-and-output mapping stored in the training set. The mean square error (MSE) during the training process is shown in Fig. 5.14.

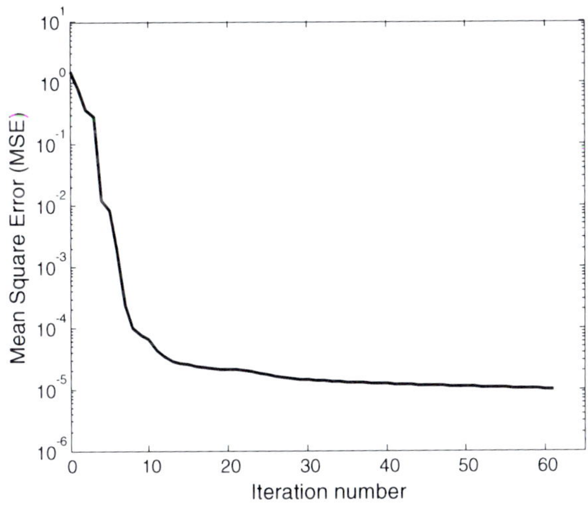

Fig. 5.14 Mean square error during the training process

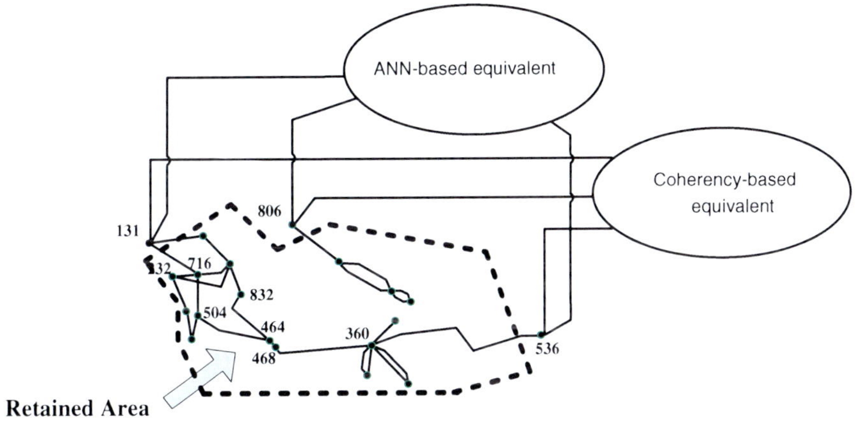

Fig. 5.15 Diagram of hybrid reduced system

It is seen in Fig. 5.14 that the training error decreases dramatically during the initial iterations. Then it reaches the threshold of 10^{-5} at the 62nd iteration. On an Intel® Core i5 Processor (2.50 GHz) PC with 3.48 GB of RAM, the entire training process is completed within 224 s. It is to be noted that the number of the neurons at each layer needs to be tuned cautiously to avoid the potential over-fitting problem. A trial-and-error method is used in order to determine an appropriate configuration for the FRNN. After being trained, the ANN-based equivalent is connected to the conventionally reduced system at the boundary buses, and the resulting hybrid reduced is shown in Fig. 5.15.

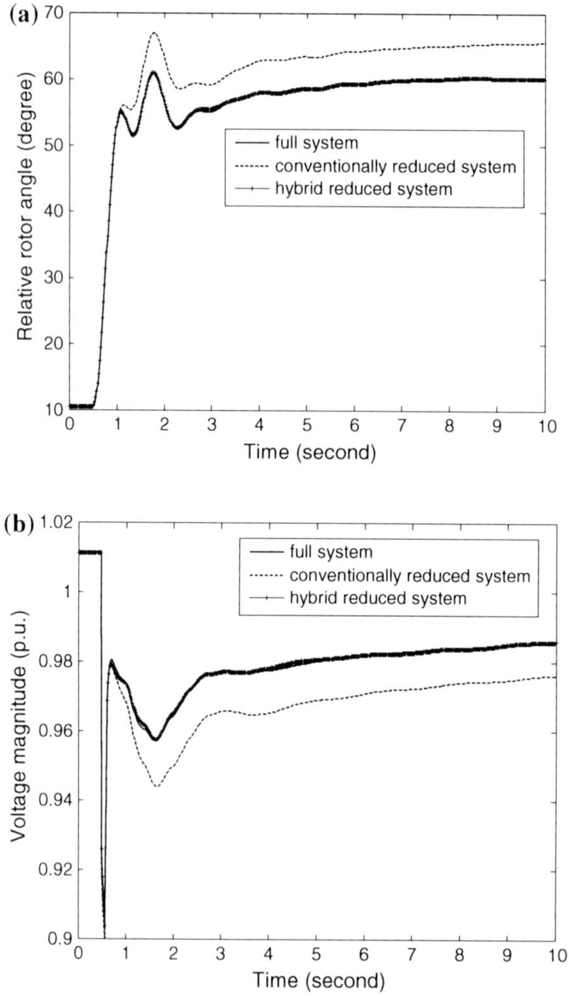

Fig. 5.16 Comparison of system responses for Case OP1_2. **a** Relative rotor angle responses of Generator 15 **b** Voltage responses at Bus 131

For Case OP1_2, the relative rotor angle response of Generator 15 and the voltage response at Bus 131 are shown in Fig. 5.16.

It can be seen in Fig. 5.16 that the response mismatches between the full system and the conventionally reduced system are compensated by the ANN-based equivalent. As expected, more accurate responses are obtained in the hybrid reduced system throughout the entire simulation duration. For the same case, the RMSEs of the generators in the retained area are compared between the conventionally reduced system and the hybrid reduced system. The results are shown in Fig. 5.17.

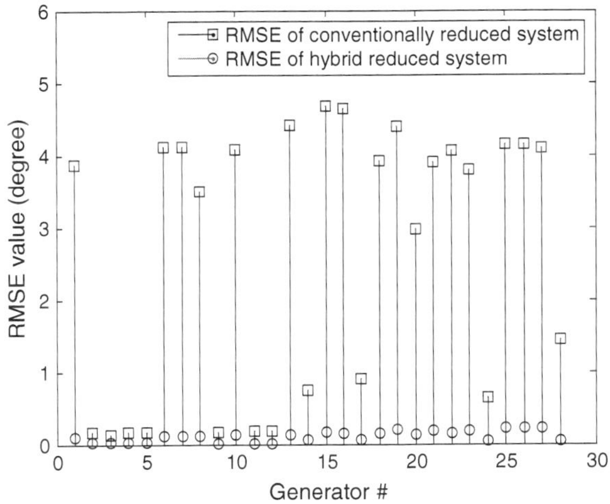

Fig. 5.17 RMSE comparison for Case OP1_2

Table 5.3 Summary of average RMSEs in the reduced systems under OP1

Case	Average RMSE (degree)	
	Conventionally reduced system	Hybrid reduced system
OP1_1	2.61	0.13
OP1_2	2.64	0.11
OP1_3	0.24	0.10
OP1_4	0.33	0.12
OP1_5	0.36	0.10
OP1_6	0.52	0.19

It is observed from Fig. 5.17 that with the supplementary power injections provided by the ANN-based equivalent, the RMSEs decrease significantly for most of the retained generators. For all the training cases, the average RMSE values in the conventionally reduced system and the hybrid reduced system are compared in Tables 5.3 and 5.4. It is seen that the ANN-based equivalent can effectively improve the accuracy of the coherency-based equivalent formed in DYNRED. Its effectiveness varies for different cases. For example, the improvement for a severe contingency in Case OP1_2 and OP2_2 is more significant. For all the trained cases, 60.6 % of the simulation execution time has been saved using the hybrid equivalent on the same PC as described before.

To validate the accuracy of the hybrid equivalent for the system disturbances that are not included in the training set, the following contingencies are tested:

- CON_7: a 3-phase fault at Bus 360 is cleared after 0.12 s by tripping the line from Bus 360 to Bus 468;

Table 5.4 Summary of average RMSEs in the reduced systems under OP2

Case	Average RMSE (degree)	
	Conventionally reduced system	Hybrid reduced system
OP2_1	3.67	0.15
OP2_2	3.69	0.12
OP2_3	0.28	0.11
OP2_4	0.37	0.13
OP2_5	0.36	0.08
OP2_6	0.52	0.14

Table 5.5 Summary of average RMSEs in the reduced systems for untrained contingencies

Case	Average RMSE (degree)	
	Conventionally reduced system	Hybrid reduced system
OP1_7	2.65	0.14
OP1_8	0.28	0.12
OP2_7	3.70	0.15
OP2_8	0.31	0.14

- CON_8: a 3-phase fault at Bus 504 is cleared after 0.10 s by tripping the line from Bus 504 to Bus 464.

It can be seen that fault clearing time in CON_7 is different from CON_1 and CON_2; while a new faulted bus is defined in CON_8. The RMSEs in different reduced system under OP1 and OP2 are summarized in Table 5.5.

It can be concluded from Table 5.5 that for the untrained contingencies, the ANN-based equivalent can still generalize its input-and-output mapping obtained from the most similar cases in the training set and achieve better equivalencing accuracy by providing essential compensation to the conventionally reduced system.

Table 5.6 Summary of average RMSEs in the reduced systems under untrained operating condition OP3

Case	Average RMSE (degree)	
	Conventionally reduced system	Hybrid reduced system
OP3_1	3.08	0.14
OP3_2	3.10	0.11
OP3_3	0.20	0.11
OP3_4	0.30	0.12
OP3_5	0.32	0.08
OP3_6	0.49	0.16
OP3_7	3.11	0.14
OP3_8	0.24	0.13

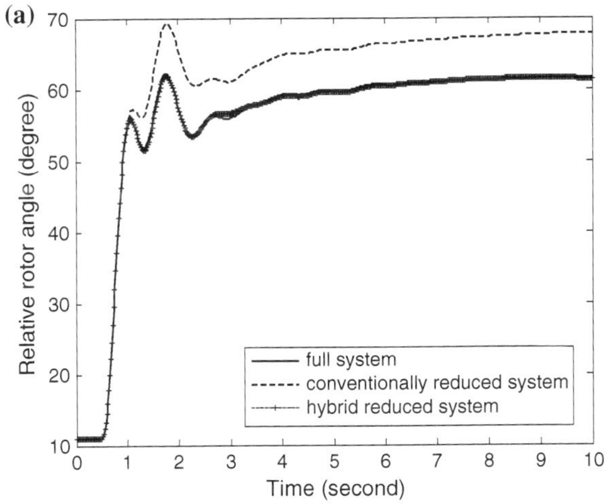

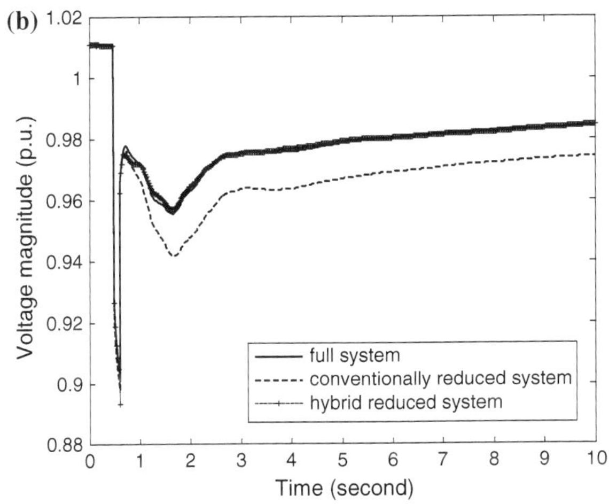

Fig. 5.18 Comparison of system responses for Case OP3_7. **a** Relative rotor angle responses of Generator 15 **b** Voltage responses at Bus 131

Furthermore, to test the performance of the hybrid dynamic equivalent for an untrained system condition, a new load condition OP3 is defined with the loads and generations in the retained area increased by 5 %. For all the contingencies, the average RMSE values in the conventionally reduced system and the hybrid reduced system are compared in Table 5.6.

It can be seen from Table 5.6 that for the new operating condition OP3, the improved equivalencing accuracy can also be obtained using the hybrid equivalent. The similarity between the trained and untrained conditions is successfully captured by the ANN-based equivalent. The same conclusion can be drawn from the relative

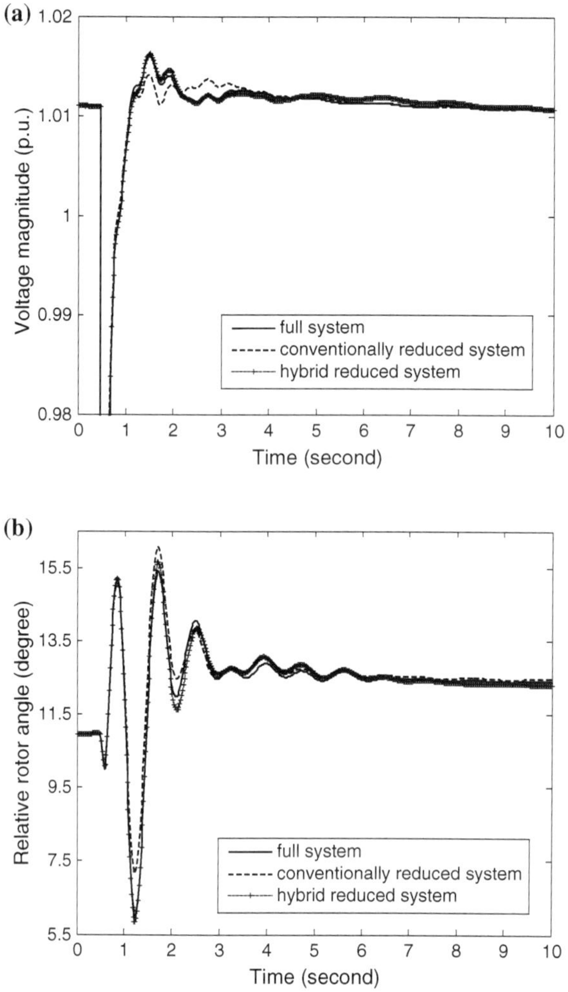

Fig. 5.19 Comparison of system responses for Case OP3_8. **a** Relative rotor angle responses of Generator 15 **b** Voltage responses at Bus 131

rotor angle response of Generator 15 and voltage magnitude at Bus 131 for Case OP3_7 and OP3_8 that are shown in Figs. 5.18 and 5.19, respectively.

Because the generalization accuracy of a neural network is dependent on its training samples, the ANN-based equivalent might not be effective for certain untrained cases. To address this issue, an extended training set that includes these unsatisfactory cases might become necessary. It is also noted that when dramatic change in the system condition occurs, a single hybrid equivalent might not be able to capture all the characteristics in the external area. To resolve this issue, an adaptive hierarchy ANN-based equivalent model shown in Fig. 5.20 will become more applicable.

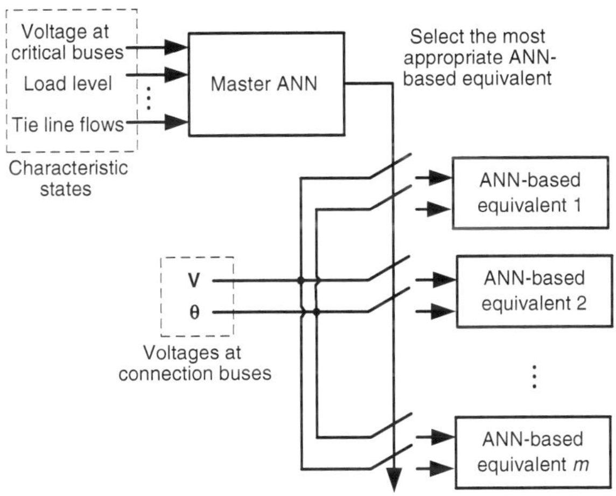

Fig. 5.20 Scheme of hierarchy ANN-based equivalents

In Fig. 5.20, a set of ANN-based equivalents are developed for different typical operating conditions. Then, a master ANN is trained to select an appropriate candidate ANN-based equivalent based on the critical information about the present system operating condition. Because each ANN-based equivalent is only trained to work within a certain range of operating conditions, a significant reduction in the computational effort can be achieved in formulating the candidate ANN-based equivalent models.

5.5 Conclusions

Power system dynamic equivalents play a critical role in on-line dynamic security assessment given the increasing size of interconnected power systems. To improve the accuracy of the conventional coherency-based dynamic equivalent, a novel hybrid dynamic equivalent has been discussed in this chapter. Unlike other research efforts on dynamic equivalents, the ANN-based equivalent in the proposed hybrid configuration is designed to compensate for the discrepancy between the full system and the coherency-based reduced system through supplementary power injections at the retained area boundary buses. The study conducted on a realistic test system shows that the training set can be efficiently formed using the trajectory sensitivity of the boundary bus voltages with respect to the power injections of the ANN-based equivalent. For the selected cases, the ANN-based equivalent is capable of providing desirable compensations to the coherency-based equivalent and improving the equivalencing accuracy.

It is to be noted that the accuracy of the proposed hybrid equivalent system can be easily improved by including more cases into the training set. Accordingly, a systematic way to determine an appropriate set of training cases needs to be investigated. For future applications, a study using the voltage response measured by PMUs as the references in determining the required power injections of the ANN-based equivalent could be considered for the application of the proposed method for on-line DSA.

References

1. S.T.Y. Lee, F.C. Schweppe, Distance measures and coherency recognition for transient stability equivalents. IEEE Trans. Power Apparatus Syst. **PAS-92**(5), 1550–1557 (1973).
2. R. Nath, S.S. Lamba, K.S.P. Rao, Coherency based system decomposition into study and external areas using weak coupling. IEEE Trans. Power Apparatus Syst. **PAS–104**, 1443–1449 (1995).
3. J.H. Chow, J.R. Winkelman, M.A. Pai, P.W. Sauer, Model reduction and energy function analysis of power system using singular perturbation techniques. in Proceedings of 25th IEEE Conference on Decision and, Control, 1206–1211 (1986).
4. P. Kundur, G.J. Rogers, D.Y. Wong, J. Ottevangers, L. Wang, Dynamic reduction. EPRI, Palo Alto, CA, Tech. Rep. TR-102234 Project 2447–01 (1993).
5. R. Podmore, Identification of coherent generators for dynamic equivalents. IEEE Trans. Power Apparatus Syst. **PAS-97**, 1344–1354 (1978).
6. R. Podmore, A. Germond, Development of dynamic equivalent for transient stability studies. EPRI EL-456 Project 763, (1977).
7. M.L. Ourari, L.A. Dessaint, V.Q. Do, Dynamic equivalent modeling of large power systems using structure preservation technique. IEEE Trans. Power Syst. **21**(3), 1284–1295 (2006)
8. S.K. Joo, C.C. Liu, L.E. Jones, J.W. Choe, Coherency and aggregation techniques incorporating rotor and voltage dynamics. IEEE Trans. Power Syst. **19**(2), 1068–1075 (2004)
9. A.M. Stankovic, A.T. Saric, M. Milosevic, Identification of non-parametric dynamic power system equivalents with artificial neural networks. IEEE Trans. Power Syst. **18**(4), 1478–1486 (2003)
10. H. Shakouri, H.R. Radmanesh, Identification of a continuous time nonlinear state space model for the external power system dynamic equivalent by neural networks. Electr. Power Energy Syst. **31**, 334–344 (2009)
11. A.M. Azmy, I. Erlich, P. Sowa, Artificial neural network-based dynamic equivalents for distribution systems containing active sources. in Proceedings of IEE Generation. Transmission, and. Distribution **151**(6), 681–688 (2004)
12. T.B. Nguyen, M.A. Pai, Dynamic security-constrained rescheduling of power systems using trajectory sensitivities. IEEE Trans. Power Syst. **18**(2), 848–854 (2003)
13. A. Zamora-Cardenas, C.R. Fuerte-Esquivel, Multi-parameter trajectory sensitivity approach for location of series-connected controllers to enhance power system transient stability. Electr. Power Syst. Res. **80**(9), 1096–1103 (2010)
14. L. Wang, G. Zhang, DYNRED enhancement project. EPRI, Palo Alto, CA, Tech. Rep. for Software Product ID 1020268, (2010).
15. F. Ma, V. Vittal, Right-sized power system dynamic equivalents for power system operation. IEEE Trans. Power Syst. **26**(4), 1998–2005 (2011)
16. F. Ma, X. Luo, V. Vittal, Application of dynamic equivalencing in large-scale power systems. in Proceedings of IEEE PES 2011 General Meeting, Detroit, MI, July 24–29, (2011).
17. S.S. Haykin, *Neural Networks: A Comprehensive Foundation* (Prentice-Hall, Englewood Cliffs, 1999)
18. Neural Network Toolbox, for use with MATLAB User's Guide, Version7, The Math Works Inc, 2010.

Chapter 6
Krylov Subspace and Balanced Truncation Methods for Power System Model Reduction

Shanshan Liu, Peter W. Sauer, Dimitrios Chaniotis and M. A. Pai

Abstract In this chapter, we discuss two mathematical approaches for model reduction of power systems which do not use coherency information. The advantages of these approaches lie in the ability to handle large systems. The Krylov and balanced truncation methods take into account system reachability and observability in obtaining reduced-order models of the external system, and thus would perform better than methods based simply on eigenvalues. In the case of balanced truncation approach, a sensitivity analysis is carried out. Through the eigenvalue information, we also establish a connection of these methods with the coherency- based approaches.

6.1 Introduction

6.1.1 Background

There is a vast literature on the model reduction of large-scale power systems with dynamics. Till the 1970s, the techniques were largely based on static models of a power system. The fundamental work of Podmore [1] in proposing coherency

S. Liu (✉)
Electric Power Research Institute, Palo Alto, CA, USA
e-mail: sliu@epri.com

P. W. Sauer
University of Illinois, Urbana-Champaign, IL, USA
e-mail: sauer@ece.uiuc.edu

D. Chaniotis
ENTSO-E, Brussels, Belgium
e-mail: dimitrios.chaniotis@entsoe.eu

M. A. Pai
University of Illinois, Urbana-Champaign, IL, USA
e-mail: mapai@uiuc.edu

J. H. Chow (ed.), *Power System Coherency and Model Reduction*,
Power Electronics and Power Systems 94, DOI: 10.1007/978-1-4614-1803-0_6,

as a basis for dynamic model reduction [2] set the stage for a physically based method. Practical software that emerged from that work has been of great value to utilities [3]. Another equally powerful method which uses the fast and slow dynamics in a large power system, proposed by Kokotović, Avramović, Winkelman, and Chow [4], is based on the concept of time-scale separation and has formed the basis for model reduction techniques, islanding, and study of oscillations. These topics are discussed in detail in the other chapters. The selective modal analysis (SMA) approach [5], which is another approach based on eigenvalue information, is discussed in Chap. 8 of this book. A somewhat related approach using electromechanical distance measure as a means for detecting coherency was proposed in [6]. Reference [7] proposed a RMS coherency measure approach based on eigensystem information. An approach which takes into account the changing pattern of coherency depending on system topology or altered operating conditions is proposed in [8]. Reference [9] uses an artificial neural network approach to coherency identification.

In contrast to the physically based coherency approaches, a new approach based on linear model reduction ideas was proposed by Chaniotis and Pai using Krylov subspace ideas [10]. The application of the balanced truncation method proposed by Liu [11] is an improvement over the Krylov-based approach for handling very large systems. Both methods use the concept of input-output properties of a linear time-invariant system at the boundary buses connecting the study area to the external system. With this brief overview, we shall discuss first the Krylov subspace approach.

6.2 Model Reduction Overview

The development of a mathematical model for a physical system can be a challenging task. A useful model must be able to capture many characteristics of the system. In power systems, the dimension of the models may easily reach the order of several thousand for applications like dynamic simulation or trajectory sensitivity analysis. Therefore, these types of analysis pose a formidable computational burden. Engineers have long treated this problem by reducing the order of the power system models using physically based reasoning. This may result in neglecting several phenomena that are assumed not to be of great significance to the response of the system. However, as systems become larger, a theoretically based formal approach is necessary. Model reduction consists of replacing the original system with one of a much smaller dimension according to the following guidelines:

- The reduced system must be an accurate representation of the original one for the analysis performed.
- The cost of generating the reduced model must be much smaller than performing the analysis using the original model.

For model reduction, power systems may be partitioned into two areas: the *study area* and the *external area* [1, 2]. The study area contains the variables of interest; therefore, it is modeled in detail. A fault is assumed to occur in the study area. The external area is important only as far as it influences the analysis in the study area. Thus, it is necessary to model only its input-output behavior at the boundary buses. Evidently, the modeling need not be detailed, and hence, model reduction may be performed in the external area.

6.3 Krylov Subspace approach

Krylov subspace methods were initially proposed for the solution of the eigenvalue problems [12, 13]. They were quickly adapted for the solution of symmetric, positive-definite linear systems of equations of the form $Ax = b$, where $A \in R^{N\times N}$ and $b \in R^{N\times 1}$ are given and $x \in R^{N\times 1}$ is the unknown vector [14, 15]. These results formed the basis for the first Krylov subspace technique, called the conjugate gradient method. Mainly because of the inherent parallelization capabilities and low storage requirement, Krylov methods offer an attractive technique. There is a great deal of interest in these methods in other areas of engineering, such as electromagnetics, where the size of A is extremely high.

The idea behind the conjugate gradient method is to seek the solution of $Ax = b$ from a Krylov subspace defined as

$$\mathcal{K}_k\{A, b\} = \left[b\; Ab\; \cdots\; A^{k-1}b\right]$$

so that the A-norm of the error (that is, $||(x-x^*)^T A(x-x^*)||_2$, where x^* is the solution) is minimized. There are many variations and improvements on the conjugate gradient method. Of these, the bi-conjugate gradient (Bi-CG) is perhaps the most interesting because it opens the field for the treatment of non-symmetric linear systems [16]. Other techniques have also been proposed, such as the quasi-minimum residual (QMR) method [17], the conjugate gradient squared (CGS) method [18], and the bi-conjugate gradient stabilized (Bi-CGSTAB) method [19].

The most robust and generic method in the family of Krylov subspace solvers is the generalized minimal residual (GMRES) method [20]. GMRES is guaranteed to converge in a finite number of iterations but its cost increases step by step. Therefore, it is usually restarted after a fixed number of iterations to reduce the cost. The idea of preconditioning was introduced in order to transform the system into one with a more favorable eigenvalue distribution for the Krylov subspace solvers. For a list of references to preconditioning techniques, see [21].

Consider the linear single-input single-output time-invariant system

$$\dot{x}(t) = Ax(t) + bu(t), \quad y(t) = c^T x(t) + du(t) \tag{6.1}$$

The vector $x(t) \in R^{N\times 1}$ is the vector of state variables, $A \in R^{N\times N}$ is the state matrix, $b \in R^{N\times 1}$ is the input vector of the system, $c^T \in R^{1\times N}$ is the output vector of the system, and d is the scalar feed-through term.

A reduced-order approximation to (6.1) takes the corresponding form

$$\dot{\hat{x}}(t) = \hat{A}\hat{x}(t) + \hat{b}u(t), \quad \hat{y}(t) = \hat{c}^T\hat{x}(t) + \hat{d}u(t) \tag{6.2}$$

where $\hat{x}(t) \in R^{m\times 1}, \hat{A} \in R^{m\times m}, \hat{b} \in R^{m\times 1}, \hat{c}^T \in R^{1\times m}, \hat{d}$ is a scalar, and $m < N$. The output $\hat{y}(t)$ approximates the true output $y(t)$. However, in general, no simple relation exists between $\hat{x}(t)$ and $x(t)$.

An important consideration is the transfer function of a system, denoted by

$$H(s) = c^T(sI - A)^{-1}b \tag{6.3}$$

where s represents the complex frequency. Without loss of generality, the feed-through term d of the original model is assumed to be zero. The transfer function of the reduced-order model, $\hat{H}(s)$, can be defined in a similar manner.

The expansion of the transfer function (6.3) around a certain interpolation point σ is given by

$$H(s) = \sum_{j=1}^{\infty} c^T\left[(\sigma I - A)^{-1}\right]^j b \cdot (s - \sigma)^{j-1}$$

The shifted moments of the system represent the value and the subsequent derivatives of the transfer function around σ and are given by

$$c^T\left[(\sigma I - A)^{-1}\right]^j b$$

The Krylov subspace-based moment matching technique approximates the system transfer function around certain interpolation points of interest $\sigma^{(k)}$ by using the Krylov subspaces generated by

$$V = \text{column span} \bigcup_{k=1}^{K} \mathscr{K}_{J_{bk}}\left\{(\sigma^{(k)}I - A)^{-1}, (\sigma^{(k)}I - A)^{-1}b\right\} \tag{6.4}$$

$$Z = \text{column span} \bigcup_{k=1}^{K} \mathscr{K}_{J_{ck}}\left\{(\sigma^{(k)}I - A^T)^{-1}, (\sigma^{(k)}I - A^T)^{-1}c^T\right\} \tag{6.5}$$

where $\sigma^{(k)}$ is the interpolation point.

Using V and Z, we develop the transformation matrices

$$T_{\text{right}} = \left[V \; V_{\perp} \right], \quad T_{\text{left}} = \left[Z \; Z_{\perp} \right], \tag{6.6}$$

where $V_{\perp}$ and $Z_{\perp}$ are matrices orthogonal to V and Z, respectively, such that T_{right} and T_{left} are square nonsingular matrices.

Applying this transformation to (6.1), we obtain

$$\begin{bmatrix} Z^T V & Z^T V_{\perp} \\ Z_{\perp}^T V & Z_{\perp}^T V_{\perp} \end{bmatrix} \begin{bmatrix} \dot{\hat{x}} \\ \dot{\hat{x}}_{\perp} \end{bmatrix} = \begin{bmatrix} Z^T A V & Z^T A V_{\perp} \\ Z_{\perp}^T A V & Z_{\perp}^T A V_{\perp} \end{bmatrix} \begin{bmatrix} \hat{x} \\ \hat{x}_{\perp} \end{bmatrix} + \begin{bmatrix} Z^T b \\ Z_{\perp}^T b \end{bmatrix} u$$

$$y = \left[c^T V \; c^T V_{\perp} \right] \begin{bmatrix} \hat{x} \\ \hat{x}_{\perp} \end{bmatrix}$$

Then, assuming $Z^T V$ is nonsingular, the leading subsystem is retained to form the reduced model

$$\dot{\hat{x}} = (Z^T V)^{-1} Z^T A V \hat{x} + (Z^T V)^{-1} Z^T b u, \quad y = c^T V \hat{x} \tag{6.7}$$

The moment preserving property of the reduced system is provided by Theorem 1.

Theorem 6.1 *[22] Given (6.4) and (6.5), then*

$$-c^T [\left(\sigma^{(k)} I - A \right)^{-1}]^{jk-1} (\sigma^{(k)} I - A)^{-1} b = \hat{c}^T [\left(\sigma^{(k)} I - \hat{A} \right)^{-1}]^{jk-1} (\sigma^{(k)} I - A) \hat{b}$$

for $j_k = 1, 2, \ldots, J_{bk} + J_{ck}$ *and* $k = 1, 2, \ldots, K$.

Power systems are rarely single-input single-output systems. Therefore, one has to extend the previous analysis to include multi-input multi-output systems. Then the Krylov subspaces are replaced by block Krylov subspaces

$$\mathscr{K}_j\{A, B\} = \text{column span} \left\{ B, AB, A^2 B, \ldots, A^{j-1} B \right\}$$

to account for the multidimensional input and output matrices B and C. Certain issues arise when block methods are applied to model reduction. The most important one is that the dimension of the reduced model is significantly larger because the block Krylov subspace has to be large in order to contain all the information generated by the individual Krylov subspaces corresponding to the columns of B and C.

The construction of the bases V and Z in transformation (6.6) is purely an implementation choice. Various approaches have been developed to construct the bases for a Krylov subspace. Well-known Krylov subspace methods are the Arnoldi, Lanczos, GMRES (generalized minimum residual), and BiCGSTAB (stabilized bi-conjugate gradient) methods. More details can be found in [21].

The algorithm used in this chapter is shown below.

Krylov-Subspace Algorithm

```
Define:
    p: the number of columns of B
    J: the number of moments matched by every interpolation point
    |σ^(0), σ^(1), ··· , σ^(K)|: interpolation points
Compute:
    B = (σ^(0) I − A)^(−1) B
    (V, R) = QR factorization of B
    m = p + 1
    for j = p : J
      for k = 1 : K
        ω = (σ^(k) I − A)^(−1) V(:, K · (j − p) + k)
        i = 1 : m − 1
          h = (ω, V(:, i))
          ω = ω − V^T(:, i) · h
        end
        if ||ω||_2 > 10^(−6)
          V(:, j + 1) = ω/||ω||_2
          m = m + 1
        end
      end
    end
```

6.4 Balanced Truncation Method

Balanced truncation is an important projection model reduction method which delivers high-quality reduced-order models by carefully choosing projection subspaces to maintain controllability and observability. It has been applied in spacecraft modeling, control system design, and other areas of large dimension modeling [23].

Rewrite the system (6.1) with multiple inputs and outputs as

$$\dot{x}(t) = Ax(t) + Bu(t), \quad y(t) = Cx(t) + Du(t) \tag{6.8}$$

where the vector $x(t) \in R^{N\times 1}$ is the vector of state variables, $A \in R^{N\times N}$ is the state matrix, $B \in R^{N\times r}$ is the input vector of the system, $C \in R^{p\times N}$ is the output vector of the system, and $D \in R^{p\times r}$ is the scalar feed-through matrix. The transfer function of the system is denoted by $H(s)$

$$H(s) = C(sI - A)^{-1}B + D \tag{6.9}$$

The balanced truncation makes use of two crucial quantities, called the reachability and observability Gramians. The reachability Gramian is defined as

$$W_r(0, t_f) = \int_0^{t_f} e^{A\tau} BB^T e^{A^T\tau} \mathrm{d}\tau \tag{6.10}$$

and the observability Gramian is defined as

$$W_o(0, t_f) = \int_0^{t_f} e^{A^T\tau} C^T C e^{A\tau} \mathrm{d}\tau \tag{6.11}$$

The balanced truncation method consists of transforming the state-space model into a balanced form, together with a truncation of those states that are both difficult to reach and difficult to observe.

An important property of this method is that asymptotic stability is preserved in the reduced-order system. Moreover, the existence of a priori error bounds [23] allows an adaptive choice of the state-space dimension of the reduced model, depending on how accurate an approximation is desired.

6.4.1 Gramians and Hankel Singular Values

Frequently, system $H(s)$ is assumed asymptotically stable, and the controllability and observability Gramians W_r and W_o are then solutions of two Lyapunov equations

$$AW_r + W_rA^T = -BB^T, \quad A^T W_o + W_oA = -C^T C \tag{6.12}$$

The square roots of the eigenvalues of the product $W_r W_o$ are so-called Hankel singular values σ of the system $H(s)$

$$\sigma_i = \sqrt{\lambda_i(W_r W_o)} \tag{6.13}$$

In many cases, the eigenvalues of $W_r W_o$ as well as the Hankel singular values σ decay very rapidly. We now consider transformation of the above system such that $x = T\hat{x},\ T \in R^{N\times N}$. We have

$$\hat{A} = T^{-1}AT, \quad \hat{B} = T^{-1}B, \quad \hat{C} = CT, \quad \hat{D} = D \tag{6.14}$$

and

$$\hat{W}_r = T^{-1}W_rT^{-T}, \quad \hat{W}_o = T^T W_o T \tag{6.15}$$

The reachable and observable system $H(s)$ is called Lyapunov balanced if

$$\hat{W}_r = \hat{W}_o = \Sigma = \mathrm{diag}(\Sigma_1, \Sigma_2) = \mathrm{diag}(\sigma_1, \ldots, \sigma_k, \sigma_{k+1}, \ldots, \sigma_N) \tag{6.16}$$

where

$$\sigma_1 > \cdots > \sigma_k > \sigma_{k+1} > \cdots > \sigma_N$$
$$\Sigma_1 = \mathrm{diag}(\sigma_1, \ldots, \sigma_k), \quad \Sigma_2 = \mathrm{diag}(\sigma_{k+1}, \ldots, \sigma_N)$$

It should be pointed out that, while the eigenvalues (or system modes) are invariant under similarity T, the eigenvalues of the Gramians are not. However, the eigenvalues of the product of the Gramians, that is, the Hankel singular values, are readily seen to be similarity invariant.

By truncating the states that are simultaneously difficult to reach and difficult to observe, which correspond to the small Hankel singular values $(\sigma_{k+1}, \ldots, \sigma_N)$, we can get the reduced-order system

$$H_r(s) = C_r\,(sI - A_r)^{-1}\,B_r + D \tag{6.17}$$

The reduced-order system satisfies

$$||H(s) - H_r(s)||_{\mathcal{H}_\infty} \leq 2 \sum_{i=k+1}^{N} \sigma_i, \quad \text{and} \quad ||H(s) - H_r(s)||_{\mathcal{H}_\infty} \geq \sigma_k \tag{6.18}$$

which equally holds if $\sum_2 = \mathrm{diag}(\sigma_n)$

A state-space balancing algorithm is described below [24].

Balanced Truncation Algorithm

1. Compute Cholesky factors of the Gramians (the Gramians themselves are not actually formed). Let L_r and L_o denote the lower triangular Cholesky factors of the Gramians W_r and W_o, i.e.,

$$W_r = L_r L_r^T, \quad W_o = L_o L_o^T \tag{6.19}$$

2. Compute the singular value decomposition

$$L_o L_o^T = U \Lambda V^T \tag{6.20}$$

3. Form the balancing transformation

$$T = L_r V \Lambda^{-1/2} \tag{6.21}$$

4. Form the balanced state-space matrices $\hat{A}, \hat{B}\cdot \hat{C}, D$.

Several other approaches exist for balancing the system, such as the Schur method and the square root method. Further details and references can be found in [23–26].

Other topics including balanced truncation for model reduction of nonlinear systems and time-varying systems are discussed in [27–29].

6.5 Application to Power Systems

In this section, we apply the Krylov subspace and balanced truncation methods to obtain linear reduced-order models of an external system. We assume that the external system is connected to study area with p tie-lines.

The second-order classic model used for the study of model reduction of the external area is

$$\begin{aligned}
\dot{\delta}_i &= \omega_i - \omega_s \\
\frac{2H_i}{\omega_s}\dot{\omega}_i &= T_{M_i} - D_i(\omega_i - \omega_s) - E_{ii}^2 G_{ii} \\
&\quad - E_i \sum_{j=1, j\neq i}^{N_G} \left(E_j G_{ij}\cos(\delta_i - \delta_j) + E_j B_{ij}\sin(\delta_i - \delta_j)\right) \\
&\quad - E_i \sum_{j=1}^{p} \left(V_j \bar{G}_{ij}\cos(\delta_i - \theta_j) + V_j \bar{B}_{ij}\sin(\delta_i - \theta_j)\right)
\end{aligned} \tag{6.22}$$

where δ_i and ω_i are the rotor angle and speed of generator i, N_G is the number of generators, and p is the number of tie-lines. H_i and D_i represent the inertia and damping coefficients, respectively, and E_i is the internal voltage of machine i. The admittance $Y_{ij} = G_{ij} + jB_{ij}$ connects machines i and j, and the admittance $\bar{Y}_{ij} = \bar{G}_{ij} + j\bar{B}_{ij}$ connects machine i and boundary bus j. A detailed description of how to deduct derive (6.22) for a power system can be found in [30].

The inputs of the external system, denoted as u, are considered to be the angles θ_i and voltage amplitudes V_i of the p connected buses belonging to the study area. The outputs of the system, denoted as y, are considered to be the angles and voltages of the p corresponding buses from the external system.

A 16-machine, 68-bus system is assumed to be the study area, and a 50-machine, 145-bus system is taken as the external area [10]. The configuration of the study and external areas is shown in Fig. 6.1.

Linearizing the system (6.22) around an equilibrium point, we obtain

$$\begin{aligned}
\begin{bmatrix} \Delta\dot{\delta} \\ \Delta\dot{\omega} \end{bmatrix} &= \begin{bmatrix} 0 & I \\ (2H)^{-1}K & -(2H)^{-1}D \end{bmatrix} \begin{bmatrix} \Delta\delta \\ \Delta\omega \end{bmatrix} + \begin{bmatrix} 0 \\ (2H)^{-1}B \end{bmatrix} \\
y &= \begin{bmatrix} C & 0 \end{bmatrix} \begin{bmatrix} \Delta\delta \\ \Delta\omega \end{bmatrix} \Delta u
\end{aligned} \tag{6.23}$$

where

$$
\begin{aligned}
K(i,j) &= E_i E_j(-G_{ij}\sin\delta_{ij}^o + B_{ij}\cos\delta_{ij}^o), \quad i,j = 1,\dots,N_G \\
K(i,i) &= E_i \sum_{j=1,j\neq i}^{N_G} E_j(G_{ij}\sin\delta_{ij}^o - B_{ij}\cos\delta_{ij}^o), \quad i = 1,\dots,N_G \\
B(i,j) &= E_i V_j(-\bar{G}_{ij}\sin(\delta_i-\theta_j) + \bar{B}_{ij}\cos(\delta_i-\theta_j)), \\
&\quad i = 1,\dots,N_G, j = 1,\dots,p \\
B(i,j+p) &= E_i(-\bar{G}_{ij}\cos(\delta_i-\theta_j) - \bar{B}_{ij}\sin(\delta_i-\theta_j)), \\
&\quad i = 1,\dots,N_G, j = 1,\dots,p \\
C(i,j) &= 1, \quad \text{if bus } i \text{ is connected through tie line } j, \\
&\quad i = 1,\dots,N_G, j = 1,\dots,p \\
C(i,j) &= 0, \quad \text{if bus } i \text{ is not connected through tie line } j, \\
&\quad i = 1,\dots,N_G, j = 1,\dots,p \\
\delta_{ij}^o &= \delta_i^o - \delta_j^o, \quad i,j = 1,\dots,N_G \\
H &= \mathrm{diag}(H_i/\omega_s), \quad D = \mathrm{diag}(D_i), \quad i = 1,\dots,N_G
\end{aligned}
$$

This formulation allows the external area to be modeled along the lines of (6.1), and the Krylov subspace model reduction methodology described in Sect. 6.3 can be applied in a straightforward manner.

6.5.1 Simulation of Single Tie-Line Case

In the first study case, the two areas in Fig. 6.1 are connected with 1 tie-line, from Bus 58 in the study area to Bus 140 in the external system. The dimension of the external system is 100 × 100. A self-clearing fault at Bus 27 in the study area is applied. The duration of the fault is 0.1 s. To compare the results of unreduced and reduced systems, we perform a full system study using the Power System Toolbox [32] without partitioning into study and external areas.

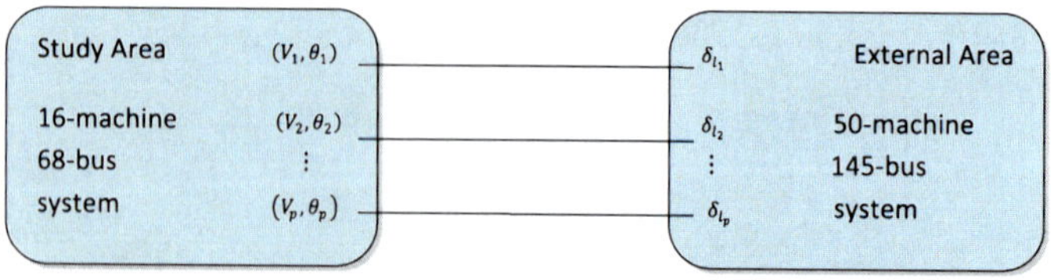

Fig. 6.1 The topology of the study and external areas

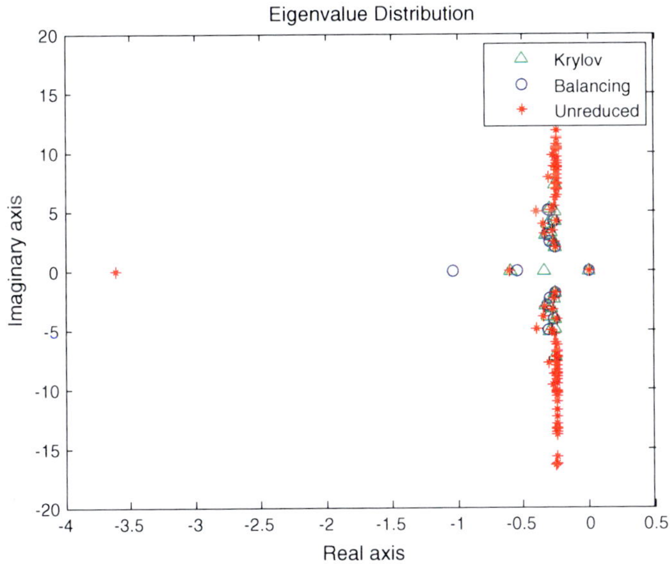

Fig. 6.2 Eigenvalue distributions of unreduced system, and reduced system vis Krylov and BT methods

In the Krylov subspace method, we start with using only one interpolation point, at zero frequency ($\sigma = 0$) [10]. The dimension of the resulting reduced system is 21×21. For the balanced truncation, the dimension of the reduced system is 13×13.

Figure 6.2 shows the eigenvalue distributions and Fig. 6.3 shows the frequency response for the full model, the Krylov subspace reduced model, and the balanced truncation reduced model. Both reduced models capture the dominant modes of the original system. Also, the frequency response suggests that Krylov subspace follows the unreduced system well for low frequencies, but is less accurate for high frequencies. This is because we use $\sigma = 0$ as the interpolation point. To improve the frequency response approximation of the Krylov subspace reduced model, we add two more frequency points $\sigma = [5, 10]$ to the interpolation list. The improved performance at high frequencies is shown in Fig. 6.4.

Figure 6.5 shows the time response simulation results of the reduced models from the Krylov subspace method and the balanced truncation method compared to the unreduced system. The labels "UnPartitioned" and "UnReduced" denote two different simulation methods. "UnPartitioned" means that both the study area and external area are simulated together. "UnReduced" means that the full study area and full external system are simulated separately: the voltages and angles of the boundary buses are used as inputs and outputs and updated each step. As expected, these two simulation methods will result in slightly different responses. In evaluating the performance of the reduced models, it is more appropriate to compare their time responses to that of the "UnReduced" model, as the reduced models are simulated separately from the study area and then reconnected. The results show not only that

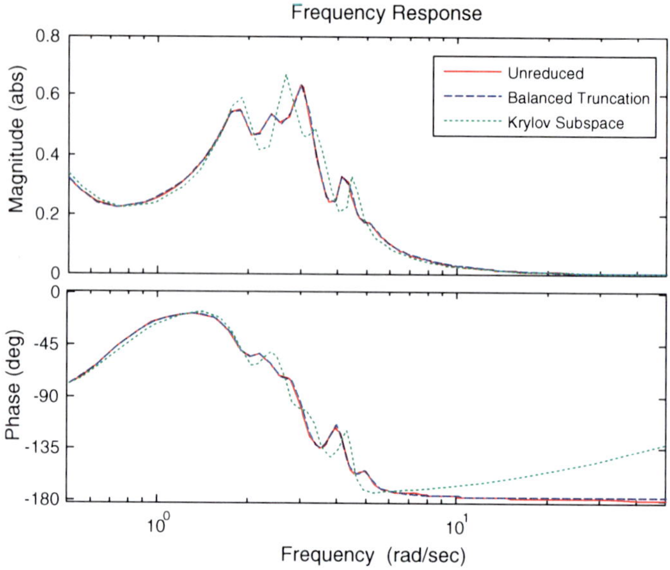

Fig. 6.3 System frequency responses—one-point interpolation

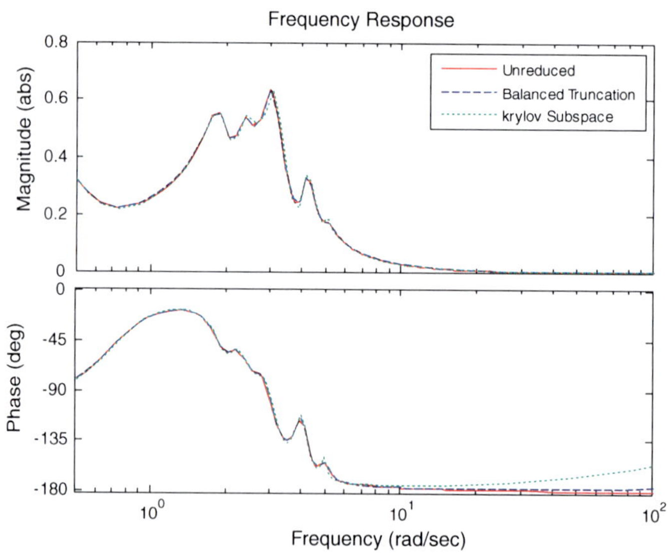

Fig. 6.4 System frequency responses—three-point interpolation

the reduced systems are good representations of the external system, but also that it is feasible to partition the system into two areas.

To quantify the accuracy of the two reduced models, we define a time response approximation error as

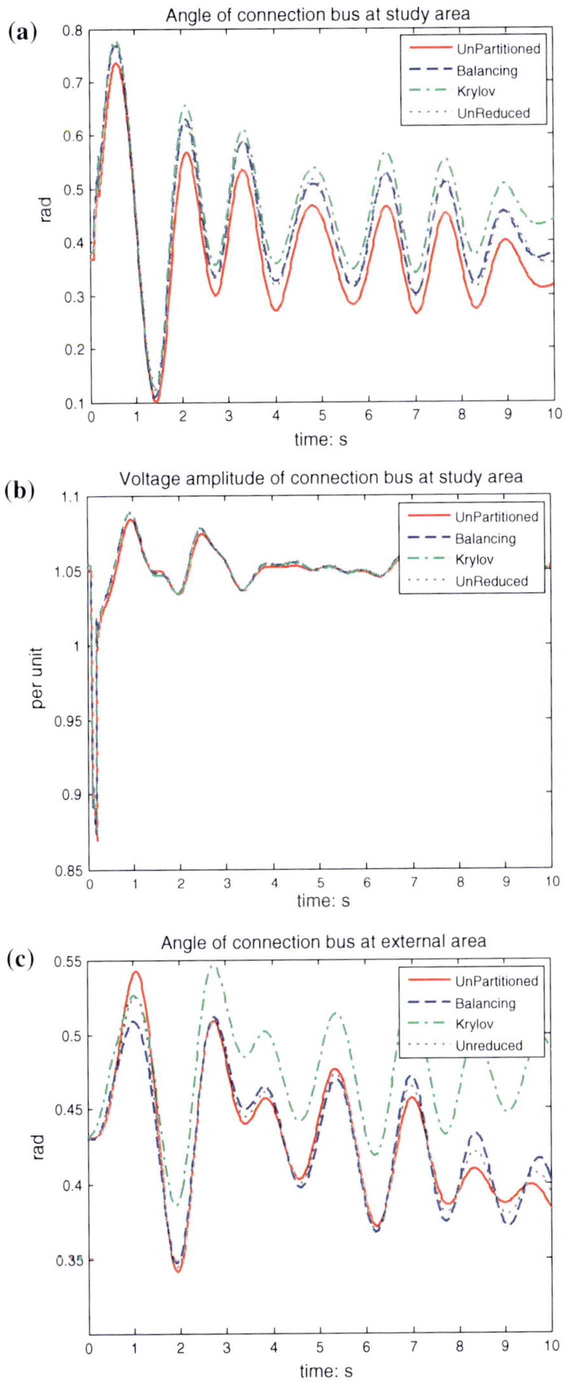

Fig. 6.5 One tie-line case

$$\varepsilon_{n_x} = \sqrt{\frac{1}{n_t}\sum_{i=1}^{n_t} |V_{n_x} - V_{n_x\text{full}}|_i^2} \tag{6.24}$$

where n_x is the boundary bus number and n_t is the number of points in the time response simulation.

For the Krylov reduced model of dimension 21, the error is $\varepsilon_{\text{Krylov}} = 0.0523$. For the balanced truncation model of dimension 13, the error is $\varepsilon_{\text{Balancing}} = 0.0125$. We can thus reach the following conclusion: The balanced truncation method can achieve a better approximation than the Krylov subspace method while using a lower dimensional system. The balanced truncation model can represent the direct input-output relation of the external system, without the redundancy in the Krylov subspace method.

The reduced system preserves most information of the unreduced system regarding the input-output relation. Partitioning the whole system into study and external areas, however, will sacrifice some accuracy.

6.5.1.1 Sensitivity Analysis

Because we perform model reduction on linearized system models, we now perform a sensitivity analysis to improve the algorithm. Sensitivity analysis (SA) is the study of how the variation (uncertainty) in the output of a mathematical model can be apportioned, qualitatively or quantitatively, to different sources of variation in the input of a model [33].

There are several possible approaches to perform SA, such as local methods, sampling-based methods, methods based on emulators, screening methods, variance-based methods, and high dimensional model representations. We use local methods for our investigation. Local methods utilize the idea of using the simple derivative of the output y with respect to a parameter α, $|\partial y/\partial\alpha)|_{x_0}$, where the subscript x_0 indicates that the derivative is taken at some fixed point in the state space (hence the "local" in the name of the class). The parameter α can be a model parameter or a state variable.

In general, the derivative can be obtained either analytically by working on the system equations or quantitatively by studying the variations of output with respect to the small variations in parameters. In this problem, our goal is to detect the system model variation, so we choose the quantitative approach to avoid working on a large number of variables and complex relationships between parameter and system model characterization, and to take advantage of computer programming.

From Figs. 6.6 and 6.7, we can see that system (6.22) is more sensitive to machine angles than bus voltages. It is also worth mentioning that, because the machine angles tend to change along the same direction, their influences on output variations might cancel each other out.

We assign an index to each parameter according to the sensitivity analysis and calculate the aggregated expected variance of the system. When such variance reaches

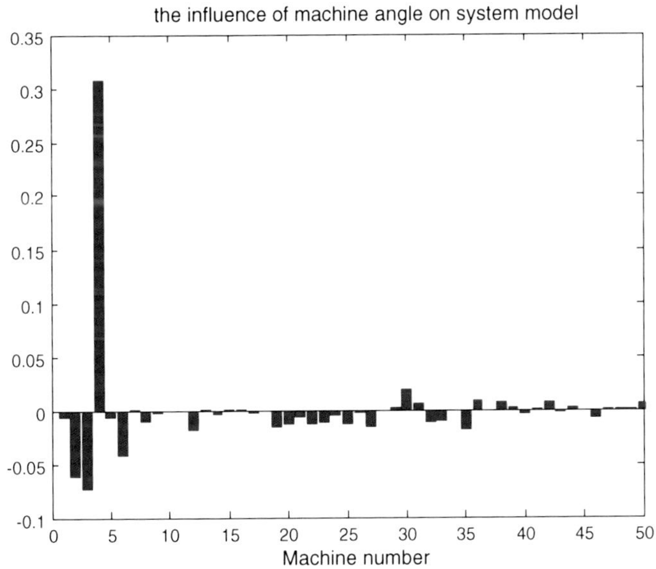

Fig. 6.6 Sensitivity analysis of machine angles

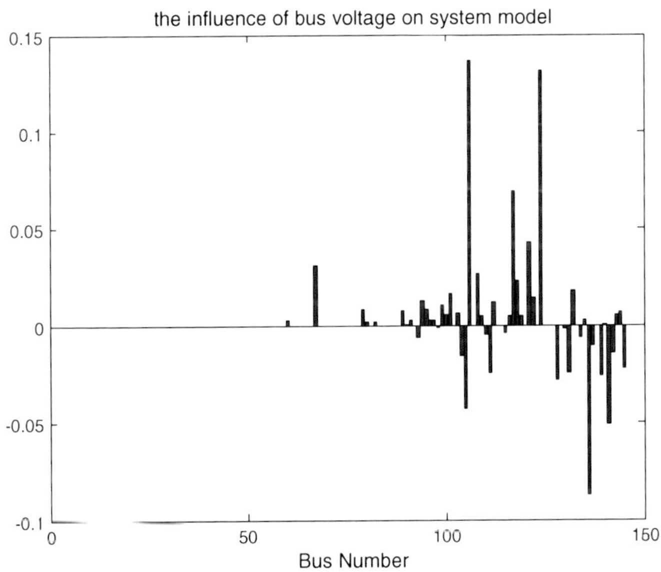

Fig. 6.7 Sensitivity analysis of bus voltages

some preset threshold, we then update the system model. The importance of such sensitivity study and system model updating is illustrated in the next section using a three tie-line case.

6.5.2 Simulation of 3 Tie-Line Case

In this section, a three tie-line case is studied, in which three tie-lines connect the study and external areas. The same fault is applied to the study area and then is cleared at 0.1 s.

The dimension of the Krylov subspace method is 27 and the interpolation points are $\sigma = [0, 5, 10, 15]$. The dimension of the balanced truncation reduced model is also 27. The Hankel singular valve error bound of the reduced model is

$$||H(s) - H_r(s)||_{\mathcal{H}_\infty} \leq 0.0195$$

Figure 6.8 shows the singular values of original unreduced system, and reduced-order models of the balanced truncation method and the Krylov subspace method, as a function of frequency. Note that the reduced-order model from the balanced truncation method approximates well the full model.

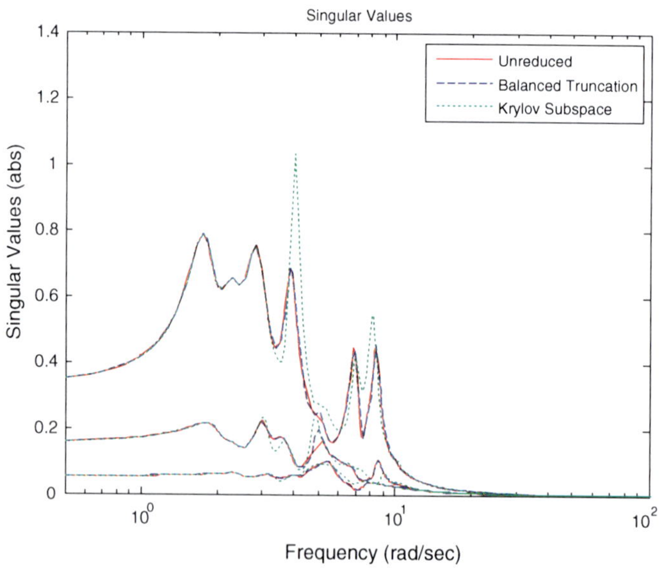

Fig. 6.8 Singular value frequency response

Figure 6.9 shows a comparison of time response simulations of the full and balanced truncation model. During the simulation, the aggregated index that represent the expected model variance have been used to decide when to update the balanced

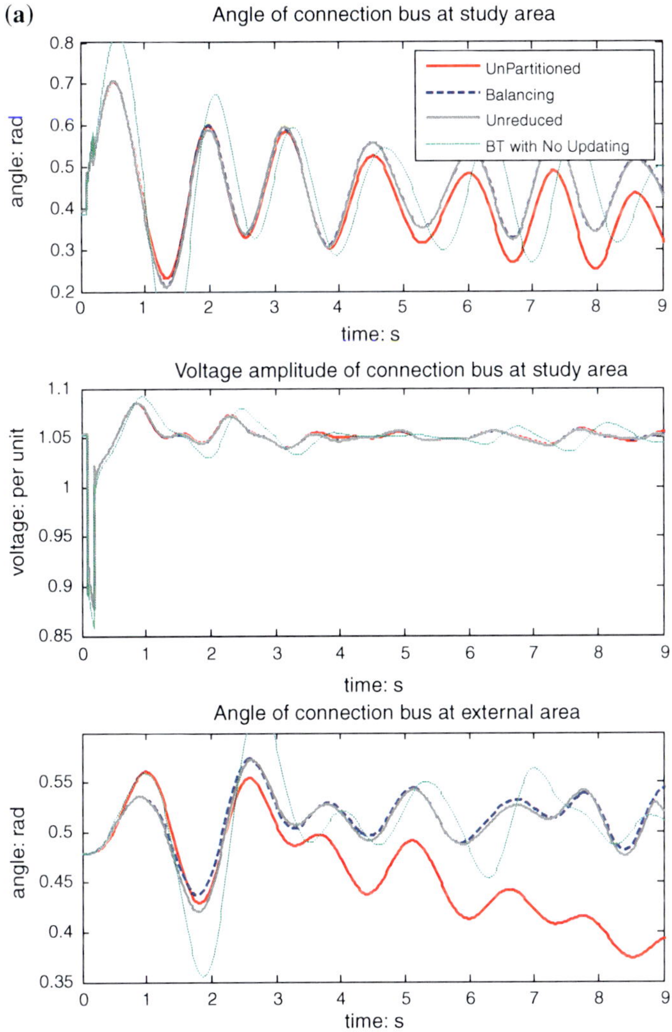

Fig. 6.9 Three tie-line case. **a** Tie-line Bus 1. **b** Tie-line Bus 2. **c** Tie-line Bus 3

truncation system model. Figure 6.9 shows that the simulation results of the balanced truncation method with no system model updating appear to have more error, whereas the simulation results of the balanced truncation method with system model updated periodically follow the simulation results of unreduced system closely. Because a linearized system is used, updating the system model based on sensitivity analysis can improve the performance, as is shown here.

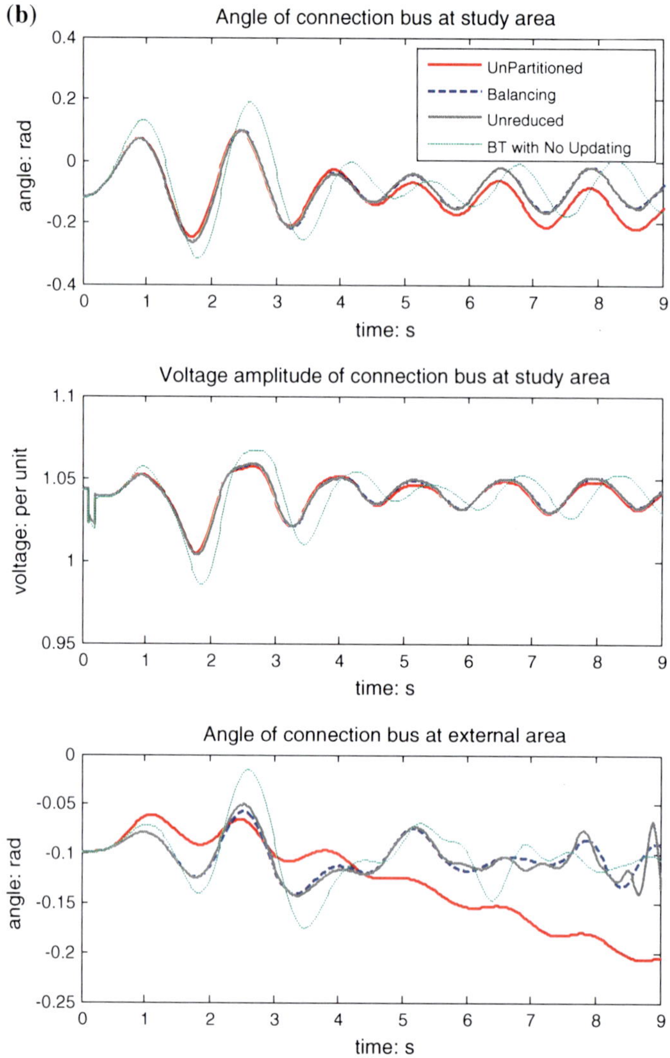

Fig. 6.9 (Continued)

6.6 Krylov Subspace Reduction and Coherency

The mechanics of the Krylov subspace model reduction dictate that the states of the unreduced system are confined to the subspace V (6.6), i.e.,

$$x = V\hat{x} \tag{6.25}$$

Therefore, the rows of V may be a good indication about the relative movement of the states of the unreduced system.

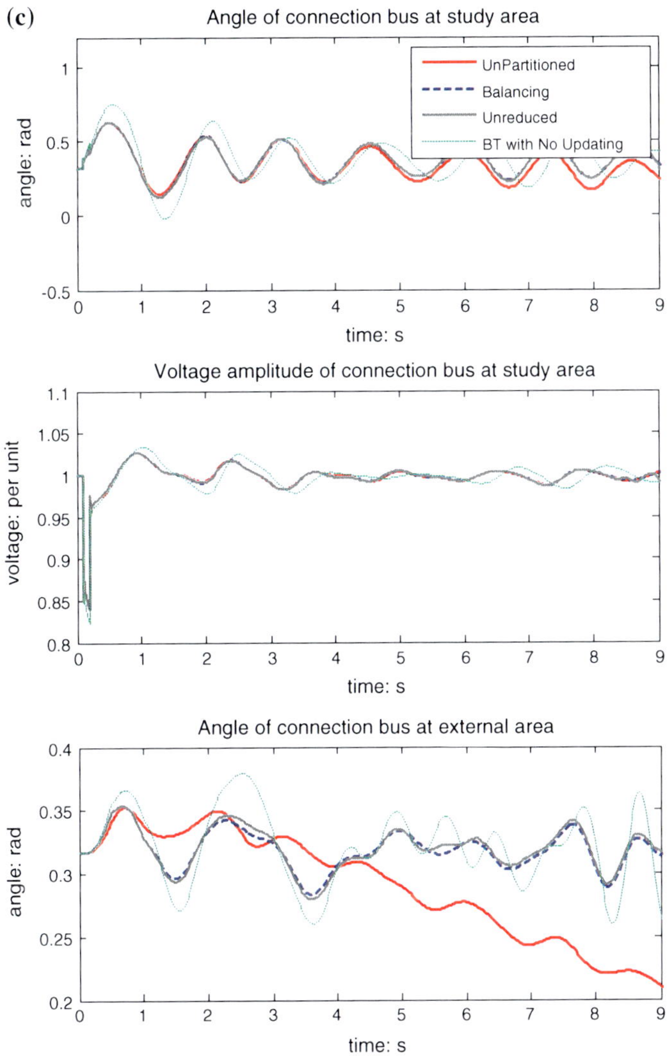

Fig. 6.9 (Continued)

Note, however, the equation above does not imply that the actual trajectories lie on subspace V, but rather that if the trajectories lie on this subspace, then the output of the reduced system matches the output of the unreduced system

$$x_i - x_j = (V(i,:) - V(j,:))\hat{x} \tag{6.26}$$

According to the definitions given in EPRIs report [2], the states i and j are coherent if the difference between x_i and x_j is small. A possible criterion to determine the closeness of x_i and x_j is the angle φ between rows $V(i,:)$ and $V(j,:)$, defined as

$$\cos\varphi = \frac{V(j,:)V(i,:)^T}{||V(j,:)||_2||V(i,:)||_2} \tag{6.27}$$

Therefore, one can define x_i and x_j as coherent if the angle φ is less than a pre-specified tolerance. Let G be the matrix

$$G(i,j) = \cos^{-1}\left(\frac{V(j,:)V(i,:)^T}{||V(j,:)||_2||V(i,:)||_2}\right) \tag{6.28}$$

Thus, the entries of G determine whether or not any two states are coherent. Equation (6.25) does not distinguish between the modes of the reduced system; it assumes that all modes may be excited. It is possible to eliminate certain high-frequency modes that are of no interest and correspond to parasitic modes by ignoring the states of the reduced system that have a relatively large participation on these modes. The small dimension of the reduced system allows a complete eigen-analysis in a robust and efficient way. Therefore, the contribution of each state to the modes of the reduced system can be identified through the use of participation factors [31]. Let v and w be the right and left eigenvectors of the reduced system corresponding to an eigenvalue λ. For complex eigenvectors, the magnitudes of each entry of v and w are considered. Then, the participation factor of state i to the eigenvalue λ is defined as

$$p_i = \frac{v_i w_i}{v^T w} \tag{6.29}$$

The participation factors can be normalized so that the largest entry is 1. The modes of interest are isolated (for example, the ones corresponding to the low frequencies) and the participation of each state of the reduced system on a certain mode can be computed from (6.29). The states that contribute the least to the modes of interest are discarded. This is equivalent to neglecting the corresponding columns of V in (6.25). Then, G is constructed using the reduced V.

To illustrate the method, the same system setting is examined for the 50-machine external system, and the coherent generators are grouped in Table 6.1 according to the values of G. Note that machines in the external areas that are not impacted by disturbances from the study area are not listed in the table. The modes retained are the ones with imaginary part less than 5 and real part greater than 1, that is, the

Table 6.1 Coherent groups for the 50-machine system

Group	Coherent generators
1	4,5
2	6,7,9,10,13,15,16,17,18,20,21,22 23,24,25,26,27,28,34
3	8,11
4	14,35,36
5	33,37,38,39
6	42,43

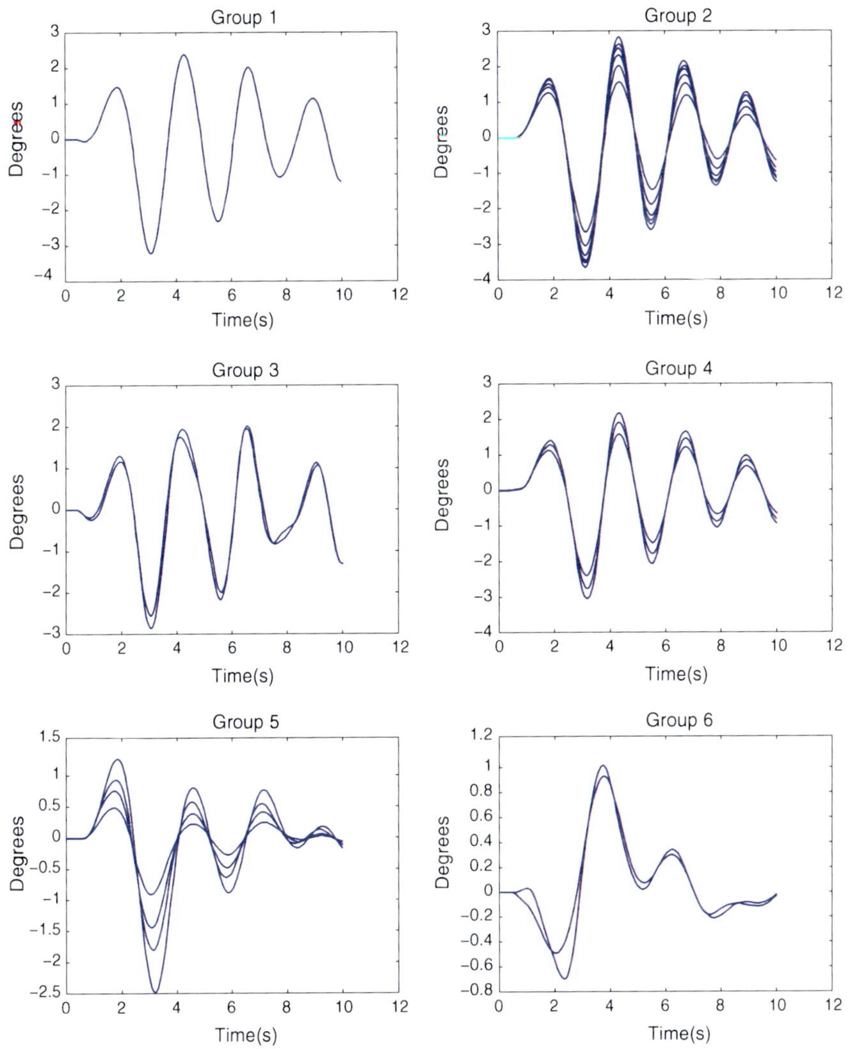

Fig. 6.10 Coherent groups

low-frequency modes. States with participation factors less than 0.3 on these modes are eliminated.

Figure 6.10 show the simulation results (using the unreduced model for the external area). The time responses in Fig. 6.10 imply that generators in Groups 1 and 2 should be termed as coherent even though our method proposes that they are not.

To test the validity of the method, one of the basic assumptions of coherency is used [2]: The coherent groups of generators are independent of the size of the disturbance. Therefore, if a fault at the same bus is considered to be cleared at 0.35 s, the coherent groups should remain the same. Figure 6.11 shows the time responses of

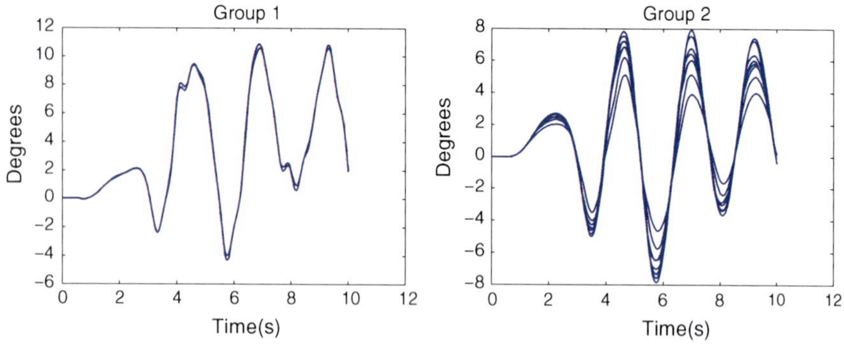

Fig. 6.11 Coherent groups

Groups 1 and 2. Clearly, the generators belonging to Groups 1 and 2 are not coherent, which confirms the results shown in Table 6.1.

6.7 Conclusions

Both Krylov subspace and balanced truncation methods are discussed in this chapter and applied to a power system for model reduction of the external system. The emphasis on input-output connections in both methods helps in removing the redundant information while preserving the relation between system inputs and outputs. Simulation results show that the reduced-order system via balanced truncation is a good approximation of the original high order system. Compared with Krylov subspace method, it represents a better approximation with lower orders. Also, the error of the balanced truncation method is bounded by Hankel singular values. Because a linearized system is used, updating the system model based on sensitivity analysis can improve the performance. In the Krylov method, a connection to the coherency approach is established.

References

1. R. Podmore, Identification of coherent generators for dynamic equivalents. IEEE Trans. Power Apparatus Syst. **PAS–97**(4), 1344–1354 (1978)
2. R. Podmore, A. Germond, Dynamic equivalents for transient stability studies. Systems control, Inc., final report prepared for EPRI project RP-763, April 1977
3. L. Wang, M. Klein, S. Yirga, P. Kundur, Dynamic reduction of large power systems for stability studies. IEEE Trans. Power Syst. **12**, 889–895 (1997)
4. P.V. Kokotović, B. Avramović, J.H. Chow, J.R. Winkelman, Coherency based decomposition and aggregation. Automatica **18**, 47–56 (1982)
5. I.J. Perez-Arriaga, G.C. Verghese, F.C. Schweppe, Selective modal analysis with applications to electric power systems, part I: Heuristic introduction. IEEE Trans. Power Apparatus Syst.

PAS–101(9), 3117–3125 (1982)
6. M.A. Pai, R.P. Adgaonkar, Electromechanical distance measure for decomposition of power systems. Electr. Power Energy Syst. **6**(4), 249–254 (1984)
7. J. Lawler, R.A. Schlueter, P. Rusche, D.L. Hackett, Modal-coherent equivalents derived from an RMS coherency measure. IEEE Trans. Power Apparatus Syst. **PAS–99**(4), 1415–1425 (1980)
8. H. You, V. Vittal, X. Wang, Slow cherency-based islanding. IEEE Trans. Power Syst. **19**, 483–491 (2004)
9. M.-H. Chang, H.-C. Wang, Novel clustering method for coherency identification using an artificial neural network. IEEE Trans. Power Syst. **9**, 2056–2062 (1994)
10. D. Chaniotis, *Krylov Subspace Methods in Power System Studies*, Ph.D. dissertation, University of Illinois at Urbana-Champaign, 2001
11. S. Liu, *Dynamic-Data Driven Real-Time Identification for Electric Power Systems*, Ph.D. dissertation, University of Illinois at Urbana-Champaign, 2009
12. W.E. Arnoldi, The principle of minimized iteration in the solution of the matrix eigenvalue. Q. Appl. Math. **19**, 17–29 (1951)
13. C. Lanczos, An iteration method for the solution of the eigenvalue problem of linear differential and integral operators. J. Res. Natl. Bur. Stan. **45**, 255–282 (1950)
14. M.R. Hestenes, E.L. Stiefel, Methods of conjugate gradients for solving linear systems. J. Res. Natl. Bur. Stan. **49**, 409–436 (1952)
15. C. Lanczos, Solution of systems of linear equations by minimized iterations. J. Res. Natl. Bur. Stan. **49**, 33–53 (1952)
16. R. Fletcher, Conjugate gradient methods for indefinite systems, in Proceedings of the Dundee Biennal Conference on Numerical Analysis 1974, New York, 1975
17. R.W. Freund, N.M. Nachtigal, An implementation of the look-ahead Lanczos algorithm for Non-Hermitian matrices: Part 2, NASA Ames Research Center, Technical report no. 90.46, 1990
18. P. Sonneved, CGS: a fast Lanczos-type solver for nonsymmetric linear systems. SIAM J. Sci. Stat. Comput. **10**, 36–52 (1989)
19. H.A. van der Vorst, Bi-CGSTAB: a fast and smoothly converging variant of Bi-CG for the solution of non-symmetric linear systems. SIAM J. Sci. Stat. Comput. **13**, 631–644 (1992)
20. Y. Saad, M.H. Schultz, GMRES: a generalized minimal residual algorithm for solving non-symmetric linear systems. SIAM J. Sci. Stat. Comput. **7**, 856–869 (1986)
21. Y. Saad, *Iterative Methods for Sparse Linear Systems*, 2nd edn. (SIAM, Philadelphia, 2003)
22. A.C. Antoulas, *Approximation of Large-Scale Dynamical Systems* (SIAM, Philadelphia, 2005)
23. S. Gugercin, A.C. Antoulas, A survey of model reduction by balanced truncation and some new results. Int. J. Control **77**(8), 748–766 (2004)
24. A. Laub, M. Heath, C. Paige, R. Ward, Computation of system balancing transformations and other applications of simultaneous diagonalization algorithms. IEEE Trans. Autom. Control **32**, 115–122 (1987)
25. B.C. Moore, Principal component analysis in linear systems: controllability, observability, and model reduction. IEEE Trans. Autom. Control **AC–26**, 17–32 (1981)
26. K. Glover, All optimal Hankel-norm approximations of linear multivariable systems and their L^∞ norms. Int. J. Control **39**, 1115–1193 (1984)
27. S. Lall, J.E. Marsden, S. Glavaski, A subspace approach to balanced truncation for model reduction of nonlinear control systems. Int. J. Robust Nonlinear Control **12**(6), 519–535 (2002)
28. S. Lall, C. Beck, Error bounds for balanced model reduction of linear time-varying systems. IEEE Trans. Autom. Control **48**, 946–956 (2003)
29. H. Sandberg, A. Rantzer, Balanced truncation of linear time-varying systems. IEEE Trans. Autom. Control **49**, 217–229 (2004)
30. P.W. Sauer, M.A. Pai, *Power System Dynamics and Stability* (Prentice Hall, Englewood Cliffs, 1998)
31. G. Rogers, *Power System Oscillations* (Kluwer, Dordrecht, 2000)
32. J. Chow, G. Rogers, Power systems toolbox, Available: http://www.ecse.rpi.edu/ chowj

33. A. Saltelli, M. Ratto, T. Andres, F. Campolongo, J. Cariboni, D. Gatelli, M. Saisana, S. Tarantola, *Global Sensitivity Analysis: The Primer* (Wiley-Interscience, New York, 2008)
34. D. Chaniotis, M. Pai, Model reduction in power systems using Krylov subspace methods. IEEE Trans. Power Syst. **20**, 888–894 (2005)

Chapter 7
Reduction of Large Power System Models: A Case Study

Kip Morison and Lei Wang

Abstract This chapter illustrates, by a practical case study, an application of the dynamic model reduction for a large power system model. This example concerns the creation of a reduced-order model from a full WECC model, to be used by BC Hydro for on-line dynamic security assessment (DSA). This model reduction was conducted using the software DYNRED. The focus of this case study is on the objective of the model reduction, the approach and procedure, the results, and the benchmarking work to ensure the suitability of the reduced models for the required applications. It is shown that, following the procedure used in the case study, a reduced model can be obtained for use with a wide range of system conditions with acceptable results and computation time saving.

7.1 Introduction

Modern interconnected electric power systems cover very large geographic areas served by many utilities or grid operators. To perform studies for such systems, it is often neither practical nor necessary to model in detail an entire interconnected system. This is particularly true for real-time control center applications for which the following two factors practically prevent the use of full system models [1]:

- The operating condition of a portion of an interconnected system may not be available to such applications at real time due to the limited observability of the SCADA/EMS for a utility or grid operator. Thus, creation of a full real-time system model is infeasible.

K. Morison (✉)
BC Hydro, 333 Dunsmuir Street, 7th Floor, Vancouver, BC V6B5R3, Canada
e-mail: kip.morison@bchydro.com

L. Wang
Powertech Labs Inc, 12388—88th Avenue, Surrey, BC V3W 7R7, Canada
e-mail: Lei.Wang@PowerTechLabs.com

J. H. Chow (ed.), *Power System Coherency and Model Reduction*,
Power Electronics and Power Systems 94, DOI: 10.1007/978-1-4614-1803-0_7,

- Real-time applications usually have rigorous speed performance requirements; for example, execution time of an application must be guaranteed so as to coordinate with other applications or operator actions. In this case, using a reduced-order model for the application is often a viable way to achieve the required speed performance.

One practice for handling the above issues is to use a reduced-order model for the large, interconnected power system. This reduced model must preserve the characteristics of the full model for a study region so that it can be used, with acceptable results, for the analysis of this study region [2].

This chapter illustrates, through a practical case study, an application of the dynamic model reduction for a large power system model. This example concerns the creation of a reduced-order model from a full WECC model, to be used by BC Hydro for on-line dynamic security assessment (DSA) [3]. This work was conducted using the software DYNRED (for DYNamic REDuction) developed in the 1990s by EPRI and recently enhanced by Powertech Labs Inc. [4]. The techniques used in DYNRED are those described in the previous chapters of this manuscript and are not repeated here. To ensure the suitability of the reduced models for the required applications, the focus of this case study will be on the objective of the model reduction, the approach and procedure, the results, and the benchmarking work.

The case study described in this chapter includes the creation of a reduced-order model that meets the technical requirements of BC Hydro. This is not, however, the final model used in BC Hydro on-line DSA system because certain business and operational considerations are not included. Such considerations are not discussed here as they are out of the scope of this work.

7.2 Objectives and Requirements

7.2.1 BC Hydro Transmission System

The transmission system of BC Hydro (Fig. 7.1a) serves the power in the province of British Columbia, Canada, and is interconnected with the power grids of the Canadian province of Alberta and the western US to form the western US/Canada power system (WECC) as shown in Fig. 7.1b.

Due to the requirements to transfer large amounts of power and the long transmission path, transient and voltage stability has often been a limiting factor for the operation of BC Hydro's power system. A viable solution to ensure system stability and security is to use an on-line DSA system which monitors a system's stability status and transfer limits periodically using real-time captured system conditions.

For the implementation of BC Hydro's on-line DSA system, one of the critical issues is to determine the model to be used for the on-line transient stability analysis. The full WECC planning models (currently sized at around 18,000 buses) cannot be

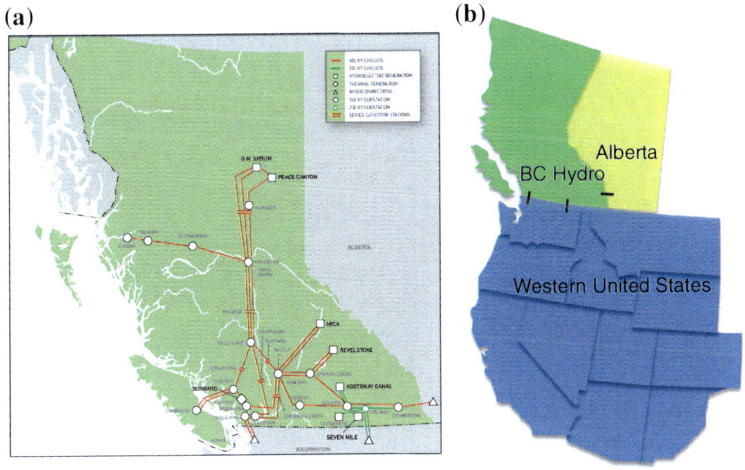

Fig. 7.1 BC hydro and WECC. **a** BC hydro transmission system. **b** WECC

used directly for reasons mentioned earlier. It was therefore decided that a reduced-order model be created and used for the on-line DSA system.

7.2.2 Model Reduction Objectives

The objectives of the model reduction work are:

1. To create an equivalent for the external systems in the US. The size of the overall system model with the reduced external model is targeted at around 7,000 buses.
2. To confirm that one reduced model can be used for all major operating scenarios with predefined performance criteria. This will allow the reduced model to be used for all real-time network applications installed in BC Hydro's EMS.

7.2.3 Performance Requirements

For the second model reduction objective, validation of the reduced model is conducted at the base system condition as well as at other operating conditions. For the base system condition, the following performance criteria were set:

1. When comparing the solved powerflow conditions (in the BC Hydro area) of the full and reduced models, the following should be sufficiently small:

    - Differences on bus voltages (>200 kV)

Table 7.1 Operating scenarios considered in model benchmarking

Scenario	Descriptions
1	Pacific DC Intertie (PDCI) is put out of service. As a result, the power carried by DC is diverted to the parallel AC transmission path
2a	BC to US transfer is set to a high value
2b	US to BC transfer is set to a high value
3	BC to Alberta transfer is set to a high value
4	A different WECC loading level is considered
5	A different transfer level is considered for a key interface in US
6	System is at the base system condition, but the fault clearance of a contingency is set to the critical clearance time (CCT) for which the system is marginally stable

- Differences on bus voltages ($\leq$200 kV)
- Differences on branch flows (MW/MVAR)
- Differences on shunt outputs (MVAR)

2. When comparing the simulations of the full and reduced models for the specified contingencies, the following quantities in the BC Hydro area should be sufficiently close (the closeness is measured by the index computed in DYNRED described later):
   - Generator rotor angles
   - Bus voltages
   - Branch flows (MW/MVAR)
3. When comparing the electromechanical oscillatory modes in the system from the full and reduced models, the main attributes (frequency and damping) should be sufficiently close.

Model benchmarking performed for the different operating scenarios is shown in Table 7.1

For simplicity of result presentation, only Scenarios 1, 2a, 3, and 6 are discussed in this chapter.

7.3 System Description

7.3.1 Full model

The full model to start with for the reduction is a standard WECC planning model. This model has

- 16,330 buses
- 1,922 generators with matching dynamic models

Table 7.2 Sizes of the full model

Area	No. of buses	No. of generators
BC [1]	2,220	290
Alberta	2,011	221
US [2]	12,099	1,411

[1] Includes separately controlled areas in British Columbia
[2] Includes all control areas in US

Table 7.2 shows the sizes of BC, Alberta, and US areas in this model.

7.3.2 Reduction Requirements

From model accuracy consideration and also because of availability of real-time data in BC Hydro's EMS, BC Hydro wants to retain the full details for the following portion of the full model in the reduced model:

- BC.
- Alberta.
- A portion of the US system surrounding BC.

The above portion of the system is referred to as the *study region*. Figure 7.2 shows this region, in which the red line is the boundary of the study region: all system details

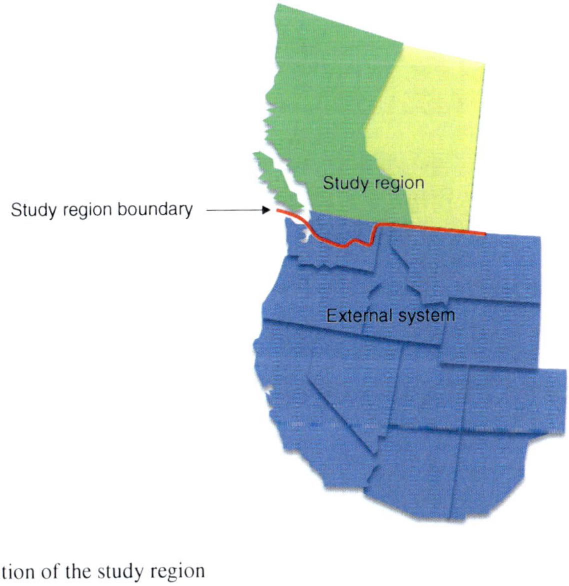

Fig. 7.2 Definition of the study region

above the red line are to be kept in the reduced model and anything below it can be reduced. The US portion in the study region includes 644 buses, so the entire study region has 4,875 buses.

7.4 Approach

7.4.1 General Procedure for a Dynamic Reduction Process

Before describing the approach used in this case study to create the reduced model, it is helpful to first review the general procedure for a dynamic reduction process. Figure 7.3 shows the flowchart of this procedure.

This process starts with a full model, including a matching powerflow and dynamics data set. This model is sanity checked to ensure its validity for the reduction, and any issues found need to be resolved by implementing appropriate fixes to data and/or models. A more important purpose of this step is to identify any characteristics in the full model (referred to as *technical reduction requirements*) that need to be preserved in the reduced model. For example, if there are critical low frequency oscillatory modes in the study region, such modes should be preserved in the reduced model, and this requires that generators significantly participating in these modes must be retained in the reduced model. The technical reduction requirements are then merged with the reduction requirements provided by the user (*user reduction requirements*), which include the specification of the study region, to form the *overall reduction requirements*.

After both the full model data and reduction requirements are finalized, they can be fed to a tool (such as the DYNRED program used in this case study) to perform the reduction with the specified parameters (the method for coherency identification and generator aggregation, etc.). At the end of this step, a reduced model with a matching powerflow and dynamics data set will be produced.

After a (draft) reduced model is created, it is critical to perform sanity checks and model validation to ensure that the reduced model not only meets the reduction requirements, but also the performance requirements. This is done by using various methods, including powerflow and time-domain simulations, at the base system condition as well as at other system conditions under which the model may be used. If the reduced model is deemed unacceptable, the reduction requirements are modified and a new reduced model is created. The process is iterated until the final reduced model is created with a set of satisfactory performance metrics.

7.4.2 Creation of the Reduced Model

Before starting reduction in this case study, it was determined that a buffer region needs to be kept in the reduced model for the following reasons:

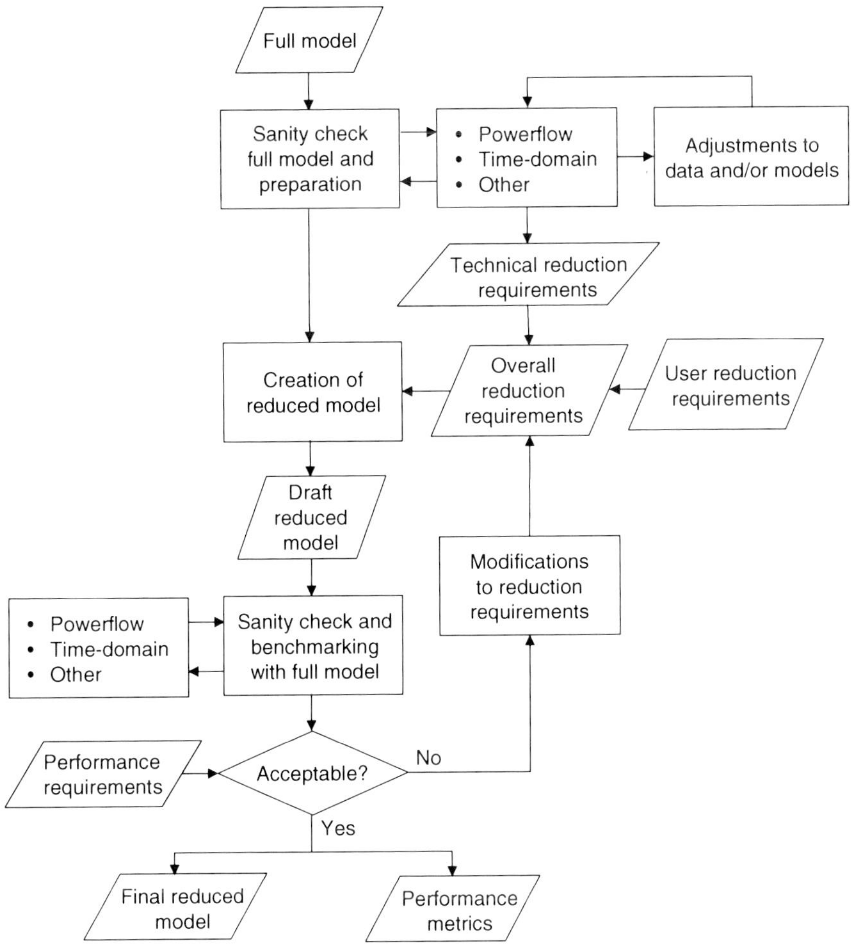

Fig. 7.3 Flowchart of a dynamic reduction process

- To facilitate the use of the reduced model in real-time applications (for example, the tie-lines connecting the study region shall be easily identified when merging the real-time system models with the external system contained in the reduced model).
- To ensure adequate system performance.
- To enable model benchmarking (for example, dispatching the full and reduced models consistently for scenarios in Table 7.2 and monitoring the system performance).

The following is the buffer region defined from the above considerations (with a total of 270 buses):

- All circuits connecting the study region.
- All 500 kV buses in the US Northwest outside of the study region.
- Key generators in the US Northwest outside of the study region.
- Buses around PDCI.
- Circuits at key interfaces in WECC.
- Generators required to preserve a 0.4 Hz inter-area mode in WECC (this was determined from eigenvalue analysis on the full model).

With the above, the portion of the system to be fully retained in the reduced model has 5,145 buses (study region + buffer region). Thus in order to achieve the target of around 7,000 buses for the reduced model, 11,185 buses in the system external to the study region and buffer region must be reduced to about 1,855 buses.

As mentioned earlier, the actual model reduction was done by the DYNRED program. The following options in DYNRED were used which appear to work well from past experience:

- Weak links method [5] was used to identify the coherent generator groups.
- Classical aggregation method was used to aggregate the generators in coherent groups.

7.4.3 Model Benchmarking

For model benchmarking using time-domain simulations, 12 contingencies (8 N-1 and 4 N-2) were considered, as shown in Fig. 7.4. These are all located on the 500 kV

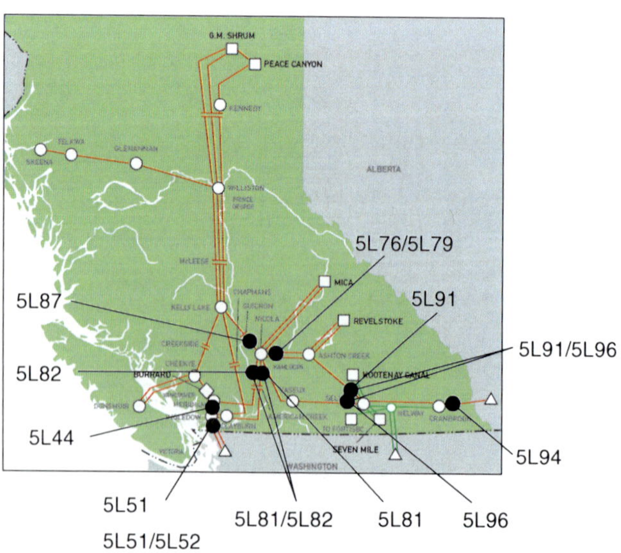

Fig. 7.4 Contingency code and locations

Table 7.3 System conditions for model benchmarking

Scenario	Description	System conditions (MW)			Actions applied
		Base case	Rating	Achieved	
1	PDCI O/S	2,500	3,100	0	PDCI is blocked to force the flow through the parallel AC system
2a	High BC-US transfer	233	3,150	2,957	• Increase generation and decrease load in BC Hydro area • Increase load in US Northwest
3	High BC-Alberta transfer	−254	1,200	819	• Increase generation in BC Hydro area • Increase load in Alberta
6	Fault clearance at CCT	N/A	N/A	N/A	Contingency 5L51 was examined for this condition

grid within the BC Hydro area. Each contingency includes a 4-cycle three-phase fault and circuit tripping to clear the fault. For some N-2 contingencies, the system is unstable. In such cases, generation and loading shedding is applied to stabilize the system.

For model benchmarking at different system conditions, system dispatches were performed to achieve the required operating conditions. The resulting system conditions and the required dispatching actions are shown in Table 7.3 for the selected scenarios discussed in this chapter.

When comparing the time-domain simulation results from the full and reduced models, one question is how to measure the closeness of two simulation trajectories. DYNRED includes a feature that produces an index to facilitate such comparison. Figure 7.5 shows typical simulation results x_f from a full model and x_r from a corresponding reduced model for one quantity x following a contingency. The Time-Domain (TD) index to measure the closeness for these two curves is defined as

$$\text{TD index} = \frac{1}{\Delta x_f T}\sqrt{\int_0^T (x_f - x_r)^2 \mathrm{d}t} \tag{7.1}$$

where T is the length of the simulation and Δx_f is the peak-to-peak deviation of the curve from the full model. This index can be viewed as a measure of the difference of the two curves (the shaded area in Fig. 7.5) with respect to the per-unitized curve from the full model.

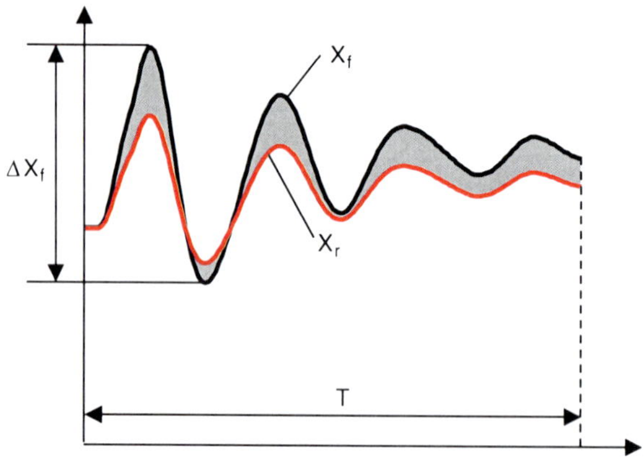

Fig. 7.5 Comparing two simulation trajectories

7.5 Results

7.5.1 Reduced Model

With the approach described above, a reduced model is created. Table 7.4 shows the summary of the reduced model.

It is seen that the size of the reduced model is within the range of the BC Hydro requirements. Note that although the reduction ratio for the number of generators is less than half at 45.9 %, the actual reduction of dynamics is higher since the equivalent generators in the external region are all represented by the 2-state classical model.

The key task in this model reduction project is to verify that the reduced model is suitable for use in stability analysis for a wide range of system conditions. This is described in the following subsections.

7.5.2 Model Benchmarking: Powerflow at the Base System Condition

The performance numbers are obtained by comparing the solved full and reduced powerflow models:

Table 7.4 Summary of the reduced model

	Number of buses	Number of generators
Full model (F)	16,330	1,922
Reduced model (R)	72,10	1,040
Reduction ratio ($\frac{F-R}{F}$)	55.8 %	45.9 %

- Differences on bus voltages (>200 kV): 0.01 %
- Differences on bus voltages (≤200 kV): 0.23 %
- Differences on branch flows (MW): 0.62 % (on branches with flows greater than 10 MW)
- Differences on branch flows (MVAR): 5.42 % (on branches with flows greater than 10 MVAR)
- Differences on shunt outputs (MVAR): 0.20 % (on shunts with outputs greater than 5 MVAR)

It is seen that most of the quantities in the reduced model are fairly close to the full model, except for the branch MVAR flows. The relatively visible differences in branch MVAR flows are primarily due to small branch reactances on these branches, and it has been identified that such visible differences occur only on a small number of branches.

7.5.3 Model Benchmarking: Time-Domain Simulations at the Base System Condition

Time-domain simulations are performed for all 12 contingencies for the full and reduced models at the base system condition. The following are the *largest* TD indices obtained for each of the four types of monitored quantities from these simulations:

- Generator relative rotor angles: 1.33 %
- Bus voltage magnitudes: 0.69 %
- Branch MW flows: 2.31 %
- Branch MVAR flows: 2.43 %

To correlate the above TD index numbers with the graphical comparison of the simulations curves, Fig. 7.6 shows the plots from which the above largest TD indices are obtained. Two conclusions can be made:

- The differences in the simulation results between the full and reduced models in Fig. 7.6 are acceptable.
- By the definition of the TD index, the differences of all other monitored quantities between the full and reduced models should be less than those in Fig. 7.6.

Therefore, it is determined that the reduced model performs satisfactorily for time-domain simulations at the base system condition.

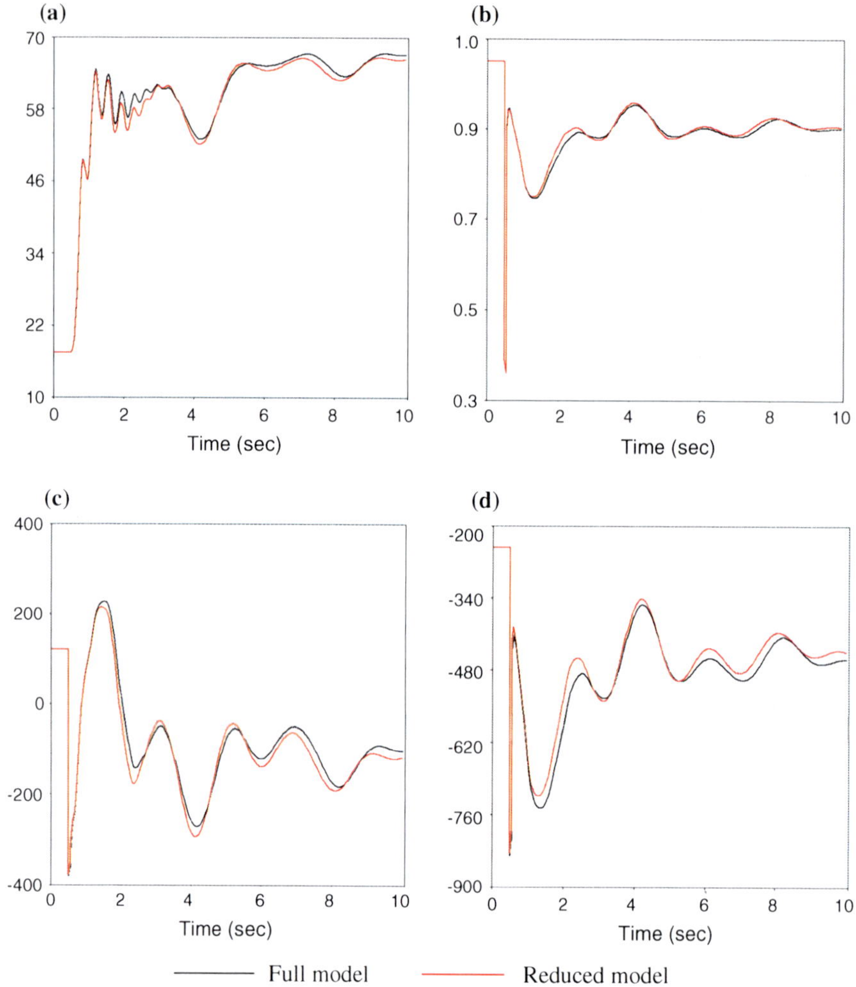

Fig. 7.6 Comparison of quantities with the largest TD indices. **a** Generator relative rotor angle. **b** Bus voltage magnitude. **c** Branch MW flow. **d** Branch MVAR flow

7.5.4 Model Benchmarking: Eigenvalues at the Base System Condition

Modes of low-frequency electromechanical oscillations in both full and reduced models between 0.2 and 1.0 Hz and with damping less than 10 % are computed. Eight modes are identified as associated with the study region (mainly in British Columbia and Alberta). Table 7.5 shows the comparison of these modes. It is clear that such modes are well preserved in the reduced model.

Table 7.5 Comparison of oscillatory modes, where f and ς are the modal frequency and damping ratio, respectively

Mode No.	Full model		Reduced model		Mode location
	f (Hz)	ς (%)	f (Hz)	ς (%)	
1	0.40	7.73	0.40	7.66	System-wide interarea mode
2	0.67	9.98	0.67	10.00	British Columbia
3	0.68	7.98	0.68	7.90	Alberta
4	0.71	5.40	0.71	5.13	British Columbia
5	0.72	9.68	0.72	9.81	Alberta
6	0.97	5.39	0.97	5.36	Alberta
7	0.95	7.55	0.95	7.55	British Columbia
8	0.97	5.39	0.97	5.36	Alberta

Table 7.6 TD indices for different system conditions

Scenario No.	Generator relative rotor angle (%)	Bus voltage magnitude (%)	Branch MW flow (%)	Branch MVAR flow (%)
1	2.09	0.88	3.81	2.95
2a	1.77	0.73	2.58	6.83
3	1.24	0.31	1.97	1.06

7.5.5 Model Benchmarking: Time-Domain Simulations at Different System Conditions

Ensuring satisfactory performance for the reduced model at different system conditions described in Table 7.3 is a more challenging task to achieve in model benchmarking. Only time-domain simulation comparisons are performed, which are considered sufficient for validating the reduced model.

Table 7.6 shows the TD indices obtained for the three scenarios (1, 2a, and 3) in Table 7.3. Results for scenario 6 are presented in the next section.

It is seen that the TD indices are, in general, in the same order as those obtained from the base system condition; thus it can be concluded that the performance of the reduced model is very close to that of the full model. A closer comparison indicates that the TD indices in Table 7.6 are marginally larger than those obtained at the base system condition, except for scenario 3. This is expected, since

- Scenarios 1 and 2a involve system dispatches applied to the external region that are different in the full and reduced models; thus the differences in system responses for these scenarios should be somehow larger than in the base system condition.
- Scenario 3 involves system dispatches only within the study region; therefore, the powerflows of full and reduced models after the dispatches should be very close. As a result, the simulation results from these two models should also be very close. The slightly smaller TD indices signify the fact that the system at this

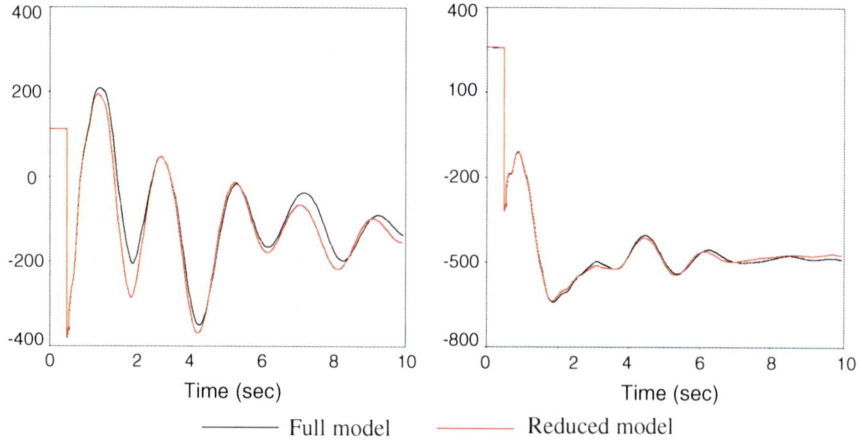

Fig. 7.7 Comparison of branch MW flows for scenarios 1 and 3

operating condition is transiently less stable than the base system condition; thus the normalizing factor (ΔX_f in Eq. 7.1) is larger.

To confirm the above points, two plots are shown in Fig. 7.7 for branch MW flows for scenarios 1 and 3. These can be compared with plot (c) in Fig. 7.6.

7.5.6 Model Benchmarking: Very Stressed System Condition

The performance of the reduced model under a much stressed system condition is tested by considering scenario 6 in Table 7.3. First, the CCT of a contingency (5L51) is obtained for both the full and reduced models, which turns out to be 13.75 cycles (consistent for both models). Then simulations are performed with fault clearance time set at 13.75 cycles (marginally stable) for both models. The TD indices obtained from these simulations are

- Generator relative rotor angles: 2.59 %
- Bus voltage magnitudes: 1.50 %
- Branch MW flows: 3.53 %
- Branch MVAR flows: 3.62 %

These are higher than the indices from the normal fault clearance cases for the obvious reason, but are still within a small boundary indicating consistent results between the two models. The plots in Fig. 7.8 confirm this.

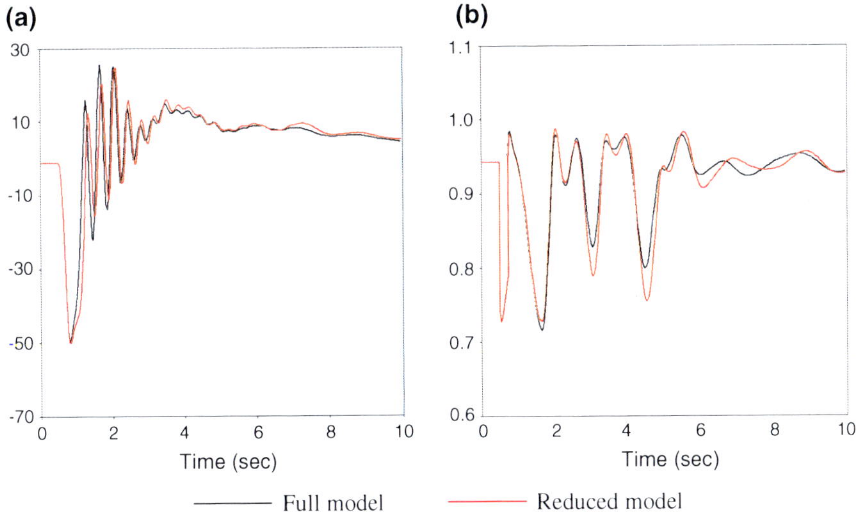

Fig. 7.8 Comparison of simulations with fault clearance at CCT. **a** Generator relative rotor angle. **b** Bus voltage magnitude

7.5.7 Computation Speed Performance

Achieving fast computation speed is one of the motivations to use a reduced model in stability analysis. A computation speed comparison shows that the time it takes to complete a typical 10 s simulation using the reduced model is approximately 52 % of the time required for the full model. Thus, the computation speed can be effectively doubled when the reduced model is used.

7.6 Conclusions

This chapter illustrates a complete dynamic reduction process through a case study in which a full WECC system model is reduced for use by BC Hydro. This process consists of the following steps:

- Specify the reduction requirements.
- Set up the performance requirements.
- Select the reduction method and parameters.
- Perform the dynamic reduction.
- Validate the reduced model.

It is shown that, following the procedure used in the case study, a reduced model can be obtained for use with a wide range of system conditions with acceptable results and computation time saving.

Acknowledgments The authors would like to thank the support given to this case study by the following personnel from BC Hydro: Dr. Ebrahim Vaahedi, Dr. Djordje Atanackovic, Dr. Michael Yao, and Mr. Guihua Wang.

References

1. J. Viikinsalo, A. Martin, K. Morison, L. Wang, F. Howell, *Transient Security Assessment in Real-time at Southern Company*. Proceedings of IEEE PSCE, Atlanta, Oct 2006
2. L. Wang, M. Klein, S. Yirga, P. Kundur, Dynamic reduction of large power systems for stability studies. IEEE Tran. Power Syst. **PWRS–12**(2), 889–895 (1997)
3. L. Wang, K. Morison, Implementation of new tools for on-line security assessment tools for reducing the risk of blackouts. IEEE Power Energ. Mag. **4**(5), 46–59 (2006)
4. EPRI DYNRED Enhancement Project, Final Report, EPRI, Palo Alto, CA, April 2010.
5. R. Nath, S.S. Lamba, K.S.R. Prakasa, Coherency based system decomposition into study and external areas. IEEE Tran. Power Syst. **PAS–104**, 1443–1449 (1985)

Chapter 8
Measurement-Based Methods for Model Reduction of Power Systems Using Synchrophasors

Aranya Chakrabortty and J. Chow

Abstract Wide-area analysis and control of large-scale electric power systems are highly dependent on the idea of *aggregation*. For example, one often hears power system operators mentioning how "Northern Washington" oscillates against "Southern California" in response to various disturbance events. The main question here is whether we can analytically construct dynamic electromechanical models for these conceptual, aggregated generators representing Washington and California, which in reality are some hypothetical combinations of hundreds of actual generators. In this chapter we present an overview of several new results on how to construct such simplified interarea models of large power systems by using dynamic measurements available from phasor measurement units (PMUs) installed at limited points on the transmission lines. Our examples of study are motivated by widely encountered power transfer paths in the Western Electricity Coordinating Council (WECC), namely a two-area radial system representing the WA-MT flow, a star-connected three-area system resembling the Pacific AC Intertie, and a generic multi-area system with more than one dominant slow mode of oscillation.

8.1 Introduction

Over the past few years, several catastrophic phenomena, such as cascade failures and blackouts in different parts of the North American power grid, have forced power system researchers to look beyond the traditional approach of analyzing power system functionalities in *steady-state*, and instead pay serious attention to their

A. Chakrabortty (✉)
North Carolina State University, Raleigh, NC, USA
e-mail: aranya.chakrabortty@ncsu.edu

J. Chow
Rensselaer Polytechnic Institute, Troy, NY, USA
e-mail: chowj@rpi.edu

J. H. Chow (ed.), *Power System Coherency and Model Reduction*,
Power Electronics and Power Systems 94, DOI: 10.1007/978-1-4614-1803-0_8,

dynamic characteristics, and in a global or *wide-area* sense. This mindset has been particularly facilitated by the recent outburst of measurement and instrumentation facilities in the context of smart power grids provided by the wide-area measurement system (WAMS) technology, which uses sophisticated digital recording devices called phasor measurement units (PMUs) to record and export GPS-synchronized high-sampling-rate (6–60 samples/s) dynamic power system data [1]. Industry platforms, such as the North American Synchrophasor Initiative (NASPI) [2], have been formed to investigate ways by which PMU measurements from different parts of the US power system can possibly be exploited to gain insight into their dynamic interdependence, which could indicate how events in one area of the grid can propagate and have a significant impact on other remote areas.

However, a major road-block to wide-area analysis of large-scale power systems is the absence of concrete mathematical models that capture the aggregated electromechanical dynamics coupling one area of the system with another. For example, one often hears power system operators mentioning how "Northern Washington" oscillates against "Southern California" in response to various disturbance events. The main question here is whether we can analytically construct dynamic electromechanical models for these conceptual, aggregated generators representing Washington and California, which in reality are some hypothetical combinations of hundreds of actual generators. For example, it is well known that a 0.25 Hz interarea swing mode exists between the north–south interconnections of the Western Electricity Coordinating Council (WECC) extending from Alberta, Canada to Baja Mexico [3], with additional 0.4–0.7 Hz modes along the pacific AC intertie and the east–west interconnection. Based on such interarea modal behavior, conceptually speaking, the map of WECC can be drawn as an equivalent mass-spring-damper model [4], as in Fig. 8.1a, showing how the electromechanical dynamics of the aggregated parts of WECC may swing against each other when a disturbance sets in. Figures 8.1b and c shows the partitioning of the WECC into a 2-area system and a 9-area system. But, again, the main question to be resolved is how can we construct an explicit dynamic model for this conceptual figure, preferably in real-time, using voltage, current, or power flow signal measurements in order to establish a prototype for the nonlinear interarea dynamics of the entire interconnection. Using conventional model-based equivalencing methods [5] would be impractical for this purpose because they are highly time-consuming and numerically challenging for large-scale nonlinear simulations. More importantly, they are dependent on the precise knowledge of the model parameters (such as inertia, transient reactance, transformer reactance) of all the thousands of generators, transmission lines, and loads constituting each aggregate area.

Motivated by these fundamental questions of current interest to the PMU research community, in this chapter we develop a framework for the identification of reduced-order dynamic models of very large-scale power systems, not by using traditional model-based methods [5], but from synchrophasor measurements available from PMUs installed only at selected points on the transfer path. Our models of study in this chapter represent prototypes of three well-known transfer paths in the WECC, namely,

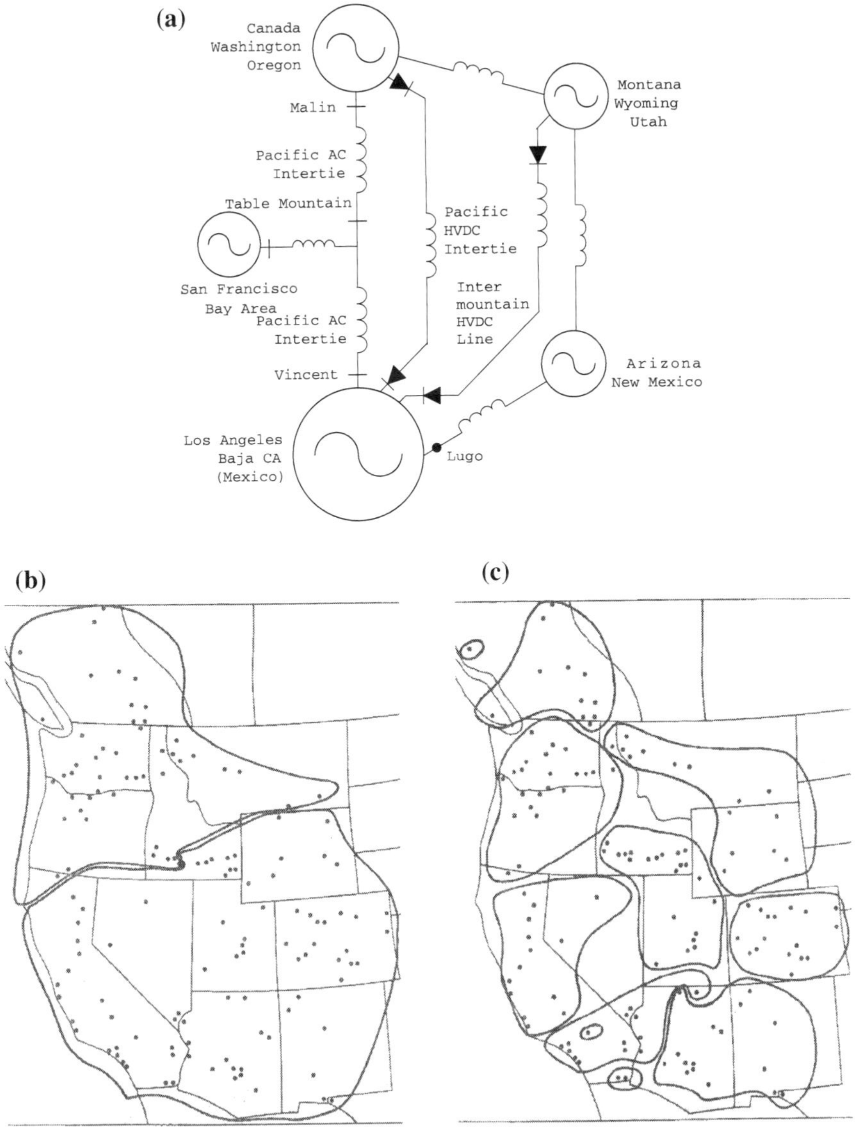

Fig. 8.1 Area aggregations in Western Electricity Coordinating Council (*WECC*). **a** Spring-mass system representation of WECC. **b** WECC 2-area partition. **c** WECC 9-area partition

1. One-dimensional models such as a two-area radial system with one dominant slow mode of oscillation (e.g., Washington-Montana transfer),
2. Two-dimensional models with algebraic nodes such as a three-area star-connected system with one or two dominant slow modes of oscillation (e.g., Pacific AC intertie),

3. Two-dimensional models with direct connectivity, i.e., a general multi-area power system with a given *interarea* topology, and two or more slow modes of oscillation.

For each of these transfer paths, we first show that the model identification reduces to a parameter estimation problem for the aggregated *intra-area* reactances and machine inertias internal to each area, and then derive analytical results showing how the voltage, phase angle, and frequency oscillations at multiple buses on the transfer path, following a small-signal disturbance, can be used to estimate these parameters. We illustrate our results with real power system disturbance events in WECC. The objective of this chapter should not, however, be confused with research on *modal identification*, the purpose of which is to estimate the eigenvalues and eigenvectors of the state matrix of the linearized power system from the measured states. Several numerical algorithms have been developed for such mode estimation from both *ringed-down* disturbance data and ambient measurements, as discussed in the seminal work of Hauer [6, 7] with established applications in wide-area monitoring in the US as well as in other countries, such as Australia [8, 9], China [10], and Denmark [11]. This chapter, on the other hand, looks farther beyond mode estimation towards identifying dynamic model parameters from the "modes".

The remainder of the chapter is organized as follows. Sections 8.2, 8.3, and 8.4 pose the interarea model estimation (IME) problem for one- and two-dimensional systems together with the validations of the respective results through PMU data analysis of WECC. Section 8.5 presents transient stability assessment using these equivalent models via energy function analysis, while Sect. 8.6 develops PMU placement methods using noisy data. Section 8.7 concludes the chapter.

8.2 Problem Formulation

Mathematical modeling of dynamic equivalents of large-scale electric power systems has seen some 40 years of long and rich research history. The foundations of this line of research were laid in the late 1970s by Chow and Kokotović, who introduced the ideas of *aggregation* and *coherency* [5], resulting in algorithms of partitioning a power network into dynamic aggregates, where each aggregate consists of a group of strongly connected generators that synchronize over a fast time-scale and, thereafter, act as a single entity, while the aggregates themselves are weakly connected to each other, and synchronize over a slower time-scale. Using singular perturbation theory, they derived analytical expressions for aggregated machine inertias and reactances in terms of the model parameters of each individual machine contained in an area. Their approach was complemented by alternative techniques of coherency such as those by Germond and Podmore [12], and de Mello, Podmore, and Stanton using circuit-theoretic approaches [13], Undrill and Turner using linear modal decomposition [14], Zaborszky et al. using enumerative clustering algorithms [15], and Nath et al. using iterative techniques to compute the coupling strengths between different areas [16]. Aggregation and coherency reduced the computational complexities of solving thousands of nonlinear equations in power system stability programs from detailed models, and was tested offline on both small-scale (such as the 48-machine NPCC system[1]) and large-scale (such as the 12,000-bus NYPP system[2]) via software programs such as *DYNEQ* and *DYNRED* [17]. However, the main limitations of the above-mentioned conventional model reduction methods are twofold: first, they are *model-based* methods, meaning that to construct reduced-order models using these methods one would need to know each constituent model explicitly; and second, in constructing an aggregate motion, most of these methods tend to capture minute details of the fast local oscillations in each area that not only increase the computational time but also may be unnecessary at times of emergency when decisions have to be made fast.

The methods proposed in this chapter are meant to circumvent these traditional barriers by developing model reduction algorithms that do not depend on individual component-level model information, and are based on measurements only. PMU measurements of voltage, current and frequency from disturbance events available from a limited number of points in the network will be utilized for this purpose. Our first system of investigation is a commonly encountered power transfer path, namely, a two-area radial power system, as shown in Fig. 8.2a, containing multiple strongly connected machines in each area with arbitrary interconnection structure. Its two-machine dynamic (interarea) model and its classical circuit representation are shown in Fig. 8.2b. The system consists of two aggregated generators G_1 and G_2, which represent coherent combinations of strongly connected machines in each respective area. Let the equivalent inertias of these aggregated machines be H_1 and H_2, respectively. The machines are connected to the high-voltage terminal buses 1 and 2 through equivalent transformers having reactances x_{T1} and x_{T2}, which, in turn, represent Thevenin equivalents of the transformer reactances in each respective area. This two-area system is useful for representing a radial transfer path in a large power system in which one coherent area is exporting power to the other coherent area. The voltage phasors at Buses 1 and 2 are given as

$$\tilde{V}_i = V_i \angle \theta_i, \quad i = 1, 2 \tag{8.1}$$

where $V\angle\theta$ denotes the polar representation $V\varepsilon^{j\theta}$. The transmission line between Buses 1 and 2 is assumed to be lossless with a reactance x_e. The line current flowing from Bus 1 to Bus 2 is $\tilde{I} = I\angle\theta_I$ with G_1 supplying power to G_2, which acts as a load. For the classical model representation, the internal voltages of the generators G_1 and G_2 are denoted as

$$\tilde{E}_1 = E_1 \angle \delta_1, \quad \tilde{E}_2 = E_2 \angle \delta_2. \tag{8.2}$$

The total reactances connecting the generator internal voltage nodes to the terminal Buses 1 and 2 are given as

[1] Northeast Power Coordinating Council.

[2] New York Power Pool.

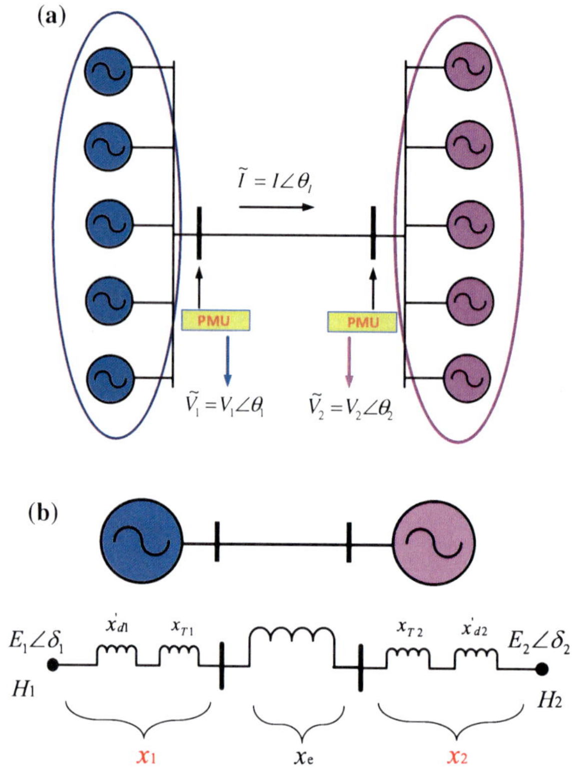

Fig. 8.2 Area aggregations in Western Electricity Coordinating Council (*WECC*). **a** Two-area power system. **b** Two-machine interarea equivalent

$$x_i = x_{Ti} + x'_{di}, \quad i = 1, 2 \tag{8.3}$$

where x'_{d1} and x'_{d2} are the equivalent direct-axis transient reactances of G_1 and G_2, respectively. The total reactance of the transfer path is, therefore, given as

$$\bar{x} = x_1 + x_e + x_2. \tag{8.4}$$

The electromechanical model of each aggregated generator, neglecting damping, can be written as [18]

$$\dot{\delta}_i = \Omega(\omega_i - \omega_s), \quad 2H_i\dot{\omega}_i = P_{mi} - P_{ei}, \quad i = 1, 2, \tag{8.5}$$

where $\Omega = 120\pi$ for a 60 Hz system, ω_s is the synchronous speed, and ω_i, P_{mi}, P_{ei} are, respectively, the angular velocity, the mechanical power input, and the electrical power output of the ith machine. All quantities are in per unit except for the phase angles which are in radians. We assume that inside each area the system configuration remains unchanged over the time-scale of the interarea mode. The two swing equations in (8.5) can then be combined to form a second-order system

$$\dot{\delta} = \Omega\,\omega, \qquad 2H\,\dot{\omega} = P_m - P_e \tag{8.6}$$

with

$$P_m = \frac{H_2 P_{m1} - H_1 P_{m2}}{H_1 + H_2}, \quad P_e = \frac{E_1 E_2}{\bar{x}} \sin(\delta) \tag{8.7}$$

where $H = H_1 H_2/(H_1 + H_2)$ is the equivalent inertia, $\delta = \delta_1 - \delta_2$, and $\omega = \omega_1 - \omega_2$. Assuming that PMU measurements of voltage, current, and bus frequency are available from Buses 1 and 2, we then pose the problem of finding the reduced model in Fig. 8.2b as follows.

Given the measured synchronized phasor variables V_1, θ_1, V_2, θ_2, I, *and* θ_I *in* Fig. 8.2a *that exhibit a few cycles of interarea oscillations, compute* E_i, δ_i, x_i, *and* H_i, $i = 1, 2$, *and* x_e *of the reduced-order two-machine system in* Fig. 8.2b *to represent the interarea dynamic behavior of the two-area power system.*

Because $\tilde{V}_1$, $\tilde{V}_2$ and $\tilde{I}$ are measured, x_e can be easily computed from $jx_e = (\tilde{V}_1 - \tilde{V}_2)/\tilde{I}$. Similarly $\tilde{E}_1$ and $\tilde{E}_2$ can be computed if x_1 and x_2 are known. Therefore, the above problem, referred to as the Interarea Model Estimation (IME) problem, reduces to the estimation of four quantities, namely x_1, x_2, H_1, and H_2. We next derive algorithms by which these four parameters can be identified from the interarea oscillations of voltage and frequency captured by the PMUs on the transfer path.

8.2.1 Reactance Estimation

Without any loss of generality, we fix our reference at the internal node of Generator 2, and assume $\delta_2 = 0$ and $\delta_1 = \delta$. The first step for this algorithm is to choose any phasor variable measured by the PMUs, for example, magnitude of bus voltages. The voltage phasor at any point P at a reactance jx away from Generator 2 (or equivalently at a distance x away if the reactance is uniformly distributed along the path), can be written as

$$\tilde{V} = a\,E_1(\cos(\delta) + j\sin(\delta)) + E_2(1 - a) \tag{8.8}$$

where $a = x/(x_1 + x_e + x_2)$ is the normalized reactance of the point P. The magnitude of $\tilde{V}(x)$, denoted as V, is, therefore,

$$V \triangleq |\tilde{V}| = \sqrt{c + 2E_1 E_2((a - a^2)\cos(\delta))} \tag{8.9}$$

where $c = (1-a)^2 E_2^2 + a^2 E_1^2$. Considering a small-signal disturbance in the system, and linearizing (8.6) and (8.9) about an equilibrium point (δ_0, $\omega_0 = 0$, V_{ss}), any change in V can be written as

$$\Delta V(a,t) = \frac{-\,E_1 E_2 \sin(\delta_0)(a - a^2)}{V(a,\delta_0)} \Delta\delta(t). \tag{8.10}$$

Of prime importance is that the Jacobian in (8.10) consists of two parts: a numerator part varying with a, and a denominator part which is the steady-state bus magnitude at the point P. From (8.10), we can write

$$V_n \triangleq \Delta V(a,t) V(a,\delta_0) = A\, a(1-a)\Delta\delta(t). \tag{8.11}$$

where $A = -\,E_1 E_2 \sin(\delta_0)$. The quantity V_n in (8.11), referred to as the *normalized voltage* is a product of two quantities, namely the change in voltage at the point P at any time instant t following the disturbance from the predisturbance equilibrium voltage $V(a,\delta_0)$ at this point, and $V(a,\delta_0)$ itself. If a PMU is located at this point P, then both of these quantities and, therefore, the normalized voltage can be calculated from the PMU measurement recordings at any fixed point of time. The RHS of (8.11) consists of the unknown constant A as well as the hypothetical state evolution $\Delta\delta(t)$, which depends only on time t and not on the spatial variable a. We refer to this as the *time–space separation* property, using which we can simply write

$$\frac{V_{n1}(a_1,t^*)}{V_{n2}(a_2,t^*)} = \frac{a_1(1-a_1)}{a_2(1-a_2)} \tag{8.12}$$

where V_{n1} and V_{n2} are, respectively, the normalized voltages at Bus 1 and Bus 2, t^* is a fixed point of time, while

$$a_1 = \frac{(x_2 + x_e)}{(x_1 + x_e + x_2)}, \quad a_2 = \frac{x_2}{(x_1 + x_e + x_2)} \tag{8.13}$$

are the normalized reactances of these two respective buses. To generate a second equation we need another measurement point, which, in other words, indicates that we must have a third PMU installed at some intermediate bus on the transfer path between Bus 1 and Bus 2 at a known distance from Bus 2. In that case, for the same time instant $t = t^*$ as before, we can use

$$\frac{V_{n3}(a_3,t^*)}{V_{n2}(a_2,t^*)} = \frac{a_3(1-a_3)}{a_2(1-a_2)} \tag{8.14}$$

where a_3 is the normalized reactance corresponding to the third bus. Equations (8.12) and (8.14) can then be solved for x_1 and x_2 using numerical algorithms. If damping is considered in (8.6), then additionally we will have to consider one more data point and form a third equation similar to (8.14).

Note 1: Even if an actual PMU is not installed between buses 1 and 2, a third voltage $\tilde{V}_3$ can still be obtained using the relation $\tilde{V}_3 = \tilde{V}_2 + jx_{23}\tilde{I}$, where x_{23} is the reactance between the third point and Bus 2 (and is assumed to be known), while $\tilde{I}$ is the line current measured by the PMU at Bus 2.

Note 2: If the transmission line has both resistance and reactance, then the spatial variable a simply needs to be defined as a ratio of the impedance of any point measured from Bus 2 to the total impedance of the transfer path. The rest of the algorithm will not change under that situation.

8.2.2 Inertia Extrapolation Algorithm

Once x_1 and x_2 have been computed, the remaining parameters to be computed are the inertias H_1 and H_2. We need two pieces of information. First, by measuring the frequency f of the swing mode in the voltage measurement (which can be done using modal decomposition algorithms), the equivalent inertia constant H can be computed from linear circuit theory as [18]

$$H = (E_1 E_2 \cos(\delta_0)\Omega)/(2\bar{x}\,(2\pi f)^2) \tag{8.15}$$

Second, to calculate H_1 and H_2 separately, we develop a companion equation by exploiting the frequencies measured at Buses 1 and 2. Neglecting losses and machine damping effects, the conservation of the total angular momentum of the two-machine system is given as

$$\begin{aligned} L = 2H_1\omega_1 + 2H_2\omega_2 &= 2\int (H_1\dot{\omega}_1 + H_2\dot{\omega}_2)\,\mathrm{d}t \\ &= \int (P_{m1} - P_{e1} + P_{m2} - P_{e2})\,\mathrm{d}t = 0 \end{aligned} \tag{8.16}$$

from which we obtain

$$\frac{H_1}{H_2} = -\frac{\omega_2}{\omega_1} \tag{8.17}$$

Hence, (8.17) can be used to solve for H_1 and H_2, provided that the estimates for the machine speeds are known. For the two-machine system, we can show that ω_1 and ω_2 can indeed be estimated from the measured frequencies ϑ_1 and ϑ_2 at Buses 1 and 2 according to the equations

$$\vartheta_1 = \frac{g_1\omega_1 + h_1(\omega_1 + \omega_2)\cos(\delta_1 - \delta_2) + k_1\omega_2}{g_1 + 2h_1\cos(\delta_1 - \delta_2) + k_1} \tag{8.18}$$

$$\vartheta_2 = \frac{g_2\omega_1 + h_2(\omega_1 + \omega_2)\cos(\delta_1 - \delta_2) + k_2\omega_2}{g_2 + 2h_2\cos(\delta_1 - \delta_2) + k_2} \tag{8.19}$$

where $g_1 = E_1^2(1-\rho_1)^2$, $h_1 = E_1E_2\rho_1(1-\rho_1)$, $k_1 = \rho_1^2E_2^2$, $g_2 = E_1^2(1-\rho_2)^2$, $h_2 = E_1E_2\rho_2(1-\rho_2)$, and $k_2 = \rho_2^2E_2^2$, with $\rho_1 = x_1/\bar{x}$ and $\rho_2 = (x_1+x_e)/\bar{x}$. Equations (8.18–8.19) can be derived simply from (8.8) by considering $\theta = \tan^{-1}(\mathrm{Im}(\tilde{V})/\mathrm{Re}(\tilde{V}))$, and then taking the time derivative to derive $\vartheta = \dot{\theta}$ as a function of a. Because the bus frequencies ϑ_1 and ϑ_2 are available from PMU measurements, we can estimate ω_1 and ω_2 using (8.18) and (8.19), calculate the ratio $\omega_2/\omega_1 = -H_1/H_2$, and solve for H_1 and H_2 using (8.15) and the relation $H = H_1H_2/(H_1+H_2)$.

8.2.3 Washington-Montana Transfer Path Modeling

We next model the east–west WECC power transfer between the aggregated generators of "Canada, Washington, Oregon" and "Montana, Wyoming, Utah" indicated in Fig. 8.1a, referred to here as *WECC transfer path* 1. The system consists of a median-size group of remote machines supplying power via a 600-mile transmission system to a load center. A disturbance initiated a 0.578 Hz oscillation across the transfer path. The event recording started 60 s before the disturbance for a total of 5 min. We assume that the transmission line is lossless. The variation of the voltage magnitudes at the two terminal buses and the midpoint over time are shown by the field measured data in Fig. 8.3a. We separate the fast and slow components of the voltage waves by bandpass filtering, as shown in Fig. 8.3b and c. The upper and lower cut-offs of the filter are chosen as 1 and 0.2 Hz, respectively, covering the interarea mode frequency. In Fig. 8.3b, we show the voltage oscillations from time $t = 60$–90 s. In this figure we can see that the oscillations are not exactly sinusoidal due to the presence of local modes in addition to the interarea mode. To extract the interarea mode, we apply the eigenvalue realization algorithm (ERA) on the three bus voltages [19]. Figures 8.4a and b shows the comparison of the interarea oscillations with the measured bus voltages at Buses 1 and 2. The oscillations are now free from the distortions due to the local mode effects, and are purely sinusoidal. Choosing a fixed time instant of $t = 75$ s, the amplitudes of the interarea oscillations at the respective buses are then measured as

$$V_{1m} = 8.746 \times 10^{-3},\ V_{2m} = 9.733 \times 10^{-3},\ V_{3m} = 0.01845 \qquad (8.20)$$

The pre-disturbance equilibrium voltages are measured as

$$V_{1ss} = 1.082,\ V_{2ss} = 1.087,\ V_{3ss} = 1.089 \qquad (8.21)$$

Therefore, the normalized amplitude of oscillations are

$$V_{1n} = 9.46 \times 10^{-3},\quad V_{2n} = 1.06 \times 10^{-3},\quad V_{3n} = 20.09 \times 10^{-3} \qquad (8.22)$$

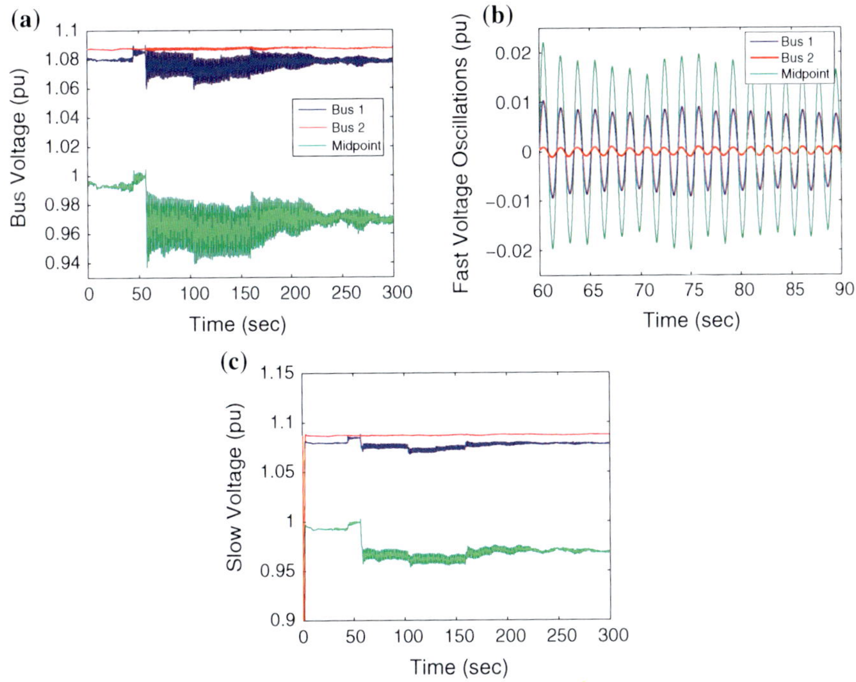

Fig. 8.3 Voltage oscillations in WECC transfer path 1. **a** Bus voltages in WECC transfer path 1. **b** Fast component of voltage magnitude. **c** Quasi-steady-state voltage magnitude

while $x_e = 0.077$ pu from least-squares estimation. Using the IME algorithm we get (in pu)

$$x_1 = 0.0121, \ x_2 = 0.0012, \ H_1 = 1050, \ H_2 = 134$$

The bus frequencies at the sending and receiving ends, their fast and slow components as well as their interarea components, used for estimating H_1 and H_2, are shown in Fig. 8.6. To verify the accuracy of the estimates, we also compare the interarea modal response of the voltage angles between the equivalent machines (as extracted via ERA) with the corresponding impulse response of the identified swing model. Figure 8.4c shows this comparison and confirms that the trajectories are sufficiently close to each other.

8.3 Star-Connected Three-Area System

Our system of interest in this section, namely a two-dimensional system with algebraic node(s), is motivated by the five-machine structure of the Pacific AC intertie model shown in Fig. 8.1a. The schematic circuit diagram of this system

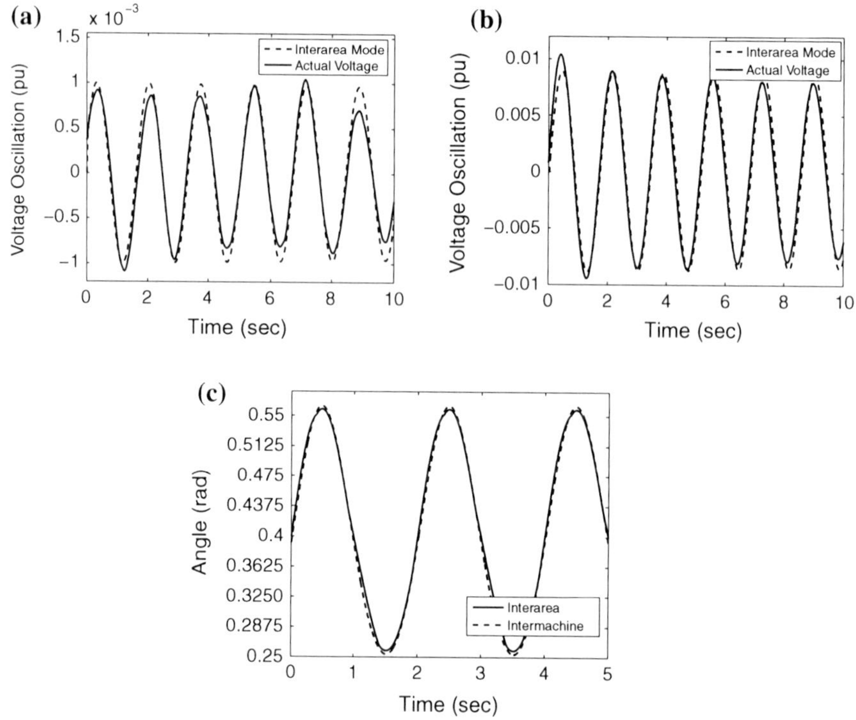

Fig. 8.4 Slow mode extraction for WECC transfer path 1 voltages. **a** Interarea oscillation at Bus 1. **b** Interarea oscillation at Bus 2. **c** Interarea versus intermachine oscillations

is shown in Fig. 8.7a. The basic problem formulation for reactance estimation for this system is similar to that in Sect. 8.2, i.e., using PMU measurements available from Buses 1, 2, 3 and 4, we need to solve for three unknown reactances, namely, $\sigma_i = x_{ei} + x_{Ti} + x'_{di}$, $i = 1, 2, 3$. Consider the star-connected three-machine equivalent of a three-area power system as shown in Fig. 8.7a. The classical model representation [18] of this three-machine system is shown in Fig. 8.7b. The system consists of three generators G_1, G_2, and G_3 with aggregated inertias H_1, H_2, and H_3 representing each coherent area [5], connected to Buses 1, 2 and 3 through transformers having equivalent reactances x_{T1}, x_{T2} and x_{T3}, respectively. The voltage phasors at Buses 1, 2, 3, and 4 are given as $\tilde{V}_i = V_i \angle \theta_i$, $i = 1, 2, 3, 4$, where $V\angle\theta$ denotes the polar representation $V\varepsilon^{j\theta}$. The transmission lines between Bus 4 and the other three buses are all assumed to be lossless, with line reactances x_{e1} between Buses 1 and 4, x_{e2} between Buses 2 and 4, and x_{e3} between Buses 3 and 4. The line current phasors shown in Fig. 8.7b are $\tilde{I}_i = I_i \angle \theta_{I_i}$, $i = 1, 2, 3$, with G_1 supplying power to G_2 and G_3, which act as loads. For the classical model representation, we denote the internal voltage phasors of the generators G_1, G_2, and G_3 as $\tilde{E}_i = E_i \angle \delta_i$, $i = 1, 2, 3$. The reactances connecting the generator internal voltages to Buses 1, 2, and 3 are given by

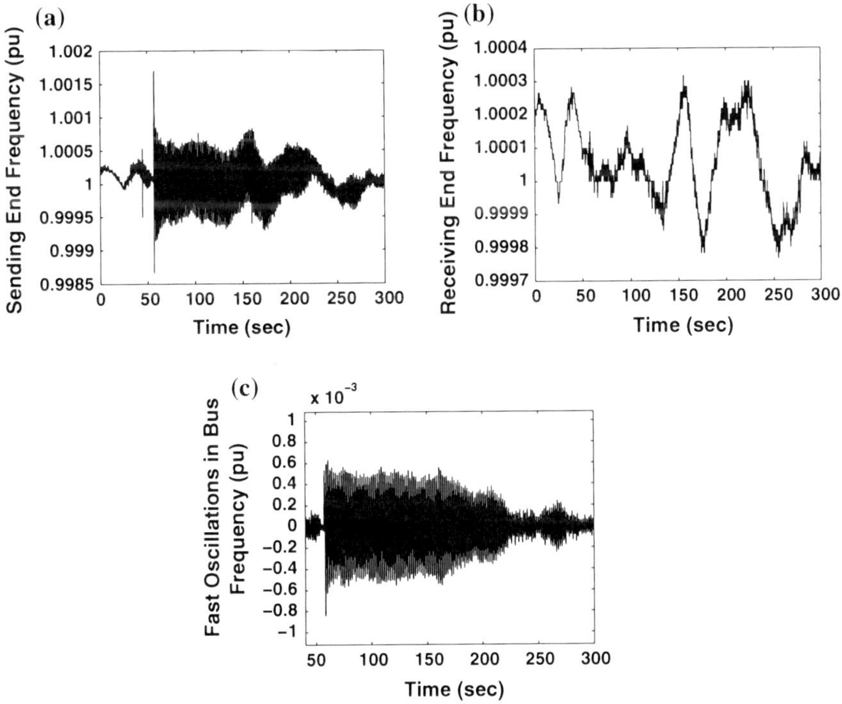

Fig. 8.5 Frequency oscillations in WECC transfer path 1. **a** Sending end frequency. **b** Receiving end frequency. **c** Fast component of sending end frequency

$$x_i = (x_{Ti} + x'_{di}), \qquad i = 1,\ 2,\ 3 \tag{8.23}$$

where x'_{d1}, x'_{d2}, and x'_{d3} are the direct-axis transient reactances of G_1, G_2, and G_3, respectively. For future use, we use the notations

$$\sigma_i = x_{ei} + x_i, \quad i = 1,\ 2,\ 3. \tag{8.24}$$

It should be noted that unlike the variant of the IME method considered in [20], here the generator G_2 is not necessarily a synchronous condenser, and, hence, there is no restrictive assumption about the equality of the voltage phase angles of Bus 2 and Bus 4.

The dynamic model of the three-machine system in Fig. 8.7b, neglecting damping, is given by

$$2H_1\,\ddot{\delta}_1 = P_{m1} - \frac{E_1 V_4}{\sigma_1}\sin(\delta_1 - \theta_4) \tag{8.25}$$

$$2H_2\,\ddot{\delta}_2 = P_{m2} - \frac{E_2 V_4}{\sigma_2}\sin(\delta_2 - \theta_4) \tag{8.26}$$

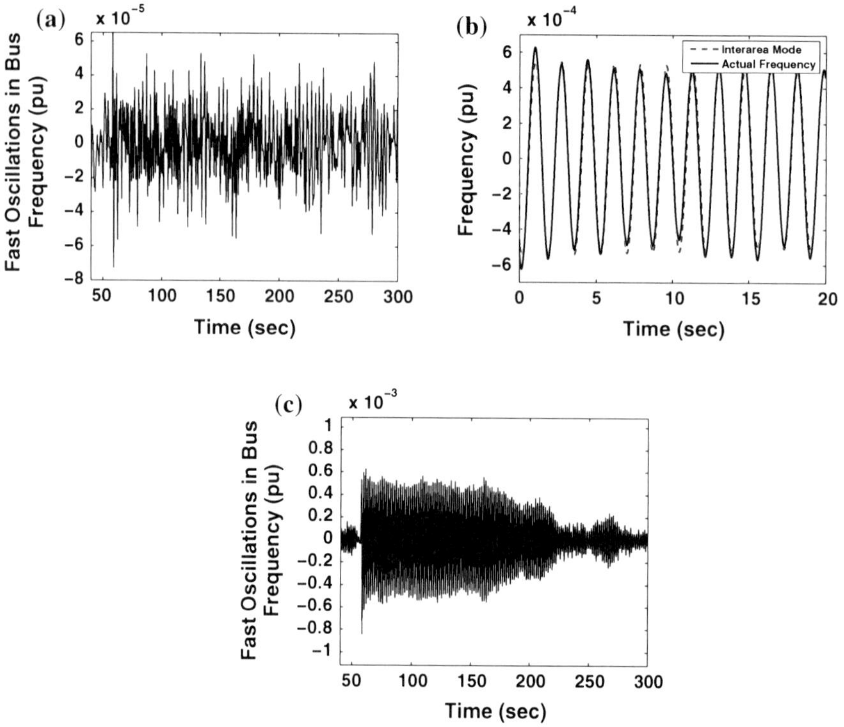

Fig. 8.6 Slow mode extraction for WECC transfer path 1 frequencies. **a** Fast component of receiving end frequency. **b** Interarea component of sending end frequency. **c** Interarea mode component of receiving end frequency

$$2H_3\ddot{\delta}_3 = P_{m3} - \frac{E_3 V_4}{\sigma_3}\sin(\delta_3 - \theta_4) \tag{8.27}$$

We assume that PMUs are located at Buses 1, 2, 3, and 4. Hence, high-sampling rate time-synchronized phasor variables $\tilde{V}_i$, $i = 1, 2, 3, 4$, and $\tilde{I}_i$, $i = 1, 2$, as a result of a disturbance are available. We pose the problem of finding the parameters of the model in Fig. 8.7b as follows.

Given the measured time-synchronized phasor variables V_1, θ_1, V_2, θ_2, V_3, θ_3, V_4, θ_4, I_1, θ_{I_1}, I_2, *and* θ_{I_2} *that exhibit a few cycles of interarea oscillations, and assuming that* E_1, E_2 *and* E_3 *are some constant values, compute* x_{e1}, x_{e2}, x_{e3}, E_1, δ_1, E_2, δ_2, E_3, δ_3, x_1, x_2, x_3, H_1, H_2 *and* H_3 *to completely characterize the dynamic behavior of the three-machine reduced system in* Fig. 8.7b.

Three of these quantities, namely, x_{e1}, x_{e2}, and x_{e3}, can readily be computed from the available bus voltages and currents using Ohm's law

$$jx_{e1} = \frac{\tilde{V}_1 - \tilde{V}_4}{\tilde{I}_1}, \quad jx_{e2} = \frac{\tilde{V}_4 - \tilde{V}_2}{\tilde{I}_2}, \quad jx_{e3} = \frac{\tilde{V}_4 - \tilde{V}_3}{\tilde{I}_3}. \tag{8.28}$$

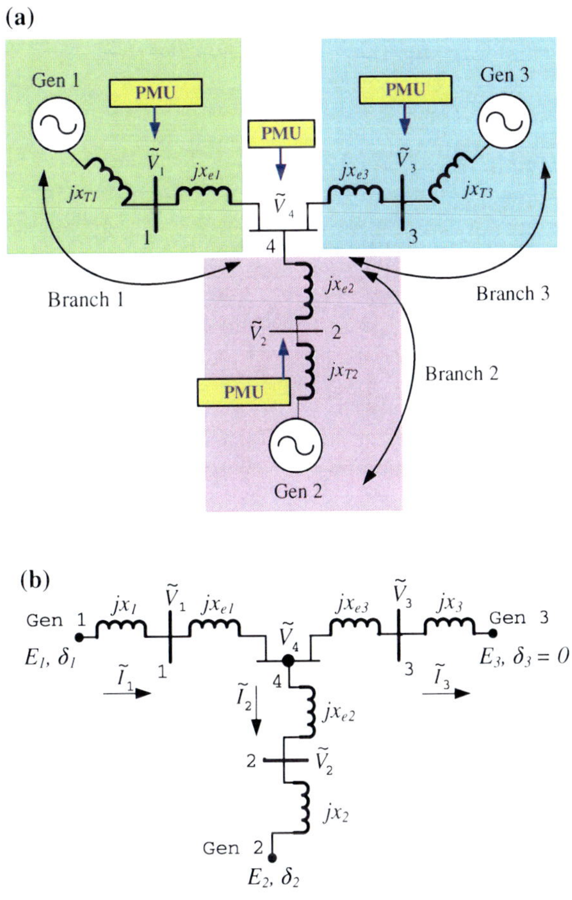

Fig. 8.7 Area aggregation of three-area power system. **a** Three-machine power system model. **b** Classical model representation

Moreover, if x_1, x_2, and x_3 are known, then the machine internal voltages can be computed from the bus voltages and the line currents. Thus, the problem reduces to the estimation of x_1, x_2, and x_3, as well as the inertias H_1, H_2, and H_3. In the following sections we develop techniques to estimate these six constant quantities.

Notations: A few notations used throughout the rest of the chapter are as follows. Subscripts R, L, and M, respectively, refer to quantities related to the right branch (between Generator 3 and Bus 4), left branch (between Bus 4 and Generator 1), and the middle branch (between Bus 4 and Generator 2) of the transfer path. The constant σ_{ij} is equal to σ_i/σ_j. The superscript i for any quantity refers to that quantity defined for Bus i, ($i = 1, 2, 3, 4$). A small change in any of the measured variables, say an angle θ, over an existing equilibrium, is denoted as $\Delta\theta$ while $\Delta\theta^{ij}$ is equal to $\Delta\theta^i/\Delta\theta^j$ where the superscript denotes the bus at which $\Delta\theta$ is measured. All quantities defined at the equilibrium are subscripted by 0.

8.3.1 Reactance Extrapolation: Branch 1

8.3.1.1 Step 1: Express $\tilde{V}_4$ as a Function of Generator Voltages

With $\tilde{E}_3$ as the reference, from circuit equations we can derive that

$$\tilde{V}_4 = \sigma(E_3 + \sigma_{31}\tilde{E}_1 + \sigma_{32}\tilde{E}_2) \tag{8.29}$$

where $\sigma \triangleq (\sigma_1\sigma_2)/(\sigma_1\sigma_2 + \sigma_2\sigma_3 + \sigma_3\sigma_1)$. We will use the expression in (8.29) for subsequent derivations in the following sections.

8.3.1.2 Step 2: Find Voltage Magnitude at an Arbitrary Point

Consider any point on the branch at a reactance jx away from Generator 3, which is taken as the reference node for this branch. The voltage phasor at this point can be written as

$$\tilde{V}_R = E_3(1 - a_3) + a_3\tilde{V}_4 \tag{8.30}$$

where $a_3 = x/\sigma_3 \in [0, 1]$. After a few calculations, it can be shown that the voltage magnitude at this point is

$$V_R(x, \delta_1, \delta_2) = \sqrt{\Sigma_R + \varpi_R(x, \delta_1, \delta_2)} \tag{8.31}$$

where Σ_R is a constant that is independent of δ_1 and δ_2,

$$\varpi(x, \delta_1, \delta_2) = \sum_{i=1}^{2} \alpha_{Ri}\cos(\delta_i) + \alpha_{R3}\cos(\delta_1 - \delta_2)$$

with $\alpha_{R1} = 2E_1E_3a_3(1 - a_3 + a_3\sigma)\sigma\sigma_{31}$, $\alpha_{R2} = 2E_2E_3a_3(1 - a_3 + a_3\sigma)\sigma\sigma_{32}$, and $\alpha_{R3} = 2E_1E_2a_3^2\sigma^2\sigma_{31}\sigma_{32}$.

8.3.1.3 Step 3: Form Normalized Voltage from a Small-Signal Perturbation

Consider a disturbance in the system, so that a small change in the voltage magnitude in (8.31), at the given point at time t, over a pre-disturbance equilibrium voltage $V_{R0}(x)$ can be written as

$$\Delta V_R(x,t) = \frac{P_{R1}(x)\Delta\delta_1(t) + P_{R2}(x)\Delta\delta_2(t)}{V_{R0}(x)} \tag{8.32}$$

where

$$P_{R1}(x) = -0.5(\alpha_{R3}\sin(\delta_{120}) + \alpha_{R1}\sin(\delta_{10}))$$
$$P_{R2}(x) = 0.5(\alpha_{R3}\sin(\delta_{120}) - \alpha_{R2}\sin(\delta_{20}))$$

and $\delta_{120} = \delta_{10} - \delta_{20}$. Next, we fix time at $t = t^*$, and denote $V_{Rn}(x) = \Delta V_R(x,t^*)V_{R0}(x)$, so that

$$V_{Rn}(x) = P_{R1}(x)\Delta\delta_1(t^*) + P_{R2}(x)\Delta\delta_2(t^*). \tag{8.33}$$

Because the normalized voltage can be measured at Buses 3 and 4 from PMU measurements following the disturbance, the quantities

$$V_n^3 \triangleq V_{Rn}(x_3), \quad V_n^4 \triangleq V_{Rn}(x_3 + x_{e3}) \tag{8.34}$$

are known. Denoting $P_{Ri}^3 = P_{Ri}(x_3)$ and $P_{Ri}^4 = P_{Ri}(x_3 + x_{e3})$, $i = 1, 2$, we can write

$$\frac{V_n^3}{V_n^4} = \frac{P_{R1}^3\Delta\delta_1(t^*) + P_{R2}^3\Delta\delta_2(t^*)}{P_{R1}^4\Delta\delta_1(t^*) + P_{R2}^4\Delta\delta_2(t^*)}. \tag{8.35}$$

The two hypothetical states $\Delta\delta_1(t^*)$ and $\Delta\delta_2(t^*)$ are unknown in (8.35). For the single interarea mode case of Sect. 8.2, this problem could be very easily bypassed as there was no term involving $\Delta\delta_2$, due to which the constant $\Delta\delta_1(t^*)$ could be cancelled in the numerator and the denominator of the right hand side of the equation, and the resulting nonlinear algebraic equation could be used toward solving for the unknown reactances. In other words, the *time–space separation property* is lost in (8.35) due to the extra interarea mode. To fix this problem, we next consider the change in the phase angles at Bus 3 and 4 as an extra degree of freedom, as follows.

8.3.1.4 Step 4: Derive the Change in Bus Phase Angles

The phase angle at any point on Branch 1, at a reactance x away from the Generator 3 node, is

$$\theta = \tan^{-1}\left(\frac{c_{R1}\sin(\delta_1) + c_{R2}\sin\delta_2}{c_{R3} + c_{R1}\cos(\delta_1) + c_{R2}\cos\delta_2}\right) \tag{8.36}$$

where $c_{R1} = E_1 a_3\sigma\sigma_{31}$, $c_{R2} = E_2 a_3\sigma\sigma_{32}$, and $c_{R3} = E_3(1 - a_3 + a_3\sigma)$. Using the fact that if $\theta = \tan^{-1}(\psi(\delta_1,\delta_2))$, then

$$\Delta\theta = \frac{1}{1+\psi^2}\left(\frac{\partial\psi}{\partial\delta_1}\Delta\delta_1 + \frac{\partial\psi}{\partial\delta_2}\Delta\delta_2\right)$$

it can be shown that a small change in the LHS of (8.36), at any time t, can be written as

$$\Delta\theta(x,t) = S_{R1}(x)\Delta\delta_1(t) + S_{R2}(x)\Delta\delta_1(t) \tag{8.37}$$

where S_{R1} and S_{R2} are given as

$$S_{R1}(x) = \frac{c_{R1}^2 + c_{R1}c_{R3}\cos(\delta_{10}) + c_{R1}c_{R2}\cos(\delta_{10}-\delta_{20})}{\vartheta_R(x)} \tag{8.38}$$

$$S_{R2}(x) = \frac{c_{R2}^2 + c_{R2}c_{R3}\cos(\delta_{20}) + c_{R1}c_{R2}\cos(\delta_{10}-\delta_{20})}{\vartheta_R(x)} \tag{8.39}$$

$$\begin{aligned}\vartheta_R(x) = c_{R1}^2 + c_{R2}^2 + c_{R3}^2 + 2(c_{R1}c_{R3}\cos(\delta_{10}) + c_{R3}c_{R2}\cos(\delta_{20})\\ + c_{R2}c_{R1}\cos(\delta_{10}-\delta_{20}))\end{aligned} \tag{8.40}$$

8.3.1.5 Step 5: Formulate a Candidate Algebraic Equation

Fixing time at $t = t^*$, the fraction of the measured changes in the phase angles at Buses 3 and 4 in terms of the functions $S_{R1}(\cdot)$ and $S_{R2}(\cdot)$ defined at these respective bus locations, can then be written as

$$\frac{\Delta\theta^3}{\Delta\theta^4} = \frac{S_{R1}^3\Delta\delta_1(t^*) + S_{R2}^3\Delta\delta_1(t^*)}{S_{R1}^4\Delta\delta_2(t^*) + S_{R2}^4\Delta\delta_2(t^*)}. \tag{8.41}$$

Selecting t^* such that $\Delta\delta_1(t^*) \neq 0$ and $\Delta\delta_2(t^*) \neq 0$, (8.35) and (8.41) yield

$$\frac{V_n^3}{V_n^4} = \frac{P_{R1}^3(\Delta\theta^{34}S_{R2}^4 - S_{R2}^3) + P_{R2}^3(S_{R1}^3 - \Delta\theta^{34}S_{R1}^4)}{P_{R1}^4(\Delta\theta^{34}S_{R2}^4 - S_{R2}^3) + P_{R2}^4(S_{R1}^3 - \Delta\theta^{34}S_{R1}^4)}. \tag{8.42}$$

The LHS of (8.42) as well as the quantity $\Delta\theta^{34}$ on the RHS are measured while the other functions on the RHS are known nonlinear functions of (x_1, x_2, x_3). Therefore, (8.42) serves as a feasible equation to solve for these three unknown reactances. The remaining two equations are constructed in the following sections using the measured phasor variables in Branches 2 and 3 of the network.

8.3.2 Reactance Extrapolation: Branch 2

Following an analysis similar to that used in Sect. 8.2.1, the normalized voltage at a point between Bus 4 and Generator 1, at a reactance x away from Bus 4,

$$V_{Ln}(x) = P_{L1}(x)\Delta\delta_1(t^*) + P_{L2}(x)\Delta\delta_2(t^*) \tag{8.43}$$

where

$$\begin{aligned}
P_{Li} &= \frac{1}{2}(-\alpha_{L3}\sin(\delta_{10}-\delta_{20}) - \alpha_{Li}\sin(\delta_{i0})), \quad i = 1,2 \\
\alpha_{L1} &= 2E_1E_3a_1(1-a_1)\sigma + 2E_1E_3(1-a_1)^2\sigma^2\sigma_{31} \\
\alpha_{L2} &= 2E_2E_3(1-a_1)^2\sigma^2\sigma_{32} \\
\alpha_{L3} &= 2E_1E_2a_1(1-a_1)\sigma\sigma_{32} + 2E_1E_2(1-a_1)^2\sigma^2\sigma_{31}\sigma_{32}
\end{aligned}$$

The equivalent of (8.42) for Branch 2 can then be written as

$$\frac{V_n^1}{V_n^4} = \frac{P_{L1}^1(\Delta\theta^{14}S_{L2}^4 - S_{L2}^1) + P_{L2}^1(S_{L1}^1 - \Delta\theta^{14}S_{L1}^4)}{P_{L1}^4(\Delta\theta^{14}S_{L2}^4 - S_{L2}^1) + P_{L2}^4(S_{L1}^1 - \Delta\theta^{14}S_{L1}^4)} \tag{8.44}$$

where the expressions for S_{L1}, S_{L2} are given as

$$\begin{aligned}
S_{Li}(x) &= \left(c_{Li}^2 + c_{Li}c_{L3}\cos(\delta_{i0}) + c_{L1}c_{L2}\cos(\delta_{10}-\delta_{20})\right)/\vartheta_L(x), \quad i = 1,2 \\
\vartheta_L(x) &= c_{L1}^2 + c_{L2}^2 + c_{L3}^2 + 2c_{L3}\sum_{i=1}^{2}(c_{Li}\cos(\delta_{i0}) + c_{L2}c_{L1}\cos(\delta_{10}-\delta_{20})) \\
c_{L1} &= a_1E_1 + (1-a_1)E_1\sigma\sigma_{31} \\
c_{L2} &= E_2(1-a_1)\sigma\sigma_{32}, \quad c_{L3} = E_3(1-a_1)\sigma
\end{aligned}$$

8.3.3 Reactance Extrapolation: Branch 3

For any point between Bus 4 and Generator 2, at a reactance x away from Bus 4, we have

$$\frac{V_n^2}{V_n^4} = \frac{P_{M1}^2(\Delta\theta^{24}S_{M2}^4 - S_{M2}^2) + P_{M2}^2(S_{M1}^2 - \Delta\theta^{24}S_{M1}^4)}{P_{M1}^4(\Delta\theta^{24}S_{M2}^4 - S_{M2}^2) + P_{M2}^4(S_{M1}^2 - \Delta\theta^{24}S_{M1}^4)} \tag{8.45}$$

where

$$\begin{aligned}
P_{Mi} &= \frac{1}{2}(-\alpha_{M3}\sin(\delta_{10}-\delta_{20}) - \alpha_{Mi}\sin(\delta_{i0})), \quad i = 1,2 \\
\alpha_{M1} &= 2E_1E_3(1-a_2)^2\sigma^2\sigma_{31} \\
\alpha_{M2} &= 2E_2E_3a_2(1-a_2)\sigma + 2E_2E_3(1-a_2)^2\sigma^2\sigma_{32} \\
\alpha_{M3} &= 2E_1E_2a_2(1-a_2)\sigma\sigma_{31} + 2E_1E_2(1-a_2)^2\sigma^2\sigma_{31}\sigma_{32}
\end{aligned}$$

with $a_2 = x/\sigma_2$, and the expressions for S_{M1} and S_{M2} are given as

$$S_{Mi}(x) = (c_{Mi}^2 + c_{Mi}c_{M3}\cos(\delta_{i0}) + c_{M1}c_{M2}\cos(\delta_{10} - \delta_{20}))/\vartheta_M(x), \;\; i = 1, 2$$

$$\vartheta_M(x) = c_{L1}^2 + c_{M2}^2 + c_{M3}^2 + 2c_{M3}\sum_{i=1}^{2}(c_{Mi}\cos(\delta_{i0}) + c_{M2}c_{M1}\cos(\delta_{10} - \delta_{20}))$$

$$c_{M1} = (1 - a_2)E_1\sigma\sigma_{31}$$
$$c_{M2} = a_2E_2 + E_2(1 - a_2)\sigma\sigma_{32}, \quad c_{M3} = E_3(1 - a_2)\sigma$$

Equations (8.42), (8.44), and (8.45) can now be used to solve for the three unknowns x_1, x_2, and x_3 using the measurements of the bus voltage magnitudes and phase angles.

8.3.4 Machine Inertia Estimation

To solve for the three inertias H_1, H_2, and H_3 we need three equations. The first two equations are given from the expressions for the frequencies of the two interarea modes. For this we consider the electromechanical swing equations for the three machines, given by (8.25–8.27). However, we should note that in these equations, the two variables V_4 and θ_4 are not constants, but functions of the generator angles according to (8.29). Considering this fact, we next linearize (8.25–8.27) about the post-disturbance equilibrium, to get a linear equation of the form

$$\ddot{\delta} = \mathscr{H}^{-1}\mathscr{A}\delta \tag{8.46}$$

where $\delta = \text{col}(\delta_1, \delta_2, \delta_3)$, $\mathscr{H} = \text{diag}(2H_1, 2H_2, 2H_3)$ and $\mathscr{A}$ is a 3×3 matrix, whose entries are all known once the three reactances x_1, x_2, and x_3 are solved for, using the method described in Sects. 8.3.1–8.3.3. To find the interarea dynamics, we next fix δ_3 as the reference angle, and define the relative angular separations as

$$\delta_{13} = \delta_1 - \delta_3, \quad \delta_{23} = \delta_2 - \delta_3. \tag{8.47}$$

Consider $\mathscr{A}_1$, $\mathscr{A}_2$ and $\mathscr{A}_3$ to be the first, second and third rows of the matrix $\mathscr{A}$ in (8.46), respectively. It can be readily seen that

$$\ddot{\delta}_{13} = (\bar{\mathscr{A}_1} - \bar{\mathscr{A}_3})\delta, \quad \ddot{\delta}_{23} = (\bar{\mathscr{A}_2} - \bar{\mathscr{A}_3})\delta \tag{8.48}$$

where $\bar{\mathscr{A}_1} = \mathscr{A}_1/(2H_1)$, $\bar{\mathscr{A}_2} = \mathscr{A}_2/(2H_2)$, $\bar{\mathscr{A}_3} = \mathscr{A}_3/(2H_3)$. Also, we have

$$\begin{bmatrix} \delta_{13} \\ \delta_{23} \end{bmatrix} = \begin{bmatrix} 1 & 0 & -1 \\ 0 & 1 & -1 \end{bmatrix} \delta. \tag{8.49}$$

Combining (8.48) and (8.49) we can write

$$\begin{bmatrix} \ddot{\delta}_{13} \\ \ddot{\delta}_{23} \end{bmatrix} = \underbrace{\begin{bmatrix} \bar{\mathscr{A}}_1 - \bar{\mathscr{A}}_3 \\ \bar{\mathscr{A}}_2 - \bar{\mathscr{A}}_3 \end{bmatrix} \begin{bmatrix} 1 & 0 & -1 \\ 0 & 1 & -1 \end{bmatrix}^+}_{\Upsilon} \begin{bmatrix} \delta_{13} \\ \delta_{23} \end{bmatrix}. \tag{8.50}$$

Denoting $\mathscr{A}_i = [a_{i1}\, a_{i2}\, a_{i3}]$, $i = 1, 2, 3$, the structure of the 2×2 matrix Υ is given as

$$\Upsilon = \begin{bmatrix} \frac{2a_{11} - a_{12} - a_{13}}{H_1} - \frac{2a_{31} - a_{32} - a_{33}}{H_3} & \frac{2a_{12} - a_{11} - a_{13}}{H_1} - \frac{2a_{32} - a_{31} - a_{33}}{H_3} \\ \frac{2a_{21} - a_{22} - a_{23}}{H_2} - \frac{2a_{31} - a_{32} - a_{33}}{H_3} & \frac{2a_{22} - a_{21} - a_{23}}{H_2} - \frac{2a_{32} - a_{31} - a_{33}}{H_3} \end{bmatrix} \tag{8.51}$$

Let the entries of this matrix be denoted as υ_{ij} where i is the row index and j is the column index. Then it follows that the eigenvalues of Υ are given as

$$\lambda_{1,2} = \frac{\upsilon_{11} + \upsilon_{22} \pm \sqrt{(\upsilon_{11} - \upsilon_{22})^2 + 4\upsilon_{12}\upsilon_{21}}}{2}, \tag{8.52}$$

and the respective eigenvectors e_1 and e_2 are given as solutions to the equations

$$\Upsilon e_1 = \lambda_1 e_1, \quad \Upsilon e_2 = \lambda_2 e_2. \tag{8.53}$$

It is clear from (8.51) that λ_1, λ_2, e_1 and e_2 are all functions of H_1, H_2, and H_3. The frequencies of the interarea oscillations can be measured from the difference signals $(\theta_1 - \theta_2)$ and $(\theta_2 - \theta_3)$ using modal analysis (such as ERA or a subspace identification algorithm). Say these frequencies (in Hz) are f_{s1} and f_{s2}, and the computed respective eigenvectors are φ_{s1} and φ_{s2}. Then we can write

$$f_{s1} = \frac{1}{2\pi}\sqrt{\lambda_1}, \quad \varphi_{s1} = e_1 \tag{8.54}$$

$$f_{s2} = \frac{1}{2\pi}\sqrt{\lambda_2}, \quad \varphi_{s2} = e_2. \tag{8.55}$$

Either (8.54) or (8.55) may be used as two equations for solving for the three unknown inertias as f_{s1}, f_{s2}, φ_{s1} and φ_{s2} are known quantities. An important point to note here is that although the three-area system under consideration contains two interarea modes of oscillation it is sufficient to consider the participation of only one of these modes in the PMU measurements for solving the EIME problem. For example, the estimation of the unknown reactances, as in (8.25), (8.33), and (8.34), involves the elimination of one of the interarea modes by considering the simultaneous measurement of voltage magnitudes and phase angles. Similarly, the estimation of the unknown inertias in (8.43–8.44) requires the modal decomposition of only one interarea mode. The resulting EIME algorithm, as summarized in Fig. 8.9, therefore, does not depend on which modal component out of the two is selected from the

PMU measurements for solving for the unknown parameters. The idea is illustrated in Fig. 8.8.

The third equation is given by the law of conservation of angular momentum, which can be simply written as

$$H_1\omega_1 + H_2\omega_2 + H_3\omega_3 = 0 \tag{8.56}$$

where ω_i is the angular speed (rad/s) of the ith generator ($i = 1, 2, 3$). These speeds are not measured, but can be estimated from the measured bus frequencies ν_1, ν_2, and ν_3 at Buses 1, 2, and 3, respectively, as discussed in [20]. The basic methodology to achieve this is to express the voltage phasor $\tilde{V}$ at any point in terms of $E_1\angle\delta_1$, $E_2\angle\delta_2$, and $E_3\angle\delta_3$ and the reactance x with respect to some chosen reference; then calculate the phasor angle

$$\theta = \tan^{-1}(\mathrm{Im}(\tilde{V})/\mathrm{Re}(\tilde{V})) \tag{8.57}$$

and compute the time derivative of θ as a function of x, ω_1, ω_2 and ω_3. For the three-machine system of Fig. 8.7b, after some calculations it can be shown that the frequency at any point on the path is given by

$$\nu(x, \delta_1, \delta_2, \delta_3) = \frac{\Pi_1(x, \delta_1, \delta_2, \delta_3)}{\Pi_2(x, \delta_1, \delta_2, \delta_3)} \tag{8.58}$$

where the functions $\Pi_1(\cdot)$ and $\Pi_2(\cdot)$ are given as

$$\begin{aligned}\Pi_1(x, \delta_1, \delta_2, \delta_3) = \omega_1 &\left[n_1^2E_1^2 + n_1n_2E_1E_2\cos(\delta_1 - \delta_2) + n_1n_3E_1E_3\cos(\delta_1 - \delta_3)\right]\\ &+ \omega_2\left[n_2^2E_2^2 + n_2n_3E_2E_3\cos(\delta_2 - \delta_3) + n_2n_1E_2E_1\cos(\delta_2 - \delta_1)\right]\\ &+ \omega_3\left[n_3^2E_3^2 + n_3n_1E_3E_1\cos(\delta_3 - \delta_1) + n_3n_2E_3E_2\cos(\delta_3 - \delta_2)\right]\end{aligned}$$

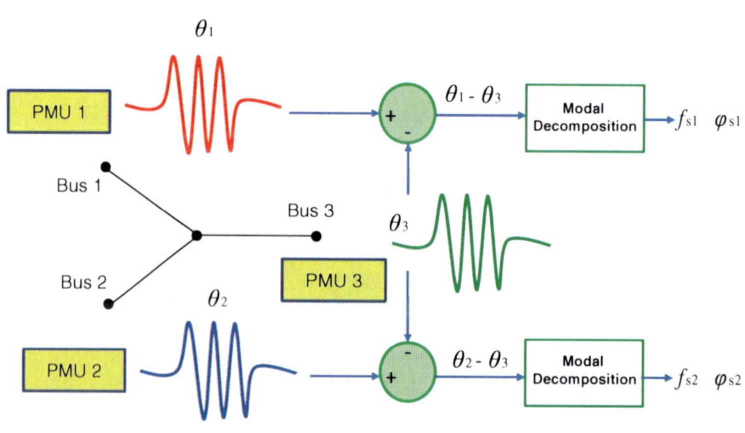

Fig. 8.8 Choice of interarea mode for computation of unknown machine inertias

A Summary of the EIME Algorithm

1. Consider the three-machine system in Figure 8.1(b). Following a disturbance in the system, measure the voltage phasors $\tilde{V}_1$, $\tilde{V}_2$, $\tilde{V}_3$, and $\tilde{V}_4$ at Buses 1, 2, 3 and 4, respectively, and the currents $\tilde{I}_1$, $\tilde{I}_2$ and $\tilde{I}_3$.
2. Calculate the line reactances x_{e1}, x_{e2}, and x_{e3} using (8.28).
3. Measure the amplitudes of oscillation in the magnitudes of $\tilde{V}_i$ $(i = 1,..,4)$ at one particular instant of time, for example, when they reach a peak simultaneously.
4. Multiply the measured amplitudes with the respective steady-state values of the voltage waveforms to get the normalized voltage amplitudes at these four buses.
5. Measure the amplitudes of oscillation in the phase angles of $\tilde{V}_i$ $(i = 1,...,4)$ at the same instant of time.
6. Solve for x_1, x_2, and x_3 using equations (8.42), (8.44), and (8.45).
7. Calculate the constant generator internal voltages using the extrapolated reactances, the bus voltages, and the line currents.
8. Calculate the inter-area swing frequency (eigenvalue) and the corresponding eigenvector for any <u>one</u> of the two inter-area modes from the measured bus voltages using modal decomposition such as ERA.
9. Use the extrapolated system parameters and the inter-area frequencies to get two equations in the machine inertias using (8.51), (8.54), and (8.55).
10. Use the measured bus frequencies ν_1 and ν_2 to estimate ω_1 and ω_2 using (8.58) and (8.62).
11. Compute H_1, H_2 and H_3 from Step 9 and equation (8.56) .

Fig. 8.9 A summary of the EIME algorithm

$$\Pi_2(x,\delta_1,\delta_2,\delta_3) = n_1^2E_1^2 + n_2^2E_2^2 + n_3^2E_3^2 + 2\Big[n_1n_2E_1E_2\cos(\delta_1-\delta_2) + n_2n_3E_2E_3\cos(\delta_2-\delta_3) + n_3n_1E_3E_1\cos(\delta_3-\delta_1)\Big]$$

with n_1, n_2, and n_3 defined as follows. The references for measuring x are fixed at Bus 4 for Branches 1 and 3, and at Generator 1 for Branch 1. Also, as defined before, a_1, a_2, and a_3 denote the normalized reactances (or equivalently distances) measured along Branches 1, 2, and 3, respectively, from their respective reference points:

1. *Branch* 1: Between Bus 4 and Generator 3

$$n_1 = \sigma\sigma_{31}(1-a_3), \quad n_2 = \sigma\sigma_{32}(1-a_3), \quad n_3 = \sigma(1-a_3)+a_3 \tag{8.59}$$

2. *Branch* 2: Between Generator 1 and Bus 4

$$n_1 = 1-a_1+a_1\sigma\sigma_{31}, \quad n_2 = a_1\sigma\sigma_{32}, \quad n_3 = a_1\sigma \tag{8.60}$$

3. *Branch* 3: Between Bus 4 and Generator 2

$$n_1 = \sigma\sigma_{31}(1 - a_2), \quad n_2 = \sigma\sigma_{32}(1 - a_2) + a_2, \quad n_3 = \sigma(1 - a_2). \tag{8.61}$$

Therefore, the frequencies at Buses 1, 2, and 3 can be expressed as

$$\nu_1 = \frac{\Pi_1(x_1, \delta_1, \delta_2, \delta_3)}{\Pi_2(x_1, \delta_1, \delta_2, \delta_3)}, \quad \nu_2 = \frac{\Pi_1(x_{e2}, \delta_1, \delta_2, \delta_3)}{\Pi_2(x_{e2}, \delta_1, \delta_2, \delta_3)},$$
$$\nu_3 = \frac{\Pi_1(x_{e3}, \delta_1, \delta_2, \delta_3)}{\Pi_2(x_{e3}, \delta_1, \delta_2, \delta_3)}. \tag{8.62}$$

Since the bus frequencies are measured (or derived by passing the bus angles through high-pass filters), ν_1, ν_2, and ν_3 are known at any chosen time instant. Also, once x_1 and x_2 are identified, E_i and δ_i $(i = 1, 2, 3)$ can be calculated for the same time instant, and hence, the three equations in (8.62) can be solved for ω_1, ω_2, and ω_3. The solution of H_1 and H_2 then follows from (8.54) or (8.55) and (8.56). The entire algorithm is described step-by-step in Fig. 8.9.

8.3.5 Reduced-Order Modeling of Pacific AC Intertie

This transfer path is a large group of generators supplying power via a 1,200-mile transmission system to a large load center with an intermediate generation cluster attached to the path, as shown in Fig. 8.1a. Figure 8.10a shows the bus voltage magnitudes at 6 buses on the transfer path, with Buses 1, 2, and 3 being the sending end, receiving end, and the intermediate generation bus, respectively. Figures 8.10b and c show the separated fast and slow components for each of the six bus voltages. As both the sending and receiving ends have a large group of generators, a significant number of swing modes contributes to the oscillations in Fig. 8.10b, only one of which is the interarea mode. The oscillation due to this mode is dominant in all six voltages. We apply the eigensystem realization algorithm (ERA) to extract the modes and their mode shapes in the time response of the voltage oscillations, starting from $t = 62$ s to $t = 80$ s. ERA shows that over this chosen time-window of 18 s, the oscillations can be approximated by a dominant interarea mode of 0.404 Hz. Figure 8.11 shows the 0.404 Hz mode components superimposed on the individual voltage magnitude oscillations at Buses 1, 2, and 3. Similar figures can be drawn for the remaining buses.

The magnitude V_{im} (in pu) of the interarea mode components of the ith bus voltage $(i = 1, 2, 3)$ can be obtained from the approximated (dotted) voltage responses in Fig. 8.11a, b, and c at a fixed point of time. Choosing this fixed time point at $t = 4$ s, where the positive peak of the second cycle occurs, we get (in pu)

$$V_{1m} = 1.897 \times 10^{-3}, \quad V_{2m} = 2.615 \times 10^{-3}, \quad V_{3m} = 5.206 \times 10^{-3} \tag{8.63}$$

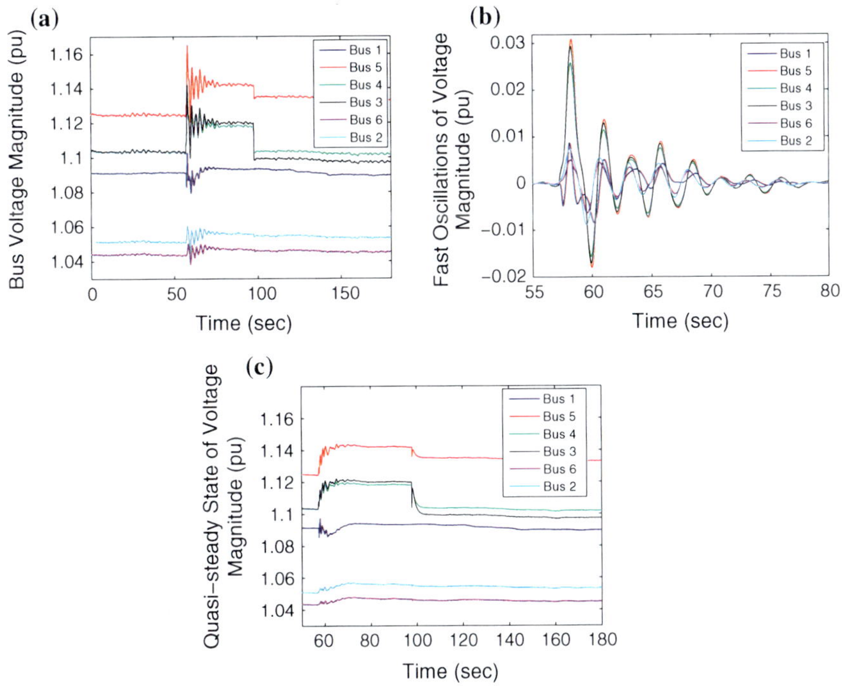

Fig. 8.10 Bus voltage magnitudes for WECC transfer path 2. **a** Bus 1 voltage. **b** Bus 2 voltage. **c** Bus 3 voltage

As the signals used for mode extraction start from $t = 62$ s in the original time response of Fig. 8.10a, the chosen fixed time-point is equal to $t = 66$ s. The quasi-steady-state values V_{iss} of the ith bus voltage ($i = 1, \ldots, 6$) are obtained from the slow parts of the voltages shown in Fig. 8.10c at the pre-disturbance time instant as

$$V_{1\text{ss}} = 1.0903, \quad V_{2\text{ss}} = 1.046, \quad V_{3\text{ss}} = 1.1234 \tag{8.64}$$

Therefore, the normalized voltage amplitudes are (in pu)

$$V_{1n} = 2.0683 \times 10^{-3}, \quad V_{2n} = 2.7353 \times 10^{-3}, \quad V_{3n} = 5.8482 \times 10^{-3} \tag{8.65}$$

Applying the IME algorithm from (8.42), (8.44), and (8.45), we obtain

$$x_1 = 0.00411, \quad x_2 = 0.00655 \tag{8.66}$$

where x_1 (in pu) is the sum of the aggregated transformer reactance and direct-axis transient reactance of the sending end equivalent generator, and x_2 (in pu) is that of the receiving end equivalent generator. The Jacobian curve is shown in Fig. 8.12.

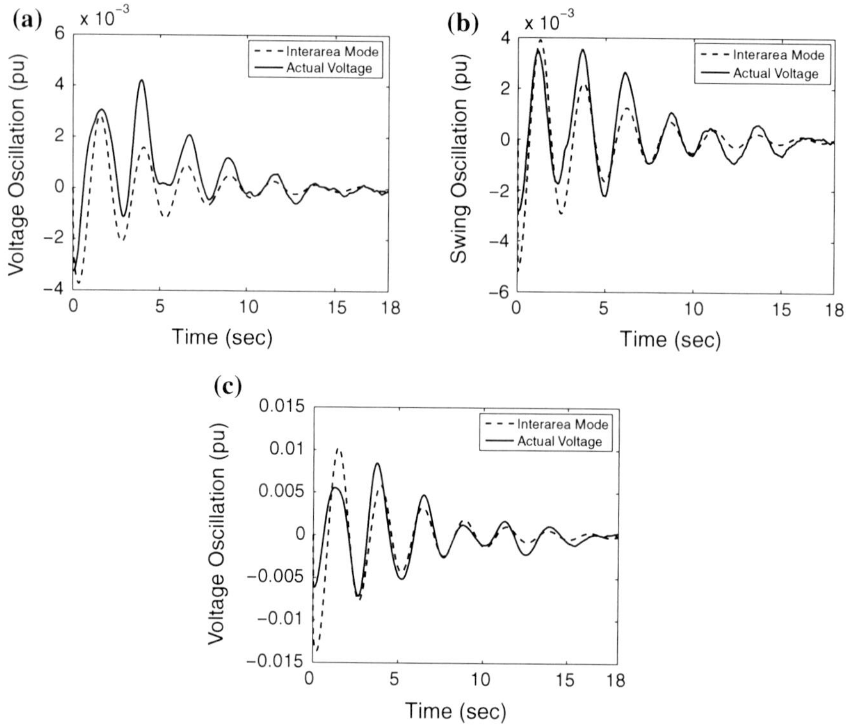

Fig. 8.11 Interarea bus voltage magnitudes for WECC transfer path 2. **a** Bus 1 voltage. **b** Bus 2 voltage. **c** Bus 3 voltage

From the curve, it can be easily seen that oscillations are much more damped on the right half of the transfer path due to the high loading effect on this side.

The inertia constant (pu) between the two dominant generators is approximately $H = 810$ pu. However, the voltage droop at the *star point* is not very significant, which indicates that the third generator does not produce a strong impact on the voltage profile. The Jacobian fit without the effect of this extra interarea mode is shown in Fig. 8.12 as the dashed curve.

8.4 Multi-Modal Interarea Equivalents

In this section we generalize the idea presented in Sect. 8.3 to a multi-area system where each area is *directly* connected to its neighboring set of areas without the existence of any algebraic bus. As before, we represent each area by an equivalent synchronous machine, and assume that the equivalent topology of the system is known. Unlike Sect. 8.3 the advantage here is that we do not need to compute the voltage phasor at any algebraic bus, and can apply the IME algorithm of Sect. 8.2

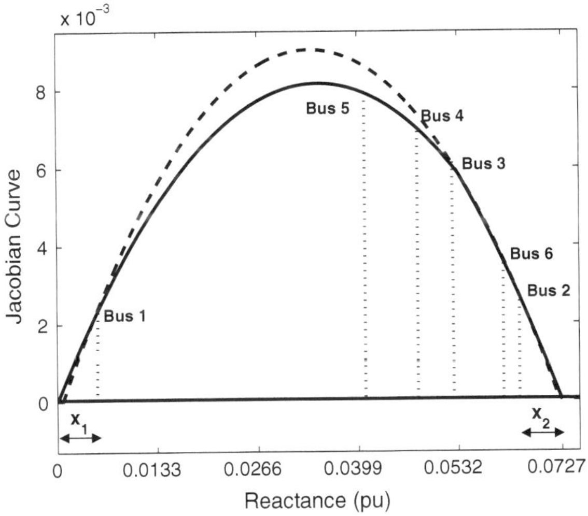

Fig. 8.12 Jacobian curves for Pacific AC Intertie—with second interarea mode (*solid*), without second interarea mode (*dashed*)

directly to every pair of connected areas in a *decentralized* fashion provided that the contribution of every slow mode is retained in the bus measurements. The idea is illustrated by the three-machine system in Fig. 8.13a, where each of the three machines G_1, G_2, and G_3, connected to each other by a ring topology, represent the equivalent of an area consisting of multiple local machines. PMUs are assumed to be installed at the terminal buses of each machine.

The swing dynamics of the equivalent system is given by

$$2H_1\ddot{\delta}_1 = P_{m1} - \frac{E_1E_2}{x_{12}}\sin(\delta_1-\delta_2) - \frac{E_1E_3}{x_{13}}\sin(\delta_1-\delta_3) \tag{8.67}$$

$$2H_2\ddot{\delta}_2 = P_{m2} - \frac{E_2E_3}{x_{23}}\sin(\delta_2-\delta_3) - \frac{E_2E_1}{x_{12}}\sin(\delta_2-\delta_1) \tag{8.68}$$

$$2H_3\ddot{\delta}_3 = P_{m3} - \frac{E_3E_1}{x_{13}}\sin(\delta_3-\delta_1) - \frac{E_3E_2}{x_{23}}\sin(\delta_3-\delta_2) \tag{8.69}$$

where P_{mi} is the effective mechanical power input to the ith area, and x_{ik} is the total reactance connecting the internal node of the ith and k^{th} equivalent machines. However, because each area may contain internal loads, we assume the line currents across each transfer path to be different, namely $\tilde{I}_1$, $\tilde{I}_2$, and $\tilde{I}_3$. All three currents are available from the respective PMU measurements. The equivalent transmission line reactances can be calculated from the bus voltage and current measurements using Ohm's law, while the total reactance between the internal node of any machine and any of the terminal buses needs to be estimated using IME. Assuming a classical model for synchronous machines, this total reactance is assumed to be the sum of the

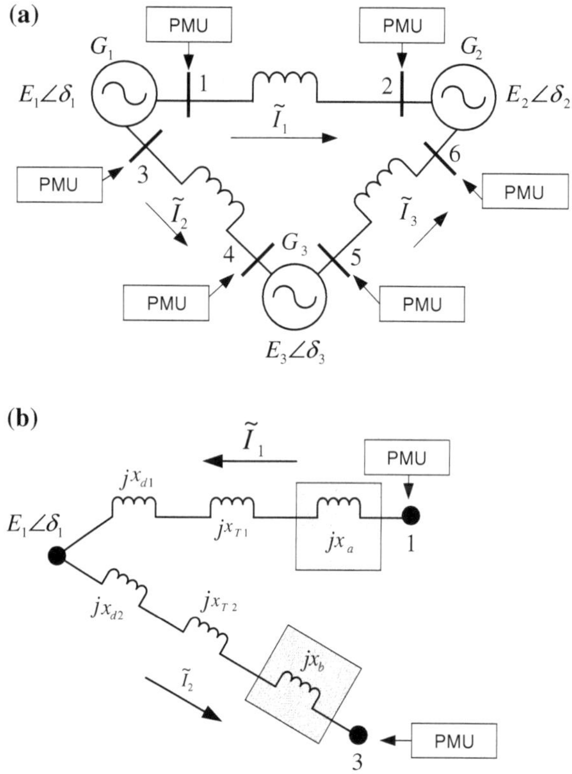

Fig. 8.13 Equivalent circuit of a multimodal power system. **a** 2-dimensional system. **b** Reactance matching

direct-axis transient reactance and the transformer reactance, namely, $(x'_{d_{ij}} + x_{T_{ij}})$, for the ith area connected to the jth terminal bus. Because Kirchoff's law holds for each transfer path independently, IME can be applied to each pair of machines to calculate this reactances irrespective of the other paths. However, two important points must be taken into consideration before applying IME:

1. Because the internal node of each machine is the point of common coupling between any two neighboring transfer paths, we must make sure that this internal voltage computed independently from the reactance estimates of each path must match with each other. For example, using Kirchoff's law at the internal node of G_1, first for transfer path 1–2 and then for the path 1–3, we get, respectively,

$$E_1 = |\tilde{V}_1 + j(x'_{d_{11}} + x_{T_{11}})\,\tilde{I}_1|, \quad E_1 = |\tilde{V}_2 + j(x'_{d_{13}} + x_{T_{13}})\,\tilde{I}_2|, \quad (8.70)$$

where, $\tilde{V}_i$ and $\tilde{I}_i$ are available from PMU measurements and the reactances are estimated independently using IME. Because the LHS may not necessarily match for both equations, we, therefore, add two fictitious reactances jx_a and jx_b on each side, and tune them till we obtain the same value of E_1. The same approach applies to the internal reactance of all other areas. Physically speaking, this may be thought of as a variable reactance that matches the internal angle of each equivalent machine at the cost of decentralized estimation of the area parameters. These fictitious reactances are illustrated in Fig. 8.13b.

2. Decentralization of IME, however, should not neglect the basic fact that the dynamics of the three machines are coupled to each other, and, therefore, all the bus measurements used for estimating the reactances and inertias must contain the contribution of each and every slow mode of oscillation. This can be easily accounted for from the fundamental principle behind coherency and aggregation [5]. For example, to obtain an analytical expression for the fast and slow oscillation dynamics, one may define the slow or aggregate variable for the kth area to be the so-called center of inertia angle for that area, namely

$$y_k \triangleq \frac{\sum_{i=1}^{n_k} H_i^k \delta_i^k}{\sum_{i=1}^{n_k} H_i^k}, \quad k = 1, 2, \ldots r \tag{8.71}$$

where δ_i^k and H_i^k are, respectively, the ith machine angle and inertia in the kth area, n_k is the total number of machines in the kth area, and r is the total number of areas. Similarly, the fast variable for the kth area can be defined as

$$z_{k,i} \triangleq \delta_i^k - \delta_1^k, \quad i = 1, 2, \ldots n_k, \quad k = 1, 2, \ldots r. \tag{8.72}$$

The time-scale separation between the fast and slow oscillations can then be expressed explicitly in the singular perturbed form

$$\frac{\mathrm{d}y}{\mathrm{d}t_s} = A_{11}y + A_{12}z, \quad \varepsilon\frac{\mathrm{d}z}{\mathrm{d}t_s} = A_{21}y + A_{22}z \tag{8.73}$$

where $\varepsilon \ll 1$ is a small parameter, and the exact expressions for the four state matrices can be found in [5]. Assuming $\varepsilon \approx 0$, the effective swing dynamics for the interarea oscillations can then be written as:

$$\frac{\mathrm{d}y}{\mathrm{d}t_s} = (A_{11} - A_{12}A_{22}^{-1}A_{21})y. \tag{8.74}$$

Because the matrix $(A_{11} - A_{12}A_{22}^{-1}A_{21})$ is not necessarily block-diagonal, it is evident that the interarea oscillation modes are not necessarily decoupled. In other words, the bus voltage oscillations available from the PMUs at each terminal bus must retain the cumulative contribution of all slow modes. This can also be seen from the time-domain representation of the ith bus voltage, namely

$$V_k = \left(\sum_{i=1}^{n-1} R_i e^{-\sigma_i t} (\sin(\omega_i t + \phi_i) + \cos(\omega_i t + \phi_i)) \right)_k \tag{8.75}$$

where n is the total number of machines in the system. However, the RHS of (8.75) contains the contribution of both local and interarea modes, only the two slowest of which will give the contribution of the interarea oscillations in our example of interest.

The net interarea component of V_k can then be written as

$$V_k^s = \Big(R_1 e^{-\sigma_1 t} (\sin(\omega_1 t + \phi_1) + \cos(\omega_1 t + \phi_1)) + R_2 e^{-\sigma_2 t} (\sin(\omega_2 t + \phi_2) + \cos(\omega_2 t + \phi_2)) \Big)_k \tag{8.76}$$

where ω_1 and ω_2 are the two slow frequencies. These two frequencies and their corresponding residues and damping factors in (8.76) can be easily computed by applying modal decomposition to the bus voltage measurement V_i, some common methods including the ERA, Prony analysis, and Matrix Pencil. Once V_k^s is extracted for $k = 1, 2, \ldots, 6$, these voltages can then be used for calculating the internal reactances of each area (for each connection) via IME in exactly the same way as in Sect. 8.2 for each pair of machines. The same idea applies to the bus frequencies for estimating the equivalent machine inertias. One interesting observation, however, is that the machines do not necessarily have unique inertias, but have as many distinct inertias as the degree of that node, each representing the effective *weight* of that area contributing to the oscillation for each respective transfer path that the machine is connected to. A schematic of the spatial variation of voltage at any fixed point of time for this three-machine system is shown in Fig. 8.14.

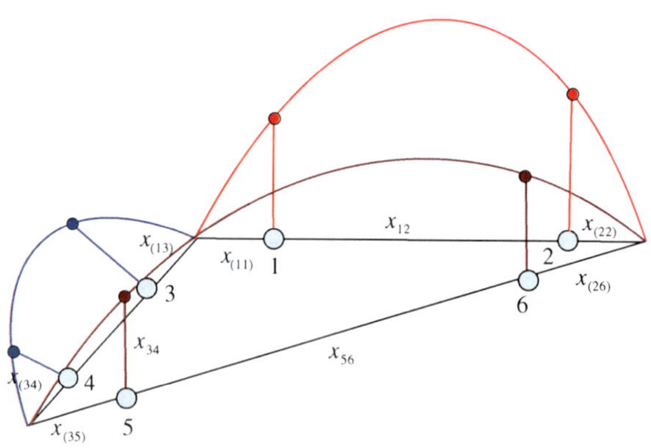

Fig. 8.14 Spatial variation of voltage for 2-dimensional system

8.5 Transient Stability Assessment Using Energy Functions

In this section we show how the reduced-order models developed in Sect. 8.2 can be exploited to formulate performance metrics for transient and damping stability assessment of two-area power systems. We develop the concepts to compute energy functions using phasor data to assess the stability margin of power transfer paths. Consider a two-machine equivalent system as in Fig. 8.2b, and let its post-fault equilibrium angle be δ_{op}. The energy function V_E of the system can be expressed in the form

$$V_E = V_{\mathrm{PE}} + V_{\mathrm{KE}} \tag{8.77}$$

where the potential energy and the kinetic energy are given by

$$V_{\mathrm{PE}} = \frac{E_1 E_2}{x_e'}\Big(\cos(\delta_{op}) - \cos(\delta) + \sin(\delta_{op})(\delta_{op} - \delta)\Big) \tag{8.78}$$

$$V_{\mathrm{KE}} = \frac{1}{2}(2H)\Omega\omega^2 = H\Omega\omega^2. \tag{8.79}$$

It should be noted that by virtue of the IME algorithm, the energy function (8.77) can be computed in terms of the machine angles and voltages extrapolated from the bus measurements. However, the bus voltages contain high-frequency local modes as well as slower interarea modes. These fast and slow components need to be separated before using the voltages to construct the energy function. We call the filtered slow component of the voltages as the quasi-steady-states $\bar{V}_1$ and $\bar{V}_2$. In real time, the post-fault equilibrium angle δ_{op} or θ_{op} is not fixed either, but rather time varying, due to turbine-generator governing and other generation and load changes. Thus we can write

$$\delta = \hat{\delta} + \delta_{\mathrm{qss}} \tag{8.80}$$

where $\hat{\delta}$ and δ_{qss} are respectively the swing component and the quasi-steady-state components of δ. We need to extract the quasi-steady-state value δ_{qss} in order to approximate the post-disturbance equilibrium angle used in the potential energy function (8.78). Based on this, we propose the transient swing energy function

$$\hat{V}_E(t) = \hat{V}_{\mathrm{KE}}(t) + \hat{V}_{\mathrm{PE}}(t) \tag{8.81}$$

to model the energy due to the dominant interarea mode, where

$$\hat{V}_{\mathrm{PE}} = \frac{\bar{E}_1 \bar{E}_2}{x_e'}\big(\cos(\delta_{\mathrm{qss}}) - \cos(\delta) + \sin(\delta_{\mathrm{qss}})(\delta_{\mathrm{qss}} - \delta)\big) \tag{8.82}$$

$$\hat{V}_{\mathrm{KE}}(t) = H\Omega\omega(t)^2 \tag{8.83}$$

where δ_{qss} is obtained in practice by bandpass filtering the δ measurement.

We next illustrate the construction of (8.81–8.83) using a disturbance event in the two-area WECC transfer path discussed in Sect. 8.2. After applying the IME algorithm to the bus measurements and extrapolating to the machine internal nodes, the time variations of the machine angular separation and frequency differences are calculated, and plotted, respectively, in Fig. 8.15a and b. The machine speed difference ω_B is mostly mono-modal, but the angle difference θ shows a distinct quasi-steady-state variation. Bandpass filtering is used to separate the oscillation and the quasi-steady-state components of δ, as shown in Fig. 8.16a and b. For the post-disturbance case, we get $x_e = 0.077$ pu from least-squares fitting, as in [21]. The equivalent machine inertia is estimated to be $H = 119$ pu. Figures 8.17a, b and 8.18 show the energy functions $\hat{V}_{\text{KE}}$, $\hat{V}_{\text{PE}}$, and $\hat{V}_E$, respectively. Note that oscillations are clearly visible in $\hat{V}_{\text{KE}}$ and $\hat{V}_{\text{PE}}$ and yet they literally disappear when $\hat{V}_{\text{KE}}$ and $\hat{V}_{\text{PE}}$ are added together to form $\hat{V}_E$. The oscillation is small-signal stable, although the damping is very low, and $\hat{V}_E$ eventually decays to a level commensurate with random perturbations on the system. If the system were negatively damped, $\hat{V}_E$ would grow. The quasi-steady-state angle δ_{qss} indicates that the sending end and receiving end of the transfer path remain synchronized, that is, transiently stable. A sudden increase in δ_{qss} indicates the loss of a portion of the transmission system or the loss of generation at the load bus, both of which would stress the transfer path. If the disturbance had caused a separation of the transfer path, δ_{qss} would grow as synchronism would be lost.

It is also worth noting that the amplitudes of oscillations in the interarea angular separation as well as frequency, following a disturbance, in two-area power systems, depend significantly on the strength of the interconnection as well as the inertia of the aggregated machines. We briefly illustrate this fact by comparing the maximum swing energy functions for two different transfer paths under two different sets of disturbances. We show that a comparison of energies of coherent machines forming interconnected transfer paths can be a good indication of the relative strengths of

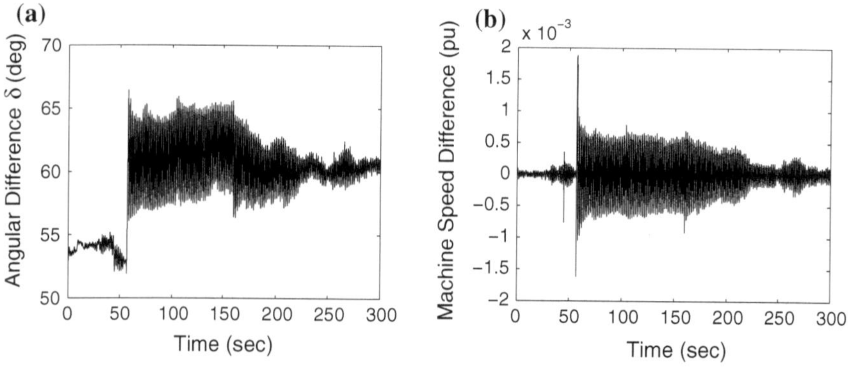

Fig. 8.15 Time variations of machine angle difference and machine speed difference in transfer path 1. **a** Machine angular difference δ (deg). **b** Machine frequency difference (pu)

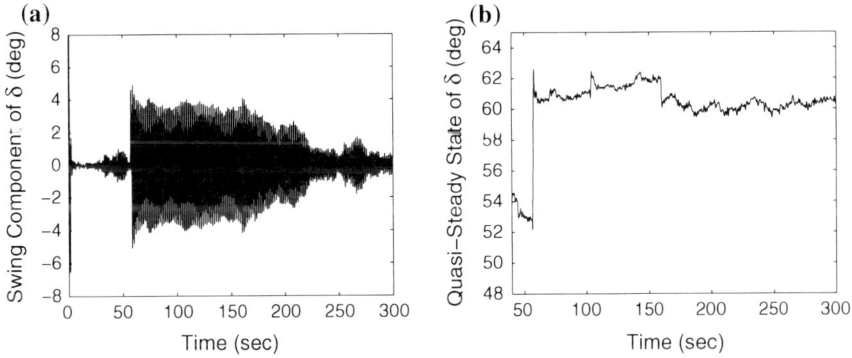

Fig. 8.16 Swing component and quasi-steady-state of machine angle difference in transfer path 1. **a** Swing component $\hat{\delta}$ on transfer path 1. **b** Quasi-steady-state δ_{qss} on transfer path 1

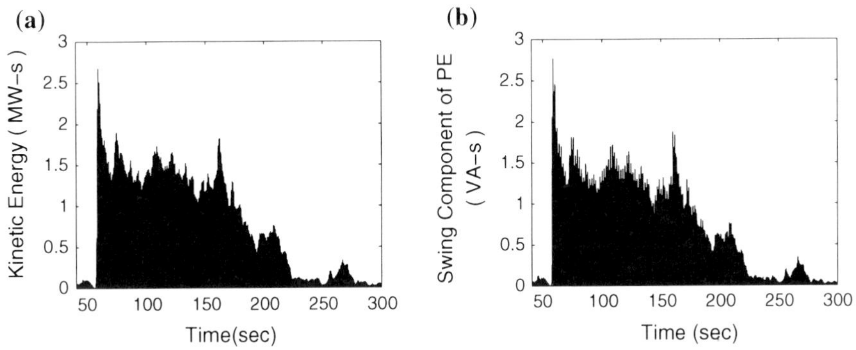

Fig. 8.17 Energy functions for transfer path 1. **a** Kinetic energy function for transfer path 1, **b** potential energy function for transfer path 1

disturbances on the transfer paths. Consider the interconnected power system with three aggregated machines as in Fig. 8.19. Generators 1 and 2 form a coherent group of machines (transfer path a) with power flowing from Generator 1 to 2, the inertia of Generator 1 being smaller than that of Generator 2. Generator 3 forms a coherent group with Generators 1 and 2 together (transfer path b), with the inertia of Generator 3 being significantly higher than that of Generators 1 and 2. The system has two interarea modes, namely a slower mode between G_3 and (G_1, G_2), and a faster mode between G_1 and G_2 with much smaller amplitude. The line reactance, or equivalently connection strength, for transfer path a is weaker compared to that for path b. We consider two sets of disturbance events, namely Events 1 and 2, which caused perturbations in transfer paths a and b, respectively. Event 1 was actually caused by a control equipment failure with possible line tripping and, hence, did not produce any significant oscillations. Event 2, on the other hand, was an earthquake event leading to significant loss of load, due to which the oscillations were more

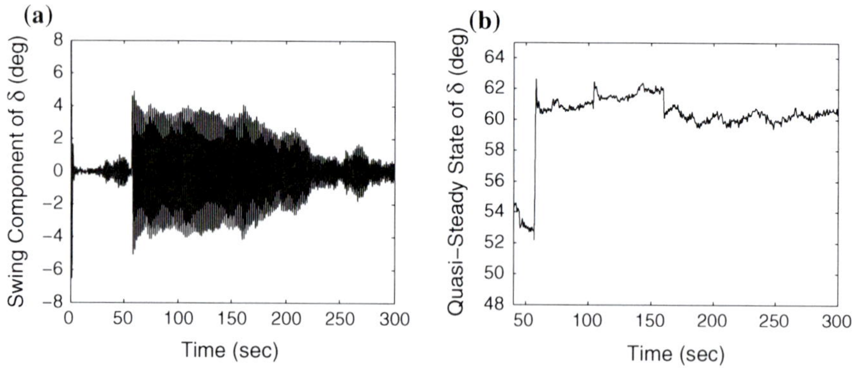

Fig. 8.18 Swing energy function for transfer path 1

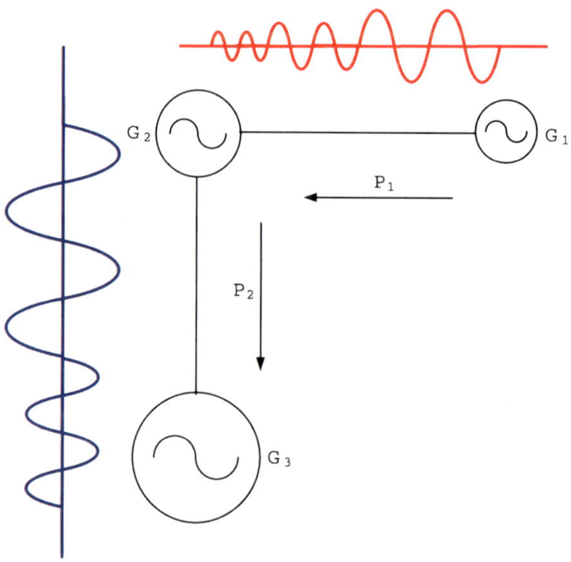

Fig. 8.19 Oscillations in two coherent areas for two events

pronounced. We next construct the swing energy functions for each transfer path for each event. Table 8.1 lists the maximum energy for each of these four cases. The numbers clearly indicate that the disturbance in transfer path *a* hardly caused any oscillations in transfer path *b* due to the high equivalent inertia of Generators 2 and 3, and the stronger connection strength between them. On the other hand, the disturbance in transfer path *b* resulted in a high energy value in transfer path *a* because of the lower machine inertia and connection strength of the latter.

Table 8.1 Swing energies of two paths for two disturbance events

Disturbance event	Maximum swing energy in path a	Maximum swing energy in path b
1	5 MW-s	0.33 MW-s
2	3.5 MW-s	17 MW-s

8.6 Equivalencing Using Noisy PMU Data

We end our discussion with a brief note on the situation where the aforesaid parameter estimation methods have to be carried out using PMU data corrupted with measurement noise. In that case, unique estimates of the model parameters are no longer available, and the problem has to be posed in terms of bounds on the estimation error. Such bounds, more commonly referred to as Cramer-Rao bounds (CRB), are widely used in the statistical signal processing literature [22]. By definition, CRB is a lower bound for the second-order moment of an unbiased parametric estimator. In this section we show an interesting fact: the CRB for estimating the *interarea* model parameters of a two-machine equivalent of the two-area radial power system of Sect. 8.2 is a function of the spatial variable a. In other words, the error in estimation depends on the location of the PMU on the transmission line. The problem, therefore, is to find the optimal location such that the estimation error is minimized.

Returning to the two-area power system model of Sect. 8.2, we linearize the model (8.6) about an initial equilibrium $(\delta_0, 0)$ where $0 < \delta_0 < 90°$, and denote the perturbed state variables as $m = \mathrm{col}(\Delta\delta, \Delta\omega)$ to obtain

$$\dot{m} = \underbrace{\begin{bmatrix} 0 & 1 \\ -\frac{E_1 E_2}{2H\bar{x}}\cos(\delta_0) & 0 \end{bmatrix}}_{A} m + \underbrace{\begin{bmatrix} 0 \\ 1 \end{bmatrix}}_{B} u \tag{8.84}$$

where u is a small disturbance input to the system. For any point P at reactance x away from Generator 2, the output matrix from (8.9) can be written as

$$C = \begin{bmatrix} \dfrac{-a(1-a)E_1 E_2 \sin(\delta_0)}{V_0} & 0 \end{bmatrix} \tag{8.85}$$

where $V_0 = \sqrt{E_2^2(1-a)^2 + E_1^2 a^2 + 2E_1 E_2 a(1-a)\cos(\delta_0)}$. After a few calculations, it can readily be shown that the discrete-time impulse response of the voltage at P is

$$V(k) = \psi(a)\,\xi(k, x_1, x_2, H_1, H_2) \tag{8.86}$$

where ξ for different values of k is listed in Table 8.2. We next stack the impulse response as

Table 8.2 Impulse response of undamped two-machine power system

k	$y(k)$
1	$\psi(a)\,K$
2	$\psi(a)\,K(1+2\cos(\sqrt{\gamma}T))$
3	$\psi(a)\,K(\cos(2\sqrt{\gamma}T))+2\cos(\sqrt{\gamma}T)(1+\cos(\sqrt{\gamma}T))$
$\vdots$	$\vdots$
n	$\psi(a)K\big(\cos((n-1)\sqrt{\gamma}T)+\left(\frac{1+\cos(\sqrt{\gamma}T)}{\sin(\sqrt{\gamma}T)}\right)$ $\sin((n-1)\sqrt{\gamma}T)\big)$

$$\mathscr{Y} = \left[\, y(1)\; y(2)\; \ldots\; y(k) \,\right]. \tag{8.87}$$

Assuming that unlimited time-series data are available, k can be any arbitrary positive integer. We partition the four unknown parameters $x_1,\ x_2,\ H_1$, and H_2 into sets $\mathbf{a} = \{x_1,\ x_2\},\ \mathbf{b} = \{H_1,\ H_2\}$ and define

$$\mathbf{H}(\mathbf{a},\,\mathbf{b}) = \partial\mathscr{Y}/\partial\mathbf{a}, \quad \mathbf{K}(\mathbf{a},\,\mathbf{b}) = \partial\mathscr{Y}/\partial\mathbf{b}. \tag{8.88}$$

Assuming that the actual measured PMU signal is

$$\tilde{y}(k) = y(k) + \tilde{n} \tag{8.89}$$

where $\tilde{n}$ is zero-mean Gaussian noise with variance σ^2, the Fisher Information Matrix for computing the error bounds is next formulated as

$$J(\mathbf{a},\mathbf{b}) = \frac{1}{\sigma^2}\begin{bmatrix} \mathbf{H}\mathbf{H}^{\mathrm{T}} & \mathbf{H}\mathbf{K}^{\mathrm{T}} \\ \mathbf{K}\mathbf{H}^{\mathrm{T}} & \mathbf{K}\mathbf{K}^{\mathrm{T}} \end{bmatrix}. \tag{8.90}$$

Because CRB is the inverse of $J(\mathbf{a},\mathbf{b})$, the tightest bound will be given by that value of a, say denoted as a^*, which maximizes the determinant of J. Moreover, as the PMU can only be placed on the transmission line, a^* must satisfy

$$a^* \in [a_1,\, a_2],\quad a_1 = \frac{x_2}{\bar{x}},\ a_2 = \frac{x_2 + x_e}{\bar{x}}. \tag{8.91}$$

Applying this procedure for the disturbance event of WECC transfer path 1 of Sect. 8.3, imposing a 10 dBW noise power on the actual measured data, the determinant of the FIM J plotted against $a \in [a_1,\, a_2]$ is shown in Fig. 8.20. From this figure, it follows that the best location P to place a PMU (or, equivalently consider a measurement from) is approximately at 40 % of the total length of the transmission line away from Bus 2, i.e., a reactance 0.0308 pu away from Bus 2, considering the output for estimation is the voltage magnitude at P. If any other variable such as the phase angle measured at P is used for estimating the parameters, then a different value of a^* will be obtained.

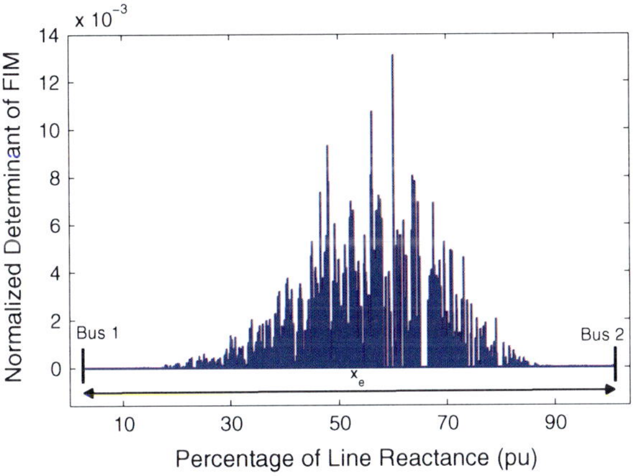

Fig. 8.20 Spatial variation of det(J)

8.7 Conclusions

In this chapter, we presented a collection of new results on model reduction of several classes of large-scale power systems using synchronized phasor data available only from a few selected points in the transmission network. The fundamental approach behind this model reduction is to formulate the reduction problem as an equivalent parameter estimation problem, which can be solved using the spatial variation of different phasor quantities from one end of the transfer path to another. The developed methods can be used to construct approximate interarea models of two representative transfer paths in the US west coast power system, each of which have been illustrated with real disturbance event data. Besides the natural benefits of model reduction, the advantage of such *dynamic equivalent* models lies in both wide-area monitoring and wide-area control. For example, these models can be directly used to construct transient energy functions operating across transfer paths, which in turn can be used as a performance metric to track the health of the interconnection following any large disturbance—in terms of damping, mode shape, rise time, settling time, etc., as shown in [21]. They will also be highly useful for efficient wide-area control designs at a global or *interarea* level. For instance, given the scale, size, and complexity of any realistic power system (e.g., WECC with roughly 2,000 generators, 11,000 transmission lines and 6,500 loads), designing PMU-based distributed controllers to shape the *interarea* responses starting from a full-order model would be highly daunting. We believe that an alternative approach of reducing such large systems into simpler chunks, and then redistributing their control efforts would give the problem a much more well-defined and less chaotic formulation. This work, however, should also be viewed as a point of departure for several future investigations. The variation of the IME model reactance as a function of power transfer levels and the correlation of fault clearing times between the IME model and the detailed model also need to be better understood.

References

1. A.G. Phadke, J.S. Thorp, M.G. Adamiak, New measurement techniques for tracking voltage phasors, local system frequency, and rate of change of frequency. IEEE Trans. Power Apparatus Syst. **102**, 1025–1038 (1983)
2. North American Synchrophasor Initiative (NASPI), www.naspi.org.
3. R.L. Cresap, J.F. Hauer, Emergence of a new swing mode in the western power system. IEEE Trans. Power Apparatus Syst. **PAS–100**(4), 2037–2045 (1981)
4. J. Ballance, B. Bhargava, G.D. Rodriguez, Use of synchronized phasor measurement system for enhancing AC-DC power system transmission reliability and capability, EIPP Meeting, Sep 2004
5. J.H. Chow, G. Peponides, P.V. Kokotović, B. Avramović, J.R. Winkelman, *Time-Scale Modeling of Dynamic Networks with Applications to Power Systems* (Springer, New York, 1982)
6. J.F. Hauer, C.J. Demeure, L.L. Scharf, Initial pesults in prony analysis of power system response signals. IEEE Trans. Power Syst. **5**(1), 80–89 (1990)
7. D.J. Trudnowski, J.W. Pierre, N. Zhou, J.F. Hauer, M. Parashar, Performance of three mode-meter block-processing algorithms for automated dynamic stability assessment. IEEE Trans. Power Syst. **23**(2), 680–690 (2008)
8. G. Ledwich, D. Geddy, P.O. Shea, Phasor Measurement Units for System Diagnosis and Load Identification in Australia, in *Proceedings of IEEE PES General Meeting*, Pittsburgh, PA, July 2008
9. D. Wilson, Oscillatory Mode Shape and Combined EMS/WAMS Data to Characterize and Locate Stability Issues, in *NASPI Meeting*, Charlotte, NC, Oct 2008
10. Q. Yang, T. Bi, J. Wu, WAMS Implementation in China and the Challenges for Bulk Power System Protection, in *Proceedings of IEEE PES General Meeting*, Tampa, FL, July 2007
11. J. Rasmussen, A.H. Nielsen, Phasor Measurement of Wind Power Plant Operation in Eastern Denmark, in *European Offshore Wind Conference & Exibition*, Berlin, Germany, 2007
12. A.J. Germond, R. Podmore, Dynamic aggregation of generating unit models. IEEE Trans. Power Apparatus Syst. **PAS–97**(4), 1060–1069 (1978). (July/Aug)
13. R.W. de Mello, R. Podmore, K.N. Stanton, Coherency-Based Dynamic Equivalents: Applications in Transient Stability Studies, in *Proceedings of PICA Conference*, New Orleans, LA, June 1975, pp. 23–31
14. J.M. Undrill, A.E. Turner, Construction of power system electromechanical equivalents by modal analysis. IEEE Trans. Power Apparatus Syst. **PAS–90**, 2049–2059 (1971)
15. J. Zaborszky, K.W. Whang, G. Huang, L.J. Chiang, S.Y. Lin, A clustered dynamic model for a class of linear autonomous system using simple enumerative sorting. IEEE Trans. Circuits Syst. **CAS–29**(11), 747–758 (1982). (Special Issue)
16. R. Nath, S.S. Lamba, K.S.P. Rao, Coherency based system decomposition into study and external areas using weak coupling. IEEE Trans. Power Apparatus Syst. **PAS–104**, 1443–1449 (1985)
17. W.W. Price, G.E. Boukarim, J.H. Chow, R. Galarza, A.W. Hargrave, B.J. Hurysz, R. Tapia, Improved Dynamic Equivalencing Software, Final Report, EPRI Project RP2447-02, 1995
18. A.R. Bergen, V. Vittal, *Power System Analysis*, 2nd edn. (Prentice Hall, NJ, 1999)
19. J.J. Sanchez-Gasca, J.H. Chow, Performance comparison of three identification methods for the analysis of electromechanical oscillations. IEEE Trans. Power Syst. **14**(3), 995–1002 (1999)
20. A. Chakrabortty, *Estimation, Analysis and Control Methods for Large-scale Electric Power Systems using Synchronized Phasor Measurements*. Ph.D Dissertation, Rensselaer Polytechnic Institute, Troy, NY, 2008
21. J.H. Chow, A. Chakrabortty, M. Arcak, B. Bhargava, A. Salazar, Synchronized phasor data based energy function analysis of dominant power transfer paths in large power systems. IEEE Trans. Power Syst. **22**(2), 727–734 (2007)
22. T. McWhorter, L.L. Scharf, Cramer-Rao bounds for deterministic modal analysis. IEEE Trans. Signal Process. **41**(5), 1847–1865 (1993)

23. A. Chakrabortty, J.H. Chow, A. Salazar, A Measurement-based Framework for Dynamic Equivalencing of Large Power Systems using WAMS, in *Proceedings of IEEE PES Conference on Innovative Smart Grid Technologies*, Jan 2010
24. N.L. Biggs, E.K. Lloyd, R.J. Wilson, *Graph Theory* (Oxford University Press, Oxford, 1976)

Chapter 9
Selective Modal Analysis

Luis Rouco, F. L. Pagola, George C. Verghese and Ignacio J. Pérez-Arriaga

Abstract Selective Modal Analysis (SMA) is a comprehensive methodology for the modeling, analysis, and control of selected parts of the dynamics of systems described by large linear time-invariant (LTI) models. The main components of SMA are sensitivity tools and algorithms for reduced-order eigenanalysis. Sensitivity tools developed within the SMA framework include participation factors, which measure the participation of the state variables in the eigenmodes and vice versa. Participation factors play a central role in SMA developments. This chapter shows the role of participation factors in the identification of dynamic patterns, design of damping controllers, and reduced-order eigenanalysis.

9.1 Introduction

Selective Modal Analysis (SMA) is a comprehensive methodology for the modeling, analysis, and control of selected parts of the dynamics of systems described by large linear time-invariant (LTI) models. Although SMA is a general approach applicable to an LTI model for any system, its motivation and main applications have come from several problems of power system dynamics (small-signal stability, subsynchronous resonance, dynamic equivalents, etc.).

L. Rouco (✉) · F. L. Pagola · I. J. Pérez-Arriaga
Universidad Pontificia Comillas, Madrid, Spain
e-mail: rouco@upcomillas.es

F. L. Pagola
e-mail: pagola@upcomillas.es

G.C. Verghese · I. J. Pérez-Arriaga
Massachusetts Institute of Technology, Cambridge, USA
e-mail: verghese@mit.edu

I. J. Pérez-Arriaga
e-mail: ipa@mit.edu

J. H. Chow (ed.), *Power System Coherency and Model Reduction*, Power Electronics and Power Systems 94, DOI: 10.1007/978-1-4614-1803-0_9,

The main components of SMA are sensitivity tools and algorithms for reduced-order eigenanalysis [1, 2]. Sensitivity tools developed within the SMA framework include *participation factors*, which measure the participation of the state variables in the eigenmodes, and vice versa. *Algorithms of reduced-order eigenanalysis* determine a set of selected eigenvalues and associated eigenvectors through iterative processes involving, at each stage, the eigenanalysis of a matrix that can be much smaller than the system state matrix. Participation factor analysis guides the design and analysis of these algorithms. Once the algorithms of reduced-order eigenanalysis have converged to the selected modes, SMA also provides a reduced-order model in terms of the relevant variables for those modes [3].

Participation factors have become widely used tools in both research and industry to identify the relationships among state variables and eigenvalues in large power system models [4, 5]. They have also attracted the attention of scholars who have tried to provide complementary [6] and alternative [7] explanations. Hence, a deeper understanding of participation factors is still worth pursuing. Participation factors play a central role in SMA developments. This chapter shows the role of participation factors in the identification of dynamic patterns, design of damping controllers, and reduced-order eigenanalysis.

Dynamic patterns are close associations between subsets of modes and of state variables. Dynamic patterns are identified using subsystem participation and shown in several examples of power system dynamic phenomena. It is well-known, for example, that in small-signal stability of power systems, poorly damped modes in the frequency range between 0.1 and 2 Hz are associated with the rotor dynamics of coupled synchronous generators, and because of this are called *electromechanical modes*. Not only do participation factors confirm such an association [2–5] but they have also helped to understand other complex models, such as those arising in the study of subsynchronous resonance [8] and of wind generators [9, 10].

Participation factors can also be seen as eigenvalue sensitivities and as residues of certain transfer functions [11]. *Eigenvalue sensitivities* are powerful tools for designing power system damping controllers (i.e., power system stabilizers) for synchronous generators [12, 13], FACTS devices [14–16], and wind generators [17]. The use of eigenvalue sensitivities with respect to controller parameters for controller design is detailed here.

The convergence of the SMA algorithms for reduced-order eigenanalysis is determined by the *participation ratio* of the relevant variables in the modes of interest. In addition to the basic algorithms, more elaborate algorithms have been developed, aimed at increasing the participation ratio of the relevant variables through variable transformations [18, 19].

The chapter is organized as follows: Sect. 9.2 defines participation factors; Sect. 9.3 shows how participation factors help to identify dynamic patterns in a wide variety of power system models; Sect. 9.4 details the use of participation factors as sensitivity tools to design power system damping controllers; and Sect. 9.5 explains the role of participation factors in the convergence of reduced-order eigenvalue analysis algorithms.

9.2 LTI Dynamic Systems and Participation Factors

This section defines participation factors and compares them with eigenvectors in mass–spring mechanical systems. Mass–spring systems exhibit the same structure and features of the classical model of power systems, which highlight the electromechanical (rotor) oscillations that are of interest in the power system small-signal stability problem. This mechanical mass–spring analogy for a power system helps to gain insight into the meaning and value of participation factors. In this section, we first introduce LTI models in state-space form for small-signal (or linearized) behavior of dynamic systems in the neighborhood of an equilibrium point. Then, eigenvalues and eigenvectors of the system state matrix are used to determine the solution of LTI systems. This then allows us to define participation factors, which are then specialized to mass–spring systems.

9.2.1 LTI Dynamic Systems

Let us start by considering a nonlinear time-invariant dynamic system with constant external inputs, written in the *explicit state-space form*, with the time-derivatives of the state variables at time t expressed only in terms of the state variables

$$\dot{\mathbf{x}}(t) = \mathbf{f}\big(\mathbf{x}(t)\big), \quad \mathbf{x}(t) \in \Re^N \tag{9.1}$$

Here $\mathbf{x}(t)$ is the vector of state variables at time t and $\mathbf{f}\,(\cdot)$ is a vector of nonlinear functions determined by the system dynamics, parameters, and constant external inputs. The time argument t will be suppressed where possible, for notational simplicity. (Nonlinear time-invariant dynamic systems can also be written in implicit state-space form, in terms not only of the state variables but also auxiliary algebraic ones. This can be a more natural way of writing models, preserving their structural properties and providing several computational advantages [20].)

Linearizing the set of nonlinear differential equations (9.1) by carrying out a first-order Taylor series expansion of $\mathbf{f}\,(\mathbf{x})$ around an operating point $\mathbf{x}_0$, where $\mathbf{f}\,(\mathbf{x}_0) = \mathbf{0}$, results in a set of LTI differential equations in state-space form

$$\Delta\dot{\mathbf{x}}(t) = \left.\frac{\partial \mathbf{f}\,(\mathbf{x})}{\partial \mathbf{x}}\right|_{\mathbf{x}_0} \Delta\mathbf{x} = \mathbf{A}\Delta\mathbf{x} \tag{9.2}$$

where

$$\Delta\mathbf{x} = \mathbf{x} - \mathbf{x}_0, \quad \mathbf{A} \in \Re^{N\times N}$$

Here $\mathbf{A}$ is the Jacobian matrix of $\mathbf{f}(\mathbf{x})$ and contains the partial derivatives of the components of $\mathbf{f}(\mathbf{x})$ with respect to the components of $\mathbf{x}$, all evaluated at $\mathbf{x} = \mathbf{x_0}$. We will refer to $\mathbf{A}$ as the state matrix of the LTI system. From now on, because our focus will be entirely on the linearized system (9.2), and for the sake of notational simplicity, $\Delta\mathbf{x}$ will be just written as $\mathbf{x}$.

The solution of the set of linear differential equations (9.2) is given in terms of the exponential of the state matrix and the initial conditions:

$$\mathbf{x}(t) = e^{\mathbf{A}t}\mathbf{x}(0) \tag{9.3}$$

where

$$e^{\mathbf{A}t} = \mathbf{I} + \mathbf{A}t + \frac{1}{2}\mathbf{A}t^2 + \cdots \tag{9.4}$$

and $\mathbf{I}$ is the identity matrix.

9.2.2 Eigenvalues and Eigenvectors

A conceptually meaningful approach to computing the exponential of the state matrix is based on the eigenvalues and eigenvectors of the state matrix. An eigenvalue λ_i of a matrix $\mathbf{A}$ and the associated right $\mathbf{v}_i$ and left $\mathbf{w}_i$ eigenvectors are defined by the conditions $\mathbf{v}_i \neq \mathbf{0}$, $\mathbf{w}_i^T \neq \mathbf{0}$ and

$$\mathbf{A}\mathbf{v}_i = \lambda_i \mathbf{v}_i \tag{9.5}$$

$$\mathbf{w}_i^T \mathbf{A} = \mathbf{w}_i^T \lambda_i \tag{9.6}$$

It is clear that the above conditions are also satisfied when any eigenvector is multiplied by an arbitrary nonzero number, so the eigenvectors are only defined to within a nonzero scale factor. It may be convenient to normalize the right eigenvector in some way that uniquely defines it, for instance, by setting its greatest (in absolute value) entry equal to 1. In the sequel it will be assumed, because many expressions will then be simpler, that the left eigenvector is normalized according to

$$\mathbf{w}_i^T \mathbf{v}_i = 1 \tag{9.7}$$

We will assume throughout that the state matrix $\mathbf{A}$ has N distinct eigenvalues (which is the typical case). The N right (and also the N left) eigenvectors are then guaranteed to be linearly independent. Reliable algorithms to compute the eigenvalues and the associated eigenvectors of a general matrix are available in a number of implementations [21, 22]. With the definitions

$$\mathbf{\Lambda} = \begin{bmatrix} \lambda_1 & \cdots & 0 \\ \vdots & \ddots & \vdots \\ 0 & \cdots & \lambda_N \end{bmatrix}, \quad \mathbf{V} = \begin{bmatrix} \mathbf{v}_1 & \cdots & \mathbf{v}_N \end{bmatrix}, \quad \mathbf{W} = \begin{bmatrix} \mathbf{w}_1^T \\ \vdots \\ \mathbf{w}_N^T \end{bmatrix} \tag{9.8}$$

we can now use (9.5–9.7) to obtain

$$\mathbf{AV} = \mathbf{V}\Lambda \tag{9.9}$$

$$\mathbf{WA} = \Lambda\mathbf{W} \tag{9.10}$$

$$\mathbf{WV} = \mathbf{I} = \mathbf{VW} \tag{9.11}$$

Note that $\mathbf{V}$ (respectively, $\mathbf{W}$) is invertible because of the independence of its columns (rows).

If the identity matrix $\mathbf{I}$ and the state matrix $\mathbf{A}$ are expressed, respectively, as $\mathbf{VW}$ and $\mathbf{V}\Lambda\mathbf{W}$, the exponential of the state matrix (9.4) becomes

$$e^{\mathbf{A}t} = \mathbf{V}\left(\mathbf{I} + \Lambda t + \frac{1}{2}\Lambda t^2 + \cdots\right)\mathbf{W} = \mathbf{V}e^{\Lambda t}\mathbf{W} \tag{9.12}$$

where $e^{\Lambda t}$ is now a diagonal matrix with the ith diagonal entry $e^{\lambda_i t}$. Substituting (9.12) in (9.3), the solution of the set of differential equations (9.2) becomes

$$\mathbf{x}(t) = \mathbf{V}e^{\Lambda t}\mathbf{W}\mathbf{x}(0) = \sum_{i=1}^{N} \mathbf{v}_i e^{\lambda_i t}\left[\mathbf{w}_i^T \mathbf{x}(0)\right] \tag{9.13}$$

The ith term in this sum is referred to as the ith *mode* of the system. Inspection of Eq. (9.13) allows one to draw the following conclusions:

- The system response is expressed as a linear combination of the N modes of the system.
- The eigenvalues of the state matrix $\mathbf{A}$ determine the system stability. An eigenvalue with a negative (positive) real part indicates an exponentially decreasing (increasing) behavior.
- The components of the right eigenvector $\mathbf{v}_i$ indicate the relative activity of each state variable in the ith mode.
- The components of the left eigenvector $\mathbf{w}_i$ weight the effect of the initial conditions in exciting the ith mode.

The dimensional units of the different components of the right eigenvectors are those of the corresponding state variables, while the dimensional units of the components of the left eigenvectors are the inverses of the dimensional units of the corresponding state variables.

9.2.3 Participation Factors

The participation factor of the jth state variable in the ith eigenmode is defined as the product of the jth components of the left and right eigenvectors associated with the ith eigenvalue

$$p_{ji} = w_{ij} v_{ji}$$

The participation factor p_{ji} can be seen as the weight of the ith mode in the jth state variable when the initial condition is the jth unit vector

$$x_j = \sum_{i=1}^{N} p_{ji} e^{\lambda_i t}$$

Participation factors are dimensionless, because the dimensions of corresponding components of the right and left eigenvectors are reciprocal to each other.

Given the normalization (9.7), the sum of the participation factors of all state variables in an eigenvalue equals 1. It also follows from the second equality in (9.11) that the sum of the participation factors of all eigenvalues in a state variable is 1. In equation form, we have

$$\sum_{j=1}^{N} p_{ji} = \sum_{i=1}^{N} p_{ji} = 1 \tag{9.14}$$

9.2.4 Participation Factors in Mass–Spring Models

Mass–spring mechanical systems and multi-machine power systems represented according to the classical model exhibit the same structure. We make use of a simple mass–spring system to illustrate participation factors and to point out how information provided by eigenvectors differs from that provided by participation factors.

An LTI mass–spring mechanical system without external forces in which each mass only moves along a line can be described by a set of linear differential equations of the form

$$\begin{bmatrix} \dot{\chi} \\ \dot{\upsilon} \end{bmatrix} = \begin{bmatrix} \mathbf{0} & \mathbf{I} \\ -\mathbf{M}^{-1}\mathbf{K} & \mathbf{0} \end{bmatrix} \begin{bmatrix} \chi \\ \upsilon \end{bmatrix}$$

where χ is the vector of mass positions, υ is the vector of mass velocities, $\mathbf{M}$ is the mass matrix, and $\mathbf{K}$ is the stiffness matrix.

If the system does not contain any position reference, a system of N_m masses exhibits $N_m - 1$ pairs of complex eigenvalues and two real eigenvalues. One of the real eigenvalues is equal to zero, as a consequence of the fact that each row of $\mathbf{K}$ sums to 0. The other real eigenvalue is also at 0 in the undamped case, but would otherwise be at some negative real value. If there is a position reference, then there are $N_m - 1$ pairs of complex eigenvalues and one real eigenvalue.

Participation factors in mass–spring systems exhibit several interesting properties. The participations of mass positions and mass velocities are identical

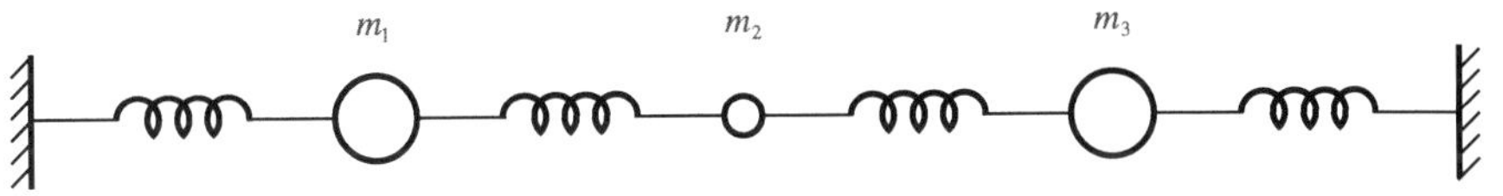

Fig. 9.1 Three-mass mass–spring system

Table 9.1 Eigenvalues of the three-masses mass–spring system

No.	λ
1,2	$0.0 \pm j0.4588$
3,4	$0.0 \pm j0.1414$
5,6	$0.0 \pm j0.0975$

Table 9.2 Right eigenvectors of the three-mass mass–spring system

Variable	λ_1	λ_3	λ_5
χ_1	0.0525	1.0000	0.9525
χ_2	−1.0000	0.0000	1.0000
χ_3	0.0525	−1.0000	0.9525
υ_1	$j0.0241$	$j0.1414$	$j0.0928$
υ_2	$-j0.4588$	$j0.0000$	$j0.0975$
υ_3	$j0.0241$	$-j0.1414$	$j0.0928$

$$p_{\chi_j,i} = p_{\upsilon_j,i}$$

and therefore

$$\sum_{j=1}^{N_m} p_{\chi_j,i} = \sum_{j=1}^{N_m} p_{\upsilon_j,i} = \frac{1}{2}$$

Furthermore, the participation factors in the ith mode can be shown to equal the peak kinetic energies of the corresponding masses when only this mode is excited.

Let us consider the mass–spring system of three masses shown in Fig. 9.1, where the ends of the outermost spring are attached to fixed supports. The stiffness coefficients of all the springs are assumed to be identical and equal to 1, while $m_1 = m_3 = 10m_2$ *and* $m_2 = 10$. The state matrix contains three pairs of purely imaginary eigenvalues, as shown in Table 9.1, because the system includes a position reference. The *shape* of each mode is indicated by the components of the right eigenvector corresponding to the position variables.

The right eigenvectors and participation factors provide much different information. Tables 9.2 and 9.3 list, respectively, the mode shapes and the participation factors of the eigenvalues with positive imaginary parts.

The mode shape components in the right eigenvector reflect the (amplitude of) motion of the associated masses in the given mode. The participation factors, on the other hand, indicate how important the various masses are to determining

Table 9.3 Participation factors of the three-mass mass–spring system

Variable	λ_1	λ_3	λ_5
χ_1	0.0131	0.2500	0.2369
χ_2	0.4739	0.0000	0.0261
χ_3	0.0131	0.2500	0.2369
υ_1	0.0131	0.2500	0.2369
υ_2	0.4739	0.0000	0.0261
υ_3	0.0131	0.2500	0.2369

the frequency and shape of the mode. Mass-weighted right eigenvectors reflecting momentum rather than energy have been proposed in power system applications to overcome the limitations of the right eigenvectors [23, 24]. Participation factors, however, emerge as more fundamental.

The mode associated with λ_5, for instance, corresponds to a *system-wide mode* in which all three masses have approximately equal motion, as indicated by the mode shape in the last column of Table 9.2. However, the frequency of the mode is mostly determined by the two large end masses (m_1 and m_3). This is brought out by the participation factors in Table 9.3, where the small mass (m_2) has a (kinetic energy and) participation factor one-tenth of those of the larger masses.

At the other extreme, the *local mode* associated with λ_1 has a frequency and shape that is largely determined by the smallest mass m_2. This is made very evident by the first column of Table 9.3, where the (kinetic energies and) participation factors for the two larger masses (m_1 and m_3) are substantially less than that of the small mass (m_2).

Finally, the mode associated with λ_3 is analogous to an *inter-area mode* in a power system and involves one end mass (m_1) swinging against the other (m_3). For this mode, the center mass plays essentially no role, as shown by its participation factor of 0, indicating no contribution to determining the mode, and its right eigenvector entries of 0, indicating no motion in this mode.

9.3 Participation Factors and Dynamic Patterns

Power system small-signal stability models are characterized by the existence of *dynamic patterns*, or close relationships between subsets of state variables and subsets of eigenvalues. For example, the poorly damped oscillations of power and frequency within the 0.1–2 Hz frequency range are well-known to be due to the electromechanical oscillations of the generator rotors.

Participation factors and related tools allow a much fuller understanding of dynamic patterns. This section illustrates how the notion of *subsystem participation* can help to identify dynamic patterns in a great variety of power system models. Dynamic patterns found in both single-machine infinite bus and multi-machine systems are first shown and discussed. Revealing dynamic patterns have also been found in the subsynchronous resonance phenomenon that occurs when a turbogenerator is connected through a series-compensated line, and in doubly fed induction generators (DFIGs) used in wind generation.

The emphasis of the section is on what information can be drawn from a wide variety of power system models using subsystem participation. Details of the procedures to build such models can be found elsewhere. Single-machine infinite bus, multi-machine, and subsynchronus resonance models can be found, for instance, in [4]. A detailed model of the DFIG model is provided in [9].

9.3.1 Subsystem Participation

The participation of a subsystem S in the ith mode, p_{Si}, is defined as the sum of the participation factors in the ith mode of all the state variables internal to subsystem S

$$p_{Si} = \sum_{j \in S} p_{ji} = \mathbf{w}_{Si}^T \mathbf{v}_{Si}$$

where $\mathbf{v}_{Si}$ and $\mathbf{w}_{Si}$ are vectors that contain only the components of the right and left eigenvectors, respectively, corresponding to the state variables that describe the subsystem S. The subsystem participation is invariant with respect to nonsingular transformations that only affect the variables involved in the subsystem.

9.3.2 Single-Machine Infinite Bus System

The simplest power system example is a synchronous generator connected to an infinite bus through a step-up transformer and a transmission line, as shown in Fig. 9.2. The standard model of a synchronous generator contains detailed representations of the rotor dynamics, the synchronous machine, the speed governor/turbine system, and the voltage regulator/excitation system.

Assuming a round-rotor synchronous machine [4], a steam turbine-governor [25] and a static excitation system [26], the explicit state-space form of the linearized model around an operating point of a synchronous generator connected to an infinite bus is of the form

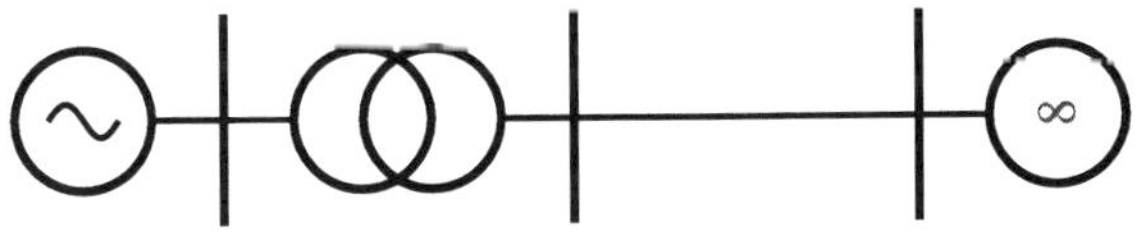

Fig. 9.2 Single-line diagram of a synchronous generator connected to an infinite bus

Table 9.4 Complex eigenvalues of the detailed model of a generator connected to an infinite bus, with the corresponding damping factor ζ and undamped natural frequency f_n

No.	λ	ζ (%)	f_n (Hz)
1,2	$-0.0858 \pm j9.6453$	0.89	1.54
3,4	$-28.4573 \pm j1.5619$	99.85	4.54

$$\dot{\mathbf{x}} = \mathbf{A}\mathbf{x} + \mathbf{B}\mathbf{u} \tag{9.15}$$

where:

$$\mathbf{x} = \begin{bmatrix} \delta \; \omega \; \psi_{fd} \; \psi_{kd} \; \psi_{kq1} \; \psi_{kq2} \; x_{\text{gov1}} \; x_{\text{gov2}} \; x_{\text{gov3}} \; x_{\text{gov4}} \; x_{\text{gov5}} \; x_{\text{exc}} \end{bmatrix}^T$$
$$\mathbf{u} = \begin{bmatrix} \omega_{\text{ref}} \; V_{\text{ref}} \end{bmatrix}^T$$

and

- δ is the rotor angle deviation with respect to synchronous rotating frame, in rad,
- ω is the rotor speed deviation from synchronous speed in pu (per unit, i.e., normalized to the nominal value),
- ψ_{fd} is the deviation from the equilibrium value of the field flux in pu,
- ψ_{kd} is the deviation from the equilibrium value of the flux in the d-axis damper winding in pu,
- ψ_{kqi} are the deviations from the equilibrium value of the fluxes in the q-axis damper windings in pu,
- $x_{\text{gov}i}$ are the deviations from the equilibrium values of the state variables that describe the speed governor/turbine system in pu,
- x_{exc} is the deviation from the equilibrium value of the state variable that describes the voltage regulator/excitation system in pu,
- ω_{ref} is the deviation from the equilibrium value of the reference of the speed governor/turbine system in pu, and
- V_{ref} is the deviation from the equilibrium value of the reference of the voltage regulator/excitation system in pu.

The model is described by 12 state variables. This linear model exhibits, for values of the parameters provided in the Appendix, two pairs of complex conjugate eigenvalues (see Table 9.4) and eight real eigenvalues (see Table 9.5). All eigenvalues lie in the left-half complex plane. Among the complex eigenvalues, there is a poorly damped complex eigenvalue pair with natural undamped frequency around 1.5 Hz; this is the so-called *electromechanical eigenvalue*. The other complex pair is characterized by high damping. The time constants of the real eigenvalues are in the range between 10 ms and 7 s.

Subsystem participations provide a clear indication of the association of eigenvalues and system components. Table 9.6 provides the magnitude of the subsystem participations for this model. It should be noted that although participation factors

Table 9.5 Real eigenvalues of the detailed model of a generator connected to an infinite bus, with the corresponding time constant τ

No.	λ	τ (sec)
5	−0.1422	7.0319
6	−1.6243	0.6156
7	−1.9867	0.5033
8	−2.8678	0.3487
9	−3.9764	0.2515
10	−9.9227	0.1008
11	−11.9528	0.0837
12	−96.8474	0.0103

Table 9.6 Subsystem participations (magnitude) of the eigenvalues of detailed model of a generator connected to an infinite bus

λ	Rotor	Machine	Turbine	Excitation
1,2	**0.9812**	0.0205	0.0084	0.0048
3,4	0.1659	**0.9136**	0.0001	0.2223
5	0.0045	0.0001	**0.9953**	0.0000
6	0.0139	**1.0508**	0.0363	0.0006
7	0.0071	0.1094	**1.1022**	0.0000
8	0.0618	0.1184	**0.8196**	0.0003
9	0.0745	0.0375	**1.1126**	0.0006
10	0.0023	0.0179	**1.0129**	0.0028
11	0.0479	**1.1326**	0.0111	0.1694
12	0.0002	0.0050	0.0000	**0.9951**

and subsystem participation are in general complex numbers, we are only paying attention to their magnitude to find the associations between state variables and eigenvalues. (The phase of the participation factors will be interpreted when the connections between eigenvalue sensitivities and participation factors are presented in Sect. 9.4.)

The conjugate pair of eigenvalues $\lambda_{1,2}$ is associated with the rotor dynamics. The conjugate pair of eigenvalues $\lambda_{3,4}$ and the real eigenvalues λ_6 and λ_{11} correspond to dynamics of the synchronous machine. Eigenvalues $\lambda_5, \lambda_7, \lambda_8, \lambda_9$, and λ_{10} correspond to the turbine dynamics. Eigenvalue λ_{12} corresponds to the excitation dynamics.

9.3.3 Multi-Machine Systems

A power system of six generators and three areas is considered to illustrate the dynamic patterns of multi-machine systems (see Fig. 9.3). The Appendix provides its data set. Although the generators are of different ratings, they are represented by the same model type as the generator connected to an infinite bus in our previous example. Hence, the linearized model of the system exhibits 72 eigenvalues.

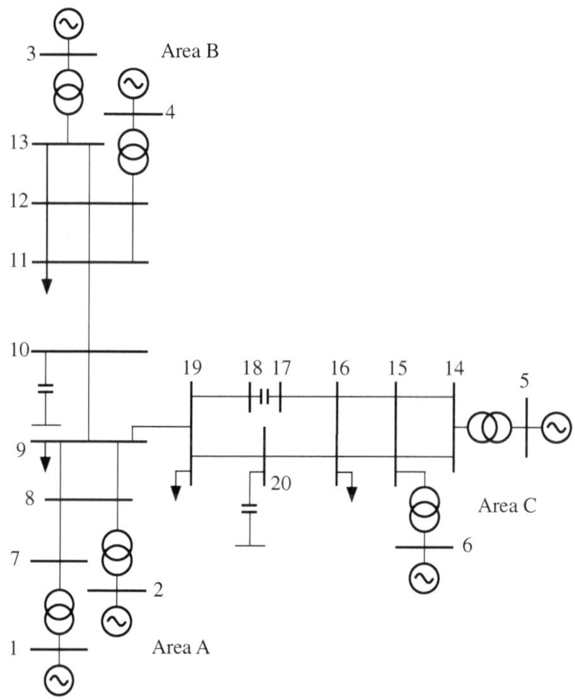

Fig. 9.3 Single-line diagram of a three-area test system

Table 9.7 Electromechanical eigenvalues of the three-area test system, with the corresponding damping factor ζ and undamped natural frequency f_n

No.	λ	ζ (%)	f_n (Hz)
1,2	$-0.6543 \pm j10.7739$	6.06	1.72
3,4	$-0.6520 \pm j10.7411$	6.06	1.71
5,6	$-0.7402 \pm j10.6533$	6.93	1.70
7,8	$-0.1727 \pm j8.2339$	2.10	1.31
9,10	$0.2177 \pm j4.6645$	−4.66	0.74

Five pairs of complex eigenvalues out of 72 eigenvalues are in the 0.1–2 Hz range (0.7–1.7 Hz); their damping is also very low (less than 10 %) and one mode is even unstable (see Table 9.7). As in the preceding example, these eigenvalues are called *electromechanical eigenvalues* because they correspond to the rotor oscillations. The mode shapes of the electromechanical eigenvalues of Table 9.7 are displayed in Table 9.8. The *shape* of each mode is determined by the right eigenvector components corresponding to generator speed deviations. The three fastest pairs of electromechanical eigenvalues ($\lambda_{1,2}$, $\lambda_{3,4}$, and $\lambda_{5,6}$) correspond to *local* oscillations within areas A, B, and C, while the two slowest eigenvalues ($\lambda_{7,8}$ and $\lambda_{9,10}$) correspond to *inter-area* oscillations.

Table 9.8 Mode shapes of the electromechanical eigenvalues of the three-area test system (The dominant positive components are in **boldface**, while the dominant negative ones are in *italics*, to convey the structure of the mode)

Generator	λ_1	λ_3	λ_5	λ_7	λ_9
1	$0.0036\angle 139°$	$0.0049\angle{-128°}$	*$0.3168\angle 179°$*	**$1.0000\angle 0°$**	**$0.7793\angle -4°$**
2	$0.0144\angle{-166°}$	$0.0073\angle{-94°}$	**$1.0000\angle 0°$**	**$0.9398\angle{-1°}$**	**$0.7919\angle -6°$**
3	$0.0010\angle -45°$	*$0.3296\angle 179°$*	$0.0035\angle{-175°}$	$0.0488\angle 161°$	*$0.7943\angle 169°$*
4	$0.0033\angle 132°$	**$1.0000\angle 0°$**	$0.0093\angle 13°$	$0.0305\angle -177°$	*$0.7474\angle 172°$*
5	*$0.3289\angle -180°$*	$0.0041\angle 130°$	$0.0132\angle 177°$	*$0.5296\angle 173°$*	**$1.0000\angle 0°$**
6	**$1.0000\angle 0°$**	$0.0142\angle -63°$	$0.0189\angle -8°$	*$0.4737\angle -179°$*	**$0.9934\angle 0°$**

Table 9.9 Generator participations (magnitude) in the electromechanical eigenvalues of the three-area test system

Generator	λ_1	λ_3	λ_5	λ_7	λ_9
1	0.0001	0.0001	**0.2345**	**0.5168**	**0.1050**
2	0.0004	0.0001	**0.7652**	**0.1501**	0.0350
3	0.0000	**0.2532**	0.0001	0.0051	**0.3560**
4	0.0000	**0.7468**	0.0000	0.0015	**0.1146**
5	**0.2437**	0.0000	0.0004	**0.2524**	**0.2945**
6	**0.7562**	0.0000	0.0000	0.0760	0.0971

Table 9.10 Generator rotor participations (magnitude) in the electromechanical eigenvalues of the three-area test system

Generator	λ_1	λ_3	λ_5	λ_7	λ_9
1	0.0001	0.0001	**0.2386**	**0.5360**	**0.1034**
2	0.0004	0.0001	**0.7890**	**0.1568**	0.0348
3	0.0000	**0.2502**	0.0000	0.0039	**0.3305**
4	0.0000	**0.7682**	0.0000	0.0008	**0.1032**
5	**0.2475**	0.0000	0.0004	**0.2426**	**0.2836**
6	**0.7700**	0.0000	0.0001	0.0693	0.0935

In the case of multi-machine power systems, not only are the generating unit component subsystems of interest, but the generating units themselves are also of interest. Table 9.9 provides the magnitude of generator participations in the electromechanical eigenvalues of the three-area test system. Only generators in the same area participate in the three fastest modes. However, most of the generators participate in the two slowest eigenvalues, which correspond to inter-area modes. Table 9.10 displays just the participations of the generator rotor dynamics in the electromechanical modes, which confirms that the selected eigenvalues are electromechanical ones.

The patterns found in small test cases are also found in very large-scale systems. Figure 9.4 shows the geographical mode shape of the slowest electromechanical eigenvalue of a linearized model of the European power system [20, 27]. The geographical mode shape displays in the geographic location of each generator the component of the right eigenvector corresponding to the generator speed deviation. The geographical mode shape of this mode shows that generators in eastern Europe oscillate against generators in western Europe. This is very similar to what happens in the three-area test system. In contrast, the generator participation is concentrated

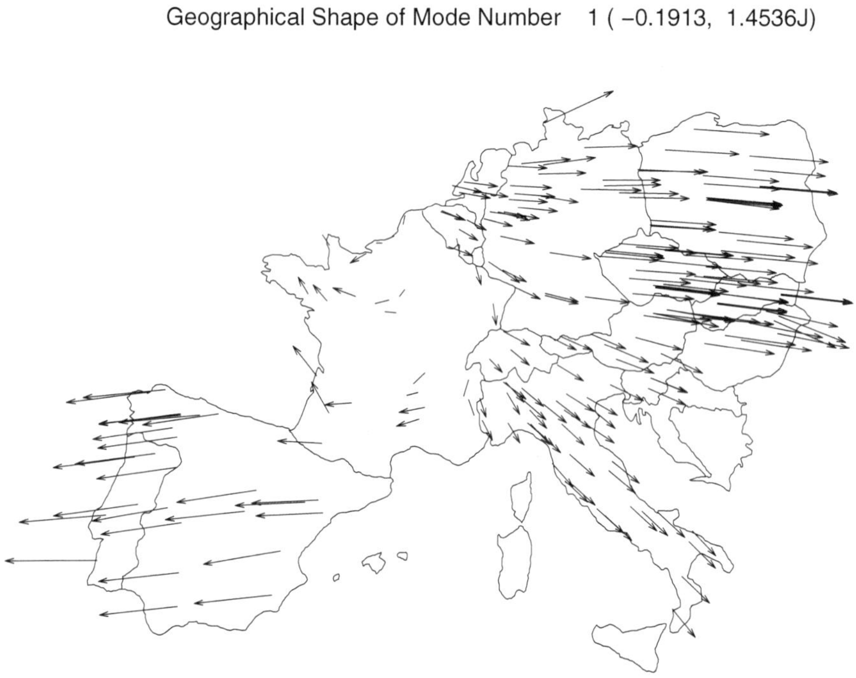

Fig. 9.4 Mode shape of the slowest electromechanical eigenvalue of a large-scale system

in a few large generators of the European system, as shown in Fig. 9.5. (The high-participation generator at the top is actually an equivalent generator representing the Danish system.)

9.3.4 Subsynchronous Resonance

Subsynchronous resonance is a small-signal phenomenon that arises in the case of turbogenerators (round-rotor generators driven by steam turbines) connected through series-compensated transmission lines [28]. The natural frequency of the series-compensated line appears at the generator rotor as two oscillations: subsynchronous and supersynchronous. Subsynchronous resonance may occur due to the interaction between the electrical subsynchronous mode and one of the torsional modes of the turbine-generator rotor.

The subsynchronous resonance phenomenon can be accurately characterized using an appropriate model of the generator connected to an infinite bus via a series-compensated transmission line, as shown in Fig. 9.6. Generator models used in power system stability studies assume that the individual masses of the turbine-generator rotor form a single rigid body, and neglect the synchronous machine stator and network dynamics. In contrast, models used in subsynchronous resonance studies

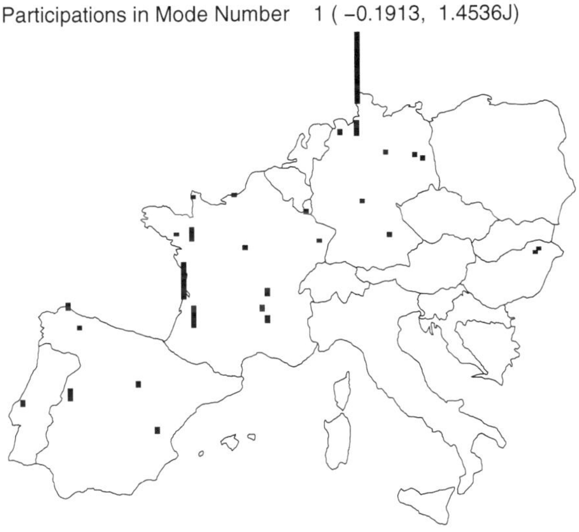

Fig. 9.5 Generator participations in the slowest electromechanical eigenvalues of a large-scale system

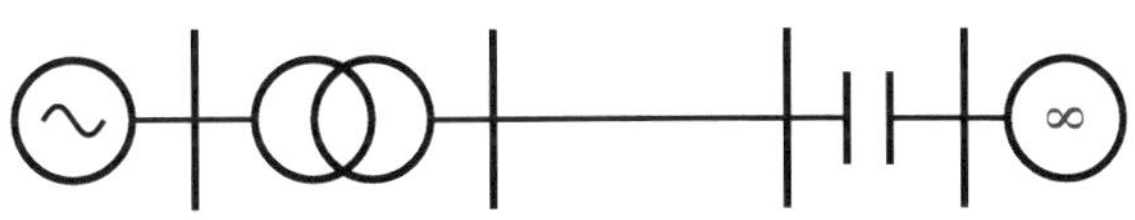

Fig. 9.6 Single-line diagram of a synchronous generator connected to an infinite bus through a series-compensated line

represent the individual rotor masses and the stator and network dynamics. Hence, the explicit state-space form of the linearized model of a synchronous generator connected to an infinite bus via a series-compensated transmission line, with the rotor represented by five masses, is written again in the form (9.15), with the following state variables representing deviations from equilibrium values:

$$\mathbf{x} = \left[\boldsymbol{\delta}^T \ \boldsymbol{\omega}^T \ \boldsymbol{\psi}^T \ \mathbf{x}_{\text{gov}}^T \ x_{\text{exc}} \ \mathbf{v}_c^T \right]^T$$

where:

$$\begin{aligned}
\boldsymbol{\delta} &= \left[\delta_{\text{hp}} \ \delta_{\text{ip}} \ \delta_{\text{lpa}} \ \delta_{\text{lpb}} \ \delta_g \right]^T \\
\boldsymbol{\omega} &= \left[\omega_{\text{hp}} \ \omega_{\text{ip}} \ \omega_{\text{lpa}} \ \omega_{\text{lpb}} \ \omega_g \right]^T \\
\psi &= \left[\psi_d \ \psi_q \ \psi_{fd} \ \psi_{kd} \ \psi_{kq1} \ \psi_{kq2} \right]^T \\
\mathbf{x}_{gov} &= \left[x_{\text{gov1}} \ x_{\text{gov2}} \ x_{\text{gov3}} \ x_{\text{gov4}} \ x_{\text{gov5}} \right]^T \\
\mathbf{v}_c &= \left[v_{cd} \ v_{cq} \right]^T
\end{aligned}$$

Table 9.11 Complex eigenvalues of the detailed model of a generator connected to an infinite bus through a series-compensated line, with the corresponding damping factor ζ and undamped natural frequency f_n

No.	λ	ζ (%)	f_n (Hz)
1,2	$-1.9343 \pm j532.2579$	0.36	84.71
3,4	$-0.0000 \pm j276.4385$	0.00	44.00
5,6	$1.4972 \pm j221.5317$	−0.68	35.26
7,8	$-0.0218 \pm j190.7095$	0.01	30.35
9,10	$-0.0192 \pm j151.6426$	0.01	24.13
11,12	$-0.0642 \pm j102.9266$	0.06	16.38
13,14	$-0.0063 \pm j9.4404$	0.07	1.50
15,16	$-28.9818 \pm j2.2753$	99.69	4.63

Table 9.12 Real eigenvalues of the detailed model of a generator connected to an infinite bus through a series-compensated line, with the corresponding time constant τ

No.	λ	τ (sec)
17	−0.1422	7.0319
18	−1.6243	0.6156
19	−1.9870	0.5033
20	−2.8709	0.3483
21	−3.9655	0.2522
22	−9.9485	0.1005
23	−11.5823	0.0863
24	−97.7062	0.0102

The model of Fig. 9.6 operating at the nominal frequency of 60 Hz is described by 24 state variables. This linear model exhibits, for values of the parameters provided in the Appendix, nine pairs of complex conjugate eigenvalues (see Table 9.11) and six real eigenvalue (see Table 9.12). Among the complex eigenvalues, there are six complex eigenvalue pairs in the range from 16 through 84 Hz that include the four torsional modes, and the supersynchronous and subsynchronous modes. A poorly damped complex eigenvalue pair with frequency around 1.5 Hz is very close to the electromechanical eigenvalue pair of Table 9.4. There is also a complex pair characterized by its high damping. The time constants of the real eigenvalues are in the range between 10 ms and 7 s. It should be noted that the operating point of the generator is the same as in the earlier single-machine infinite bus example, except that now the complexity of the model has been increased. Hence, not only are the two complex pairs of Table 9.4 found in Table 9.11 but also all the real eigenvalues of Table 9.5 can be found in Table 9.12.

Subsystem participations are again suggested to find the associations between eigenvalues and system components. The following subsystems can be identified in this model of a generator connected to an infinite bus through a series-compensated line: turbine-generator rotor, machine stator, machine rotor, speed governor-turbine, voltage regulator-excitation system, and series capacitor. Table 9.13 provides the magnitude of the subsystem participations. The table indicates that:

- The complex conjugate pairs $\lambda_{1,2}$ and $\lambda_{5,6}$ characterize the supersynchronous and subsynchronous oscillations in which machine stator circuits interact with the series capacitor.

Table 9.13 Subsystem participations (magnitude) of the eigenvalues of the detailed model of a generator connected to an infinite bus through a series-compensated line

λ	Rotor	Machine stator	Machine rotor	Turbine	Excitation	Series capacitor
1,2	0.0001	**0.5001**	0.0018	0.0000	0.0001	**0.5001**
3,4	**1.0000**	0.0000	0.0000	0.0000	0.0000	0.0000
5,6	0.0150	**0.5076**	0.0114	0.0000	0.0007	**0.5032**
7,8	**1.0077**	0.0045	0.0002	0.0000	0.0000	0.0033
9,10	**1.0028**	0.0019	0.0002	0.0000	0.0000	0.0010
11,12	**1.0036**	0.0027	0.0008	0.0000	0.0001	0.0010
13,14	**0.9778**	0.0065	0.0121	0.0087	0.0043	0.0014
15,16	0.1008	0.0090	**0.9402**	0.0001	0.1828	0.0124
17	0.0045	0.0000	0.0001	**0.9954**	0.0000	0.0000
18	0.0139	0.0024	**1.0527**	0.0359	0.0006	0.0001
19	0.0069	0.0001	0.1079	**1.1010**	0.0000	0.0000
20	0.0603	0.0004	0.1155	**0.8235**	0.0003	0.0000
21	0.0707	0.0001	0.0349	**1.1062**	0.0007	0.0002
22	0.0018	0.0005	0.0370	**1.0312**	0.0049	0.0004
23	0.0608	0.0061	**1.1172**	0.0274	0.1515	0.0070
24	0.0003	0.0001	0.0064	0.0000	**1.0066**	0.0000

- The complex conjugate pairs $\lambda_{3,4}$, $\lambda_{7,8}$, $\lambda_{9,10}$, $\lambda_{11,12}$, and $\lambda_{13,14}$ correspond to the turbine-generator rotor dynamics.
- The complex conjugate pair $\lambda_{15,16}$ and real eigenvalues λ_{22} and λ_{23} corresponds to the synchronous machine dynamics.
- The real eigenvalues λ_{17}, λ_{19}, λ_{20}, λ_{21}, and λ_{22} correspond to the turbine dynamics.
- The real eigenvalue λ_{24} corresponds to the excitation dynamics.

9.3.5 Wind Generators

Wind power generation is mostly based on DFIGs due to their ability to operate at the maximum efficiency point of the wind turbine and their continuous reactive power control capability. The model of a DFIG comprises models of the induction machine and the self-commutated converters with the associated vector controllers. Figure 9.7 displays the control scheme of a DFIG. The rotor windings are fed by a voltage-sourced three-phase converter with pulse-width modulation that provides a variable-frequency three phase voltage system. The variation of the frequency of rotor currents results in a variation of the rotor speed. Assuming that the stator frequency f_1 is constant, a variation of the rotor frequency f_2 results in a change of rotor speed as $\omega = (1 - f_2/f_1)$.

The electronic converter is built of two converters coupled through a dc-link capacitor. The rotor-side converter is used to control either the torque or the rotor

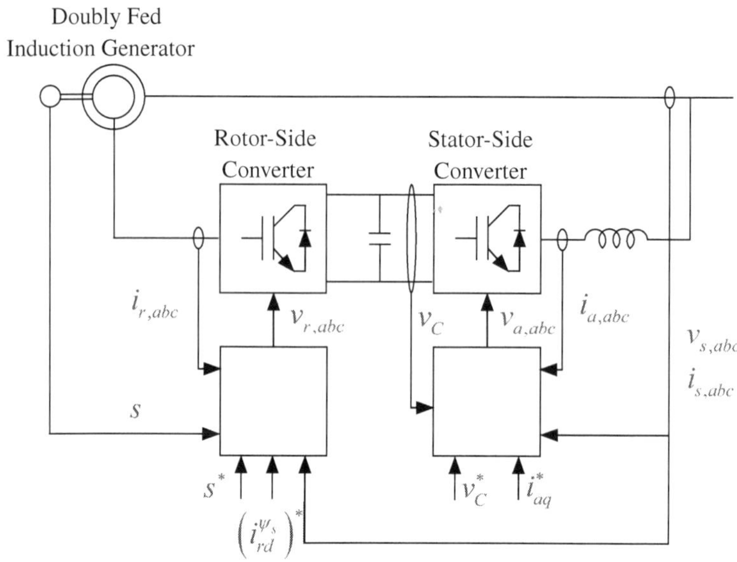

Fig. 9.7 Control scheme of a doubly fed induction generator

speed and the rotor reactive power [29]. The component in the direct axis of the rotor current in a reference system aligned with the stator flux is the excitation current and controls the generator reactive power. The usual strategy is to set it to zero. The component in the quadrature axis of rotor current in a reference system aligned with stator flux is the torque current and allows control of the electromagnetic torque and the rotor speed. The stator-side converter is used to control the overall generator reactive power and the capacitor voltage. The component in the direct axis of the stator converter current determines the active power through the stator converter and controls the capacitor voltage. The component in the quadrature axis of the stator converter current determines the reactive power provided by the stator-side converter.

Rotor current components and stator-side converter currents are controlled, respectively, by four PI controllers. Capacitor voltage and speed are controlled, respectively, by two PI controllers. The PI controllers have been designed in such a way that the equivalent second-order system exhibits the selected natural frequency and damping. The four inner PI current controllers have been designed at the natural undamped frequency of 25 rad/s whereas the two outer PI controllers have been designed at the natural undamped frequency of 2.5 rad/s. The selected damping is 70 % in all controllers.

The explicit linearized state-space form of the model of a DFIG connected to an infinite bus through a transformer and a transmission line (9.15) is written in terms of the following state and input variables:

$$\mathbf{x} = \left[\, \psi_{sd}\ \psi_{sq}\ \psi_{rd}\ \psi_{rq}\ \psi_{ad}\ \psi_{aq}\ x_{a1}\ x_{a2}\ x_{a3}\ v_C^2\ x_{r1}\ x_{r2}\ x_{r3}\ s \,\right]^T$$
$$\mathbf{u} = \left[\, i_{aq}^*\ v_C^*\ \left(i_{rd}^{\psi_s}\right)^*\ s^* \,\right]^T$$

where

- ψ_{sd} and ψ_{sq} are the machine stator flux deviations in the d- and q-axis, respectively,
- ψ_{rd} and ψ_{rq} are the machine rotor flux deviations in the d- and q-axis, respectively,
- ψ_{ad} and ψ_{aq} are the stator-side converter flux deviations in the d- and q-axis, respectively,
- x_{ai} are the deviations of the state variables that describe each of the PI controllers of the stator-side converter,
- v_C^2 is the deviation of the square of the capacitor voltage,
- x_{ri} are the deviations of the state variables that describe each of the PI controllers of the rotor-side converter,
- s is the deviation of the machine slip,
- i_{aq}^* is the deviation of the reference of the stator-side converter current in the q-axis,
- v_C^* is the deviation of the reference of the capacitor voltage,
- $\left(i_{rd}^{\psi_s}\right)^*$ is the deviation of the reference of the rotor-side converter current in the d-axis, and
- s^* is the deviation of the reference of the machine slip.

The detailed linearized model of a DFIG connected to an infinite bus is described by 14 state variables. Table 9.14 contains the eigenvalues of this linear model for the data set detailed in the Appendix. It exhibits seven complex pairs. One pair is close to the grid frequency (60 Hz). There are four pairs close to 4 Hz (25 rad/s) and two pairs close to 0.4 Hz (2.5 rad/s). The damping of these six complex pairs is 70 % (or close to that). Both the natural undamped frequency and damping of these four pairs of complex eigenvalues correspond to the design values of the PI controllers. It is worth noting that the dynamics of DFIGs used in wind generation are largely determined by the converter controllers. The poorly damped oscillation found in synchronous generators does not arise in DFIGs.

Subsystem participations are also suggested in this case to find the associations between eigenvalues and system components. The machine stator together with the two state variables of each control loop are proposed as the subsystems. Table 9.15 displays the magnitudes of the subsystem participations. The subsystem with the highest participation in each eigenvalue is highlighted. The computed subsystem participations show that

- the pair $\lambda_{1,2}$ is associated with the stator flux dynamics,
- the pairs $\lambda_{3,4}$ and $\lambda_{9,10}$ correspond to the control loops of the stator current components in the q- and d-axis, respectively,
- the pairs $\lambda_{5,6}$ and $\lambda_{7,8}$ are related to the control loop of the rotor current components in the q- and d-axis, respectively,
- the pair $\lambda_{11,12}$ is related to the control loop of the capacitor voltage, and

Table 9.14 Eigenvalues of the detailed model of a DFIG connected to an infinite bus, with the corresponding damping factor ζ and undamped natural frequency f_n

No.	λ	ζ (%)	f_n (Hz)
1, 2	$-12.9119 \pm j375.8664$	3.43	59.86
5, 6	$-17.5000 \pm j17.8536$	70.00	3.98
9,10	$-17.1593 \pm j17.7982$	69.41	3.93
13,14	$-15.8447 \pm j18.4654$	65.12	3.87
17,18	$-15.5088 \pm j15.6042$	70.49	3.50
21,22	$-1.9909 \pm j2.0335$	69.96	0.45
25,26	$-1.8362 \pm j1.8458$	70.52	0.41

Table 9.15 Subsystem participations (magnitude) of the eigenvalues of the detailed model of a DFIG connected to an infinite bus

λ	ψ_{sd}, ψ_{sq}	ψ_{rd}, x_{r2}	ψ_{rq}, x_{r1}	ψ_{ad}, x_{a2}	ψ_{aq}, x_{a1}	s, x_{r3}	v_C^2, x_{a3}
1,2	**1.0030**	0.0015	0.0016	0.0000	0.0000	0.0000	0.0000
3,4	0.0000	0.0000	0.0000	0.0000	**1.0000**	0.0000	0.0000
5,6	0.0013	0.1396	**1.0874**	0.0000	0.0000	0.0107	0.0000
7,8	0.0017	**1.1288**	0.1341	0.0000	0.0000	0.0886	0.0000
9,10	0.0000	0.0000	0.0000	**1.1574**	0.0000	0.0000	0.1577
11,12	0.0000	0.0000	0.0000	0.1574	0.0000	0.0000	**1.1574**
13,14	0.0000	0.0361	0.0000	0.0000	0.0000	**1.0355**	0.0000

- the pair $\lambda_{13,14}$ is associated with the control loop of the rotor speed.

Subsystem participations have proved to be valuable tools to identify the dynamic patterns that arise in a wide variety of linearized models of power system dynamic phenomena. Common knowledge of well-known models is confirmed, and better understanding of new models is achieved. It is also the case in many instances that dynamic patterns identified in linearized models (and especially if these patterns persist across multiple operating points) can then be traced back to similar patterns in the underlying nonlinear model. Some exploration of participation factors in the setting of limit cycles in nonlinear power system models may be found in [30].

9.4 Participation Factors and Design of Damping Controllers

The most effective and inexpensive approach to improving the damping of electromechanical oscillations is the installation of power system stabilizers in synchronous generator excitation systems. Stabilizers vary the reference of the automatic voltage regulator, using the speed deviation as an input (other inputs like electrical power or a combination of speed deviation and electrical power are also used) and thereby modulate the electrical torque applied by the synchronous machine to the rotor. The natural oscillation of the rotor is reduced by introducing, in effect, a forced oscillation of opposite sign.

Power system damping controllers have also been incorporated into other devices such as HVDC links, Static Var Compensators (SVCs), and Thyristor Controlled Series Capacitors (TCSCs). The damping controller of an HVDC link can modulate either the active power or the reactive power at each terminal. The damping controller of an SVC varies the bus shunt susceptance whereas the damping controller of a TSC modulates the effective line series reactance.

The design of power system damping controllers has attracted much attention from both engineers and researchers. Many methods have been proposed and can be found in the technical literature. A widely used approach combines frequency response with eigenvalue analysis (see for instance, [4, 5, 31]).

Eigenvalue sensitivities are powerful tools for designing power system damping controllers and for placing them (see, for instance [15, 16, 32, 33]). The first-order eigenvalue sensitivity provides an estimate of the eigenvalue shift when a controller parameter is changed. This section shows that participation factors can also be seen as eigenvalue sensitivities with respect to state matrix parameters. Moreover, powerful expressions for eigenvalue sensitivities with respect to feedback controller parameters are provided. Two applications of eigenvalue sensitivities to stabilizer design are shown: design of a single controller for damping one mode, and coordinated design of multiple controllers for damping several modes.

9.4.1 Sensitivities to State Matrix Parameters

The first-order sensitivity of the eigenvalue λ_i with respect to a parameter q of the state matrix $\mathbf{A}$ can be computed as [34]

$$S_{i,q} = \frac{\partial \lambda_i}{\partial q} = \mathbf{w}_i^T \frac{\partial \mathbf{A}}{\partial q} \mathbf{v}_i$$

Participation factors can be seen as eigenvalue sensitivities. In fact, the participation of the jth variable in the ith eigenvalue is the sensitivity of the ith eigenvalue with respect to the jth diagonal term of the state matrix

$$\begin{aligned}
\frac{\mathrm{d}\lambda_i}{\mathrm{d}a_{jj}} &= \mathbf{w}_i^T \frac{\partial \mathbf{A}}{\partial a_{jj}} \mathbf{v}_i \\
&= \begin{bmatrix} w_{i1} \cdots w_{ij} \cdots w_{iN} \end{bmatrix} \begin{bmatrix} 0 \cdots 0 \cdots 0 \\ \vdots \ddots \vdots \ddots \vdots \\ 0 \cdots 1 \cdots 0 \\ \vdots \ddots \vdots \ddots \vdots \\ 0 \cdots 0 \cdots 0 \end{bmatrix} \begin{bmatrix} v_{1i} \\ \vdots \\ v_{ji} \\ \vdots \\ v_{jN} \end{bmatrix} \\
&= w_{ij} v_{ji} = p_{ji}
\end{aligned}$$

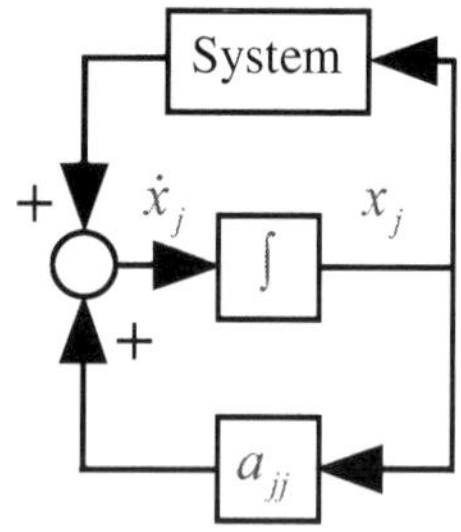

Fig. 9.8 Feedback around the jth state variable

Fig. 9.9 Closed-loop system in transfer function form

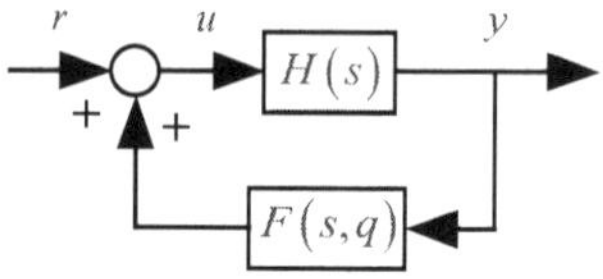

It is interesting to note that the jth diagonal term of the state matrix can be seen as a static feedback around the j-state variable (see Fig. 9.8).

Participation factors are complex numbers. So far, we have only made use of their magnitudes to identify the relationships among state variables and eigenvalues in power system models. The interpretation of the participation factors as eigenvalue sensitivities provides an explanation of both the magnitude and phase of participation factors: they, respectively, indicate the magnitude and direction of the eigenvalue shift.

9.4.2 Sensitivity to Feedback Controller Parameters

Much of our interest in eigenvalue sensitivity computation comes from needing to determine the effect of feedback controller parameters on the closed-loop eigenvalues. The preceding sensitivity results apply directly, provided that the matrix $\mathbf{A}$ describes the complete closed-loop system. However, it is typically the case that feedback controller parameters appear in many terms of the closed-loop matrices, and in complicated ways. It is therefore helpful to develop an alternative perspective for this situation, emphasizing transfer function properties. This point of view also lends valuable insight for control design.

Consider the feedback system of Fig. 9.9. The plant to be controlled is represented by the transfer function $H(s)$, and the controller is modeled by the transfer function $F(s,q)$.

The sensitivity of a pole (eigenvalue) λ_i of the closed-loop transfer function $y(s)/r(s)$ with respect to a parameter q of the controller transfer function $F(s,q)$ is the product of the residue of the closed-loop transfer function $y(s)/r(s)$

corresponding to the pole λ_i and the partial derivative of the controller transfer function with respect to the parameter q for $s = \lambda_i$

$$S_{i,q} = \frac{\partial \lambda_i}{\partial q} = R_i \left. \frac{\partial F(s,q)}{\partial q} \right|_{s=\lambda_i} \tag{9.16}$$

Recall that the residue R_i is the coefficient of $1/(s - \lambda_i)$ in the partial fraction expansion of $y(s)/r(s)$. The sensitivity expression (9.16) shows a useful separation of the effect of the feedback path, which determines the residue of the closed-loop transfer function $y(s)/r(s)$, and the effect of the parameter q in the controller transfer function for $s = \lambda_i$.

The computation of the two components of the sensitivity expression (9.16) exhibits different degrees of difficulty. The partial derivative of the controller transfer function $F(s,q)$ with respect to parameter q can be computed easily. In contrast, the residue of the closed-loop transfer function corresponding to the eigenvalue λ_i is more difficult to compute. A transfer function representation of the system for this purpose can be unwieldy for complex systems. It is more preferable to use a state-space representation for the closed-loop system.

A hybrid formulation can be useful, with a state-space model for the plant $H(s)$ and the closed-loop system in order to evaluate the residue, while preserving the transfer function model for the feedback in order to compute its derivative with respect to the parameter. The feedback system of Fig. 9.9 can be represented in this hybrid form as in Fig. 9.10 when the plant is represented in explicit state-space form and there is no direct link between the input and the output. In this case, the closed-loop residue corresponding to the eigenvalue λ_i can be computed by taking into account only the components of the right and left eigenvectors corresponding to the state variables of the plant

$$R_i = \left(\mathbf{c}_1^T \mathbf{v}_{1,i}\right) \left(\mathbf{w}_{1,i}^T \mathbf{b}_1\right) \tag{9.17}$$

It should be noted the residue expression (9.17) is obtained using the input vector, output vector and eigenvectors of the closed-loop state-space model. Of course, in the simplest case, when there is no feedback present at first, the residue is computed using the input vector, output vector and eigenvectors of the open-loop state-space model; this case arises frequently.

9.4.3 Design of a Single Damping Controller for Damping One Mode

The first-order eigenvalue sensitivity provides an estimate of the eigenvalue shift when a controller parameter is changed. The estimated eigenvalue λ_i^e after the variation in the parameter can be computed as

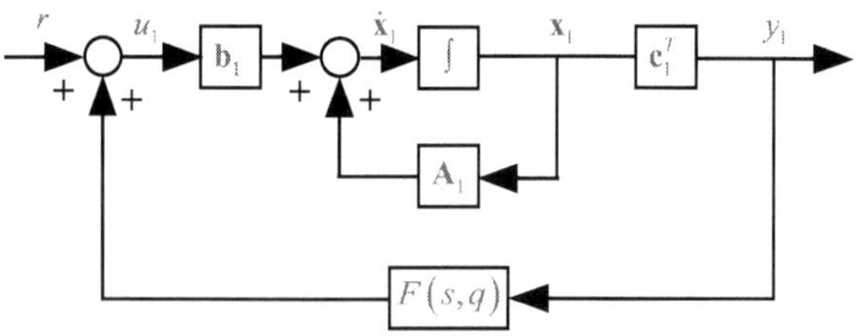

Fig. 9.10 System in hybrid form with no direct relation between the input and the output of the part represented in state space form

$$\lambda_i^e = \lambda_i + S_{i,q}\Delta q$$

The same equation can be used to compute the increment Δq in the parameter that will produce a desired eigenvalue λ_i^d. This can be equal (and usually at least close) to the estimated eigenvalue, but some approximations can produce a larger difference. After the parameter is actually changed, a new eigenanalysis is performed and the true value of the eigenvalue λ_i^t (hopefully not too far away from λ_i^e) is obtained.

If the controller transfer function is in the lead compensation form, the eigenvalue sensitivity approach to designing a single damping controller to damp a mode comprises two steps: the design of the phase compensation network of the controller and the computation of the controller gain. The phase compensation network of the controller is designed so that the phase of the eigenvalue sensitivity with respect to gain is set to (approximately) 180° at the nominal eigenvalue. In this way, an increase in gain will increase damping. The controller gain is then determined so that the decrease in the real part of the eigenvalue is equal to the desired value.

The eigenvalue sensitivity of the ith eigenvalue with respect to the variation of the gain of the controller is

$$S_{i,K_S} = \frac{\partial \lambda_i}{\partial K_S} = R_i \left. \frac{\partial F(s, K_S)}{\partial K_S} \right|_{s=\lambda_i}$$

where R_i is the residue of the transfer function between the output and the input of the controller associated to the ith mode and $F(s, K_S)$ is the transfer function of the controller

$$F(s, K_S) = K_S \left(\frac{1 + sT_{S1}}{1 + s\alpha T_{S1}} \right)^{N_s} \frac{sT_{S5}}{1 + sT_{S5}} \tag{9.18}$$

where N_s stages are assumed for the phase compensation network.

Transfer function (9.18) corresponds to the stabilizer model of Fig. 9.11 when $N_s = 2$, $T_{S1} = T_{S3}$, and $T_{S2} = T_{S4} = \alpha T_{S1}$. This is a function of the lead time constant T_{S1} because the time constant of the wash-out filter T_{S5} is assumed to be given. Combining the preceding equations, we get

$$S_{i,K_S}(T_{S1}) = R_i \left(\frac{1+\lambda_i T_{S1}}{1+\lambda_i \alpha T_{S1}} \right)^{N_s} \frac{\lambda_i T_{S5}}{1+\lambda_i T_{S5}}$$

As previously mentioned, the first step will be to set the phase of the sensitivity to 180°. The sensitivity for $T_{S1} = 0$ is

$$S_{i,K_S}(T_{S1}=0) = R_i \frac{\lambda_i T_{S5}}{1+\lambda_i T_{S5}}$$

and its argument is

$$\psi = \arg\left[S_{i,K_S}(T_{S1}=0)\right]$$

It should be noted that the wash-out filter usually adds a small phase lead to the phase of the residue, but hardly affects the magnitude of the residue.

The total phase lead, φ, should now be set to

$$\varphi = \pi - \psi \tag{9.19}$$

and the phase lead ϕ of each stage is

$$\phi = \frac{\pi - \psi}{N_s} \tag{9.20}$$

The task now is to find values of α and T_{S1} that produce the desired phase shift

$$\phi = \arg\left[\frac{1+\lambda_i T_{S1}}{1+\lambda_i \alpha T_{S1}}\right]$$

Texts that describe frequency response methods for the design of controls and compensations, see for example [35], provide sets of figures and formulas that are strictly applicable for purely imaginary eigenvalues. They can be used as a good approximation when the real part is small enough, and then either the natural frequency (imaginary part) or the undamped natural frequency (magnitude) are adequate. We will use the natural frequency ω_{di} of the original mode $\lambda_i = \sigma_i + j\omega_{di}$.

Some easy-to-use formulas are given now. The filtering ratio is determined as

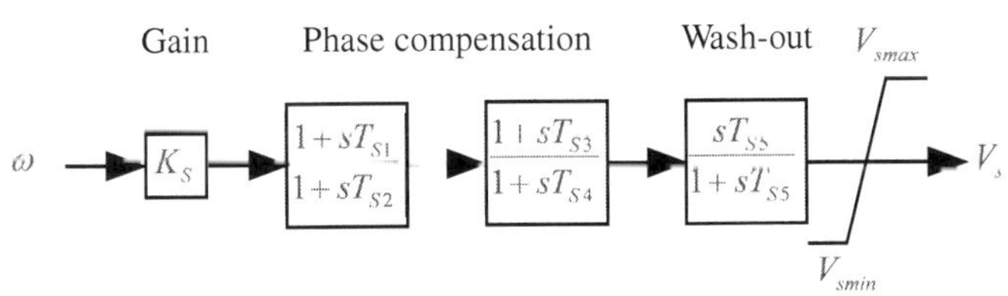

Fig. 9.11 Stabilizer model

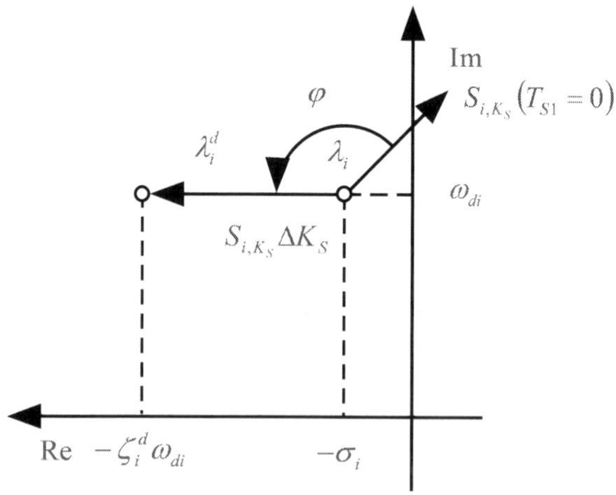

Fig. 9.12 Geometric interpretation of the two-step approach to design a damping controller to damp a mode

$$\alpha = \frac{1 - \sin\phi}{1 + \sin\phi} \tag{9.21}$$

and the time constant of the numerator is determined as

$$T_{S1} = \frac{1}{\sqrt{\alpha}\omega_{di}}$$

Once the controller phase compensation has been designed, the controller gain is computed to give the desired shift in the real part of the eigenvalue

$$K_S = \frac{\mathrm{Re}\left[\Delta\lambda_i\right]}{\mathrm{Re}\left[S_{i,K_S}\right]} = \frac{\mathrm{Re}\left[\lambda_i^d - \lambda_i\right]}{\mathrm{Re}\left[S_{i,K_S}\right]}$$

Figure 9.12 shows a geometric interpretation of the eigenvalue sensitivity approach for designing a single damping controller. The starting point is the eigenvalue sensitivity with respect to a variation of the controller gain when $T_{S1} = 0$. In Fig. 9.12, a phase lead of φ is provided by the phase compensation network. Once the phase of the eigenvalue sensitivity S_{i,K_s} is set to (approximately) 180°, the gain K_s is determined to achieve the real part of the desired eigenvalue λ_i^d.

9.4.4 Stabilizer Design for a Single-Machine Connected to an Infinite Bus

The sensitivity approach is applied here to the design of a power system stabilizer for the generator represented by the detailed model presented in Sect. 9.3, connected to an infinite bus. A stabilizer that uses the rotor speed as input is considered. The stabilizer is aimed at improving the damping of the electromechanical eigenvalue from 0.89 to 15 %. The original and the desired electromechanical eigenvalues are, respectively,

$$\lambda = -0.0858 + j9.6453, \ \lambda^d = -1.4468 + j9.6453$$

The time constant of the wash-out filter is assumed to be 5 s. The residue of the transfer function ω / V_{ref} corresponding to the electromechanical mode is

$$R_{\omega/V_{ref}} = 0.2007\angle 130.11^\circ$$

The sensitivity of the eigenvalue with respect to the gain of the stabilizer assuming no phase compensation is

$$S_{K_S}(T_{S1} = 0) = 0.2007\angle 131.30^\circ$$

As previously mentioned, the small effect of the wash-out filter on both the magnitude and the phase of the eigenvalue sensitivity can be seen. The total required phase lead is $\varphi = 180^\circ - 131.30^\circ = 48.70^\circ$. Assuming two equal lead stages, the phase to be compensated by each one is $\phi = 24.35^\circ$. The lead ratio that provides this phase lead is $\alpha = 0.4161$. The time constant of the numerator of the lead compensation is $T_{S1} = 0.1607$ s, using the damped natural frequency of the electromechanical eigenvalue $\omega_d = 9.65$ rad/s. After designing the phase compensation, the stabilizer gain can be determined to attain the desired eigenvalue λ^d. The value of the stabilizer gain that results is $K_S = 2.8220$ pu. The full set of parameters of the designed power system stabilizer in lead compensation transfer function form is contained in Table 9.16.

The eigenvalue sensitivity provides an estimate of the electromechanical eigenvalue that results when the power system stabilizer is incorporated. The estimated and the true electromechanical eigenvalues are, respectively, $\lambda^e = -1.4468 + j9.6453$ and $\lambda^t = -1.5589 + j9.4030$. Both the estimated and the true eigenvalues are very close, which validates the approximation of the eigenvalue shift provided by the eigenvalue sensitivity. The damping of the final electromechanical eigenvalue is 16.36 %, which is greater than the desired value. Figure 9.13 compares the time response of the linearized model without and with stabilizer in response to a step at the reference of the excitation system. The generator speed deviation is displayed. Figure 9.13 confirms the performance of the designed stabilizer.

Table 9.16 Parameters of a stabilizer for a single generator connected to an infinite bus

K_S (pu)	$T_{S1} = T_{S3}$ (s)	$T_{S2} = T_{S4}$ (s)	T_{S5} (s)
2.8220	0.1607	0.0669	5.0

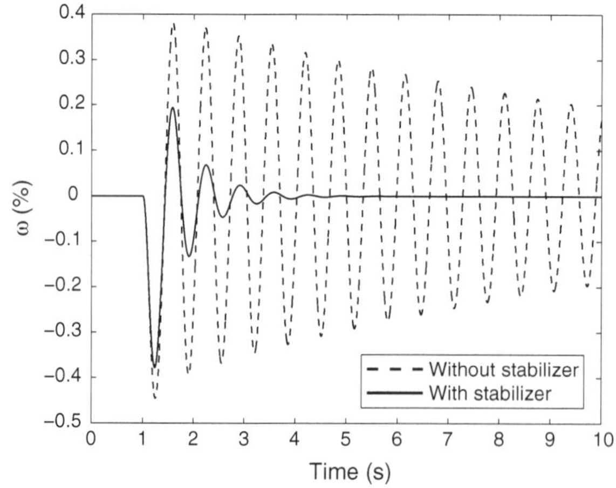

Fig. 9.13 Comparison of the time response of the linearized model of a synchronous generator connected to an infinite bus without and with a stabilizer, following a step at the reference of the excitation system

9.4.5 Coordinated Design of Multiple Damping Controllers for Damping Several Modes

The coordinated approach to the design of damping controllers addresses the problem of designing multiple controllers to improve the damping of several modes. The eigenvalue sensitivity approach to designing a single damping controller can be extended to design multiple damping controllers.

The two-step approach to coordinated design of multiple controllers to damp several modes is considered when the controller transfer function is in the lead compensation form. The first step comprises the independent design of the phase compensation network of each controller whereas the second step performs the coordinated computation of the gains of all controllers. The phase compensation network of each controller is first designed so that the phase of the sensitivities (for all significant eigenvalues) is close to 180°. Then the gains of all controllers are determined so that the damping of all eigenvalues is greater than a selected bound.

9.4.5.1 Design of the Controller Phase Compensation Networks

The procedure given in Sect. 9.4.3 can be extended to multiple eigenvalues. Because it is no longer possible to tune the sensitivity for just one eigenvalue, the idea is to use average phases to adjust the parameters of each controller.

Assume the transfer function of the jth controller to be

$$F\left(s, K_{Sj}\right) = K_{Sj}\left(\frac{1 + sT_{S1j}}{1 + s\alpha_j T_{S1j}}\right)^{N_{sj}} \frac{sT_{S5j}}{1 + sT_{S5j}} \tag{9.22}$$

The sensitivities of the eigenvalues with respect to changes in the gains of the controllers are

$$S_{i,K_{Sj}}\left(T_{S1j}\right) = R_{ij}\left(\frac{1 + \lambda_i T_{S1j}}{1 + \lambda_i \alpha_j T_{S1j}}\right)^{N_{sj}} \frac{\lambda_i T_{S5j}}{1 + \lambda_i T_{S5j}}$$

The sensitivities for $T_{S1j} = 0$ are given by

$$S_{i,K_{Sj}}\left(T_{S1j} = 0\right) = R_{ij}\frac{\lambda_i T_{S5j}}{1 + \lambda_i T_{S5j}}$$

An average phase ψ_j will be defined for each controller as the argument of the sum of the sensitivities (this gives more weight to more significant sensitivities)

$$\psi_j = \arg\left[\sum_{i=1}^{N_e} S_{i,K_{Sj}}\left(T_{S1j} = 0\right)\right]$$

where N_e is the total number of considered eigenvalues.

The total and partial phase lead, according to the number of stages N_{sj} and the filtering ratio of each controller α_j, is determined from the average phase ψ_j by using Eqs. (9.19), (9.20), and (9.21).

For the determination of the time constant T_{S1j} of the transfer function (9.22), an optimization procedure (in just one variable) is used, taking advantage of the fact that the cosine when the argument is minimum (-1) is 180°. A weighting factor β_{ij} gives more importance to more significant sensitivities. It is normalized so that in the ideal case (all sensitivities set to 180°), the value of the function $G\left(T_{S1j}\right)$ will be -1. The optimization problem is defined as

$$\min G\left(T_{S1j}\right) = \min \sum_{i=1}^{N_e} \beta_{ij} \cos\left(\arg\left[S_{i,K_{Sj}}\left(T_{S1j}\right)\right]\right) \tag{9.23}$$

where the weights are determined from the residues:

$$\beta_{ij} = \frac{\left|R_{ij}\right|}{\sum_{i=1}^{N_e}\left|R_{ij}\right|}$$

9.4.5.2 Computation of the Controller Gains

Once the phase of the sensitivities is close to 180°, the gains of the controllers are determined to shift the eigenvalues to yield the desired damping. The gains of the controllers are determined by solving a linear programming problem. The objective function is to minimize the control action. The control action is expressed as the sum of the gains, weighted by the sensitivities:

$$\min \sum_{j=1}^{N_c} \gamma_j K_{Sj}$$

where N_c is the total number of controllers being designed and

$$\gamma_j = \sum_{i=1}^{N_e} \left| S_{i,K_{Sj}} \right|$$

The weighting function γ_j helps to normalize different modeling practices (for instance, different units).

The constraints are the minimum values of the real parts of the eigenvalues and the lower and upper bounds of the gains

$$\sum_{j=1}^{N_c} \mathrm{Re}\left[S_{i,K_{Sj}}\right] K_{Sj} \le \mathrm{Re}\left[\lambda_i^d - \lambda_i\right], \quad i = 1, \ldots, N_e, \quad 0 \le K_{Sj} \le K_{Sj}^{\max}$$

Should the phase of the eigenvalue sensitivity be 180°, the imaginary part of the desired eigenvalue would remain constant and the real part would be defined by the desired eigenvalue damping.

9.4.6 Coordinated Design of Power System Stabilizers in a Multi-Machine System

Both approaches to the coordinated design of multiple damping controllers to damp several modes are illustrated in the design of the power system stabilizers of all generators in the three-area test system. The aim is the improvement of the damping of the five electromechanical modes to at least 15 %.

The phase compensation of each generator is independently designed. For each generator, the average phase ψ_j of eigenvalue sensitivities is determined and the number of phase compensation stages is selected. Then the filtering ratio is obtained. The time constant is subsequently determined by solving the nonlinear optimization problem. The results are provided in Table 9.17.

The average phase of the eigenvalue sensitivities is between 120° and 150° (this is related to the fact that all generators are equipped with static excitation systems). The selected number of stages of the phase compensation networks was always two. Therefore, the phase ϕ to be provided by each stage is between 10° and 30°. The filtering ratio is in all cases around 0.5. The performance of the overall process is quite satisfactory because the value of $G(T_{S1})$ after incorporating the lead compensation is close to -1.

The solution of the linear programming problem, assuming that the upper limits of the stabilizer gains is 15 pu, provides the values of the gains (see Table 9.18). Among six generator candidates, only three generators have nonzero gains. This means that only three generators are needed to provide the required damping. All eigenvalues were computed after incorporating the designed stabilizers in the system model. All eigenvalues are stable. The performance of the overall approach is checked by comparing the original eigenvalues, the estimated, and the new eigenvalues after incorporating the designed stabilizers (see Table 9.19).

It is not always easy to match the eigenvalues that result from incorporating the stabilizers with the original ones. A simple approach consists of computing the root locus as the stabilizer gain increase from zero to the nominal values. Figure 9.14 shows the locus of the electromechanical eigenvalues. Figure 9.15 details the locus of the electromechanical eigenvalues corresponding to the local oscillations.

The comparison of the estimated and the new eigenvalues confirms the accuracy of the eigenvalue sensitivities. The damping of the original and the new eigenvalues can also be compared (see Table 9.20). The damping of the local modes is greater than the selected bound in three out of four modes, whereas the damping of the inter-area modes is a bit lower. This is due to two circumstances: (1) the stabilizer gains are mainly determined to improve the damping of the inter-area modes, and (2) the eigenvalue shift is only approximated by the first-order eigenvalue sensitivities. Figure 9.16 compares the time response of the linearized system without and with stabilizers in response to a step at the reference of the excitation system of generator 1. Differences of the speed deviation of two generators ($\omega_1 - \omega_3$ and $\omega_1 - \omega_5$) are displayed because they show the inter-area oscillations. Figure 9.16 confirms the performance of the designed stabilizers.

Table 9.17 Two-step design of speed deviation stabilizers in the three-area test system

Generator	ψ (deg)	N_s	ϕ (deg)	α	T_{S1} (sec)	T_{S2} (sec)	$G(T_{S1})$
1	167.43	2	20.55	0.480	0.1025	0.0492	−0.9760
2	124.17	2	27.91	0.362	0.1082	0.0392	−0.9956
3	132.25	2	23.88	0.424	0.1118	0.0474	−0.9714
4	119.10	2	30.45	0.327	0.1163	0.0381	−0.9969
5	134.33	2	22.83	0.441	0.1047	0.0461	−0.9607
6	118.60	2	30.70	0.324	0.1139	0.0369	−0.9920

Design of the phase compensation networks

Table 9.18 Two-step design of speed deviation stabilizers in the three-area test system: Gain computation

Generator	K_S (pu)
1	10.7947
2	0
3	15
4	0
5	15
6	0

Table 9.19 Two-step design of speed deviation stabilizers in the three-area test system

λ	λ	λ^e	λ^t
1,2	$-0.6543 \pm j10.7739$	$-2.3297 \pm j11.3061$	$-2.4175 \pm j10.3965$
3,4	$-0.6520 \pm j10.7411$	$-2.5028 \pm j11.1790$	$-2.0809 \pm j10.6809$
5,6	$-0.7402 \pm j10.6533$	$-1.6522 \pm j10.9335$	$-1.3028 \pm j9.9813$
7,8	$-0.1727 \pm j8.2339$	$-2.1569 \pm j7.9729$	$-2.6864 \pm j7.9305$
9,10	$0.2177 \pm j4.6645$	$-0.6997 \pm j4.3800$	$-0.6261 \pm j4.4179$

Original, estimated, and true eigenvalues

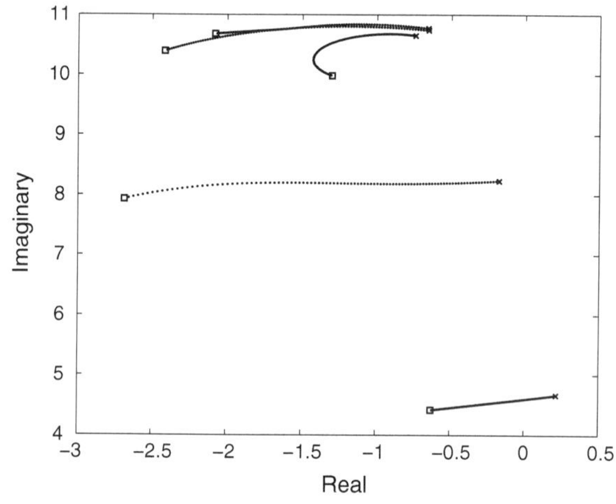

Fig. 9.14 Locus of the electromechanical eigenvalues as the stabilizer gains increase from zero to the nominal values

9.5 Participation Factors and Reduced-Order Eigenanalysis

Accurate characterization of power system electromechanical oscillations requires detailed models of the power system. The typical model of a generating unit is described by 10 state variables. This means that the state matrix of a power system of 100 generators is on the order of 1000×1000. As the state matrix of a realistic linearized model of a power system does not exhibit any special computational property (symmetry, sparsity, etc.) that can be taken advantage of, a matrix of that size

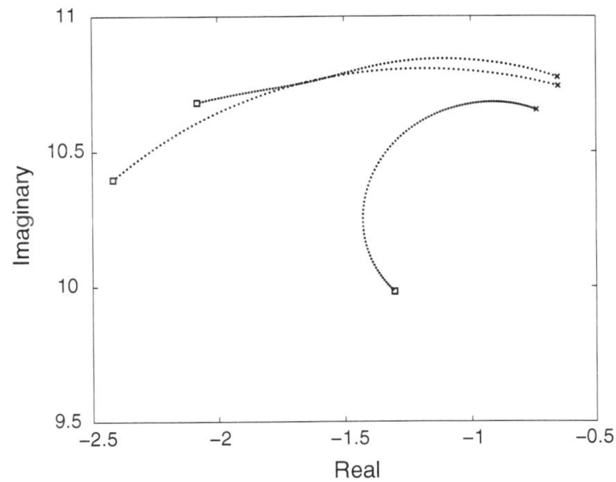

Fig. 9.15 Locus of the local electromechanical eigenvalues as the stabilizer gains increase from zero to the nominal values

Table 9.20 Two-step design of speed deviation stabilizers in the three-area test system: Original, estimated and true eigenvalue damping (%)

λ	ζ	ζ^e	ζ^t
1,2	6.06	20.18	22.65
3,4	6.06	21.85	26.77
5,6	6.93	14.94	12.94
7,8	2.10	26.11	32.08
9,10	−4.66	15.77	14.03

is close to the limits of the eigenvalue analysis algorithms for general matrices based on the QR transformation [22]. The SMA algorithms of reduced-order eigenanalysis can be applied to the small-signal stability problem because:

- Only a few modes are of interest, which are those that characterize the poorly damped oscillations.
- The modes of interest are associated with a subset of state variables that describe the dynamics of the rotors of the generators.

This section presents the role of participation factors in the convergence of SMA algorithms of reduced-order eigenanalysis. The basic idea of SMA algorithms is presented first. Then the core algorithms and the algorithms with variable transformations are detailed. The application of the algorithms to the small-signal stability problem of power systems is illustrated throughout the section.

9.5.1 Basic Idea

Let us recall the set of LTI differential equations (9.2) that describe an undriven LTI dynamic system. SMA algorithms for reduced-order eigenvalue analysis take advantage of the close association between subsets of modes and subsets of state

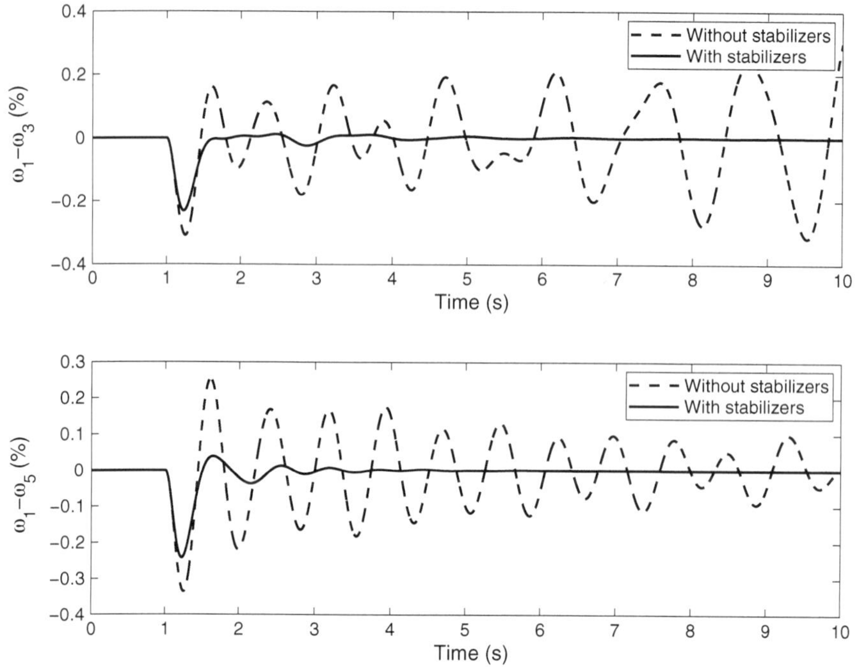

Fig. 9.16 Comparison of the time response of the linearized model of the three-area test system without and with stabilizers, following a step at the reference of the excitation system of generator 1

variables that typically exist in most physically motivated LTI systems. We have already referred to such associations as *dynamic patterns*.

In general, an intimate understanding of the physical system is what leads to the identification of such dynamic patterns. If, however, the results of a complete eigenanalysis are available, the dynamic patterns can be identified more systematically by using the participation factors, as has been shown in Sect. 9.3. In particular, the study of small test cases has shown that the electromechanical modes together with the variables that describe the rotor dynamics constitute a distinct dynamic pattern.

Once the dynamic patterns and the modes of interest have been identified, the state variables can be separated, for the modes of interest, into two subsets: the *relevant* variables $\mathbf{r}$, and the *less relevant* variables $\mathbf{z}$. According to this partitioning of the state variables, the original system (9.2) can be rewritten as

$$\begin{bmatrix} \dot{\mathbf{r}} \\ \dot{\mathbf{z}} \end{bmatrix} = \begin{bmatrix} \mathbf{A}_{11} & \mathbf{A}_{12} \\ \mathbf{A}_{21} & \mathbf{A}_{22} \end{bmatrix} \begin{bmatrix} \mathbf{r} \\ \mathbf{z} \end{bmatrix}, \quad \mathbf{r} \in \Re^{n}, \quad \mathbf{z} \in \Re^{N-n} \tag{9.24}$$

The system (9.24) can be represented in the block diagram form of Fig. 9.17. The basic idea of the SMA algorithms for the determination of a subset of the system eigenvalues is to iteratively build a reduced-order system expressed in terms of the

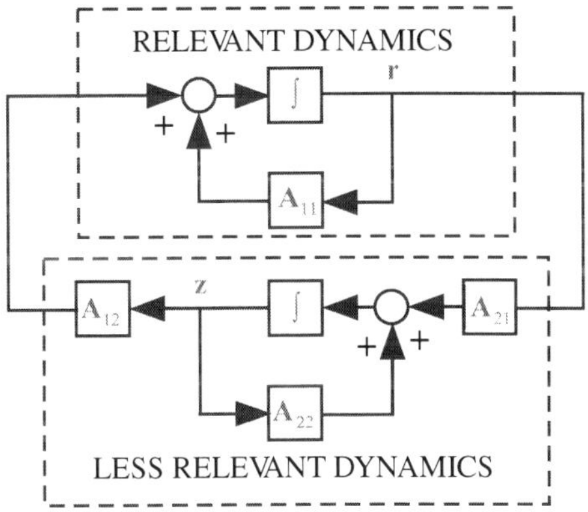

Fig. 9.17 Relevant and less relevant dynamics

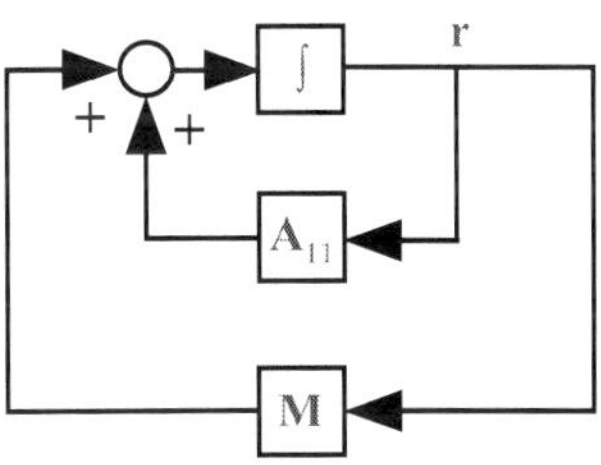

Fig. 9.18 Reduced-order system

relevant variables for the desired subset of eigenvalues, incorporating the effects of the less relevant variables by approximate static equivalents that will be iteratively refined. The core of the relevant dynamics is characterized by the $\mathbf{A}_{11}$ matrix whereas the effects of the less relevant dynamics are accounted for by a correction matrix $\mathbf{M}$, yielding

$$\dot{\mathbf{r}} = (\mathbf{A}_{11} + \mathbf{M})\,\mathbf{r} = \mathbf{A}_R\mathbf{r} \tag{9.25}$$

The system of equations (9.25) can be made to possess modes of interest, by a proper choice of $\mathbf{M}$, as we shall presently show. Although we use the same symbol $\mathbf{r}$ in (9.25) as in (9.24) for notational simplicity, we actually get the same functions of time if and only if $\mathbf{M}$ and the initial conditions are correctly chosen. The system (9.25) can be represented in the block diagram form of Fig. 9.18.

Let us assume that an eigenvalue λ_1 of $\mathbf{A}$ and its associated right and left eigenvectors, $\mathbf{v}_1$ and $\mathbf{w}_1$, respectively, are exactly known. The definition of the right eigenvector applied to the system written in the partitioned form (9.24) is

$$\begin{bmatrix} \mathbf{A}_{11} & \mathbf{A}_{12} \\ \mathbf{A}_{21} & \mathbf{A}_{22} \end{bmatrix} \begin{bmatrix} \mathbf{v}_{\mathbf{r}1} \\ \mathbf{v}_{\mathbf{z}1} \end{bmatrix} = \begin{bmatrix} \mathbf{v}_{\mathbf{r}1} \\ \mathbf{v}_{\mathbf{z}1} \end{bmatrix} \lambda_1 \tag{9.26}$$

It follows from (9.26) that the part of the right eigenvector corresponding to the relevant variables satisfies

$$\left[\mathbf{A}_{11} + \mathbf{A}_{12}\left(\lambda_1 \mathbf{I} - \mathbf{A}_{22}\right)^{-1} \mathbf{A}_{21}\right] \mathbf{v}_{\mathbf{r}1} = \mathbf{v}_{\mathbf{r}1} \lambda_1 \quad (9.27)$$

assuming that λ_1 is not an eigenvalue of $\mathbf{A}_{22}$.

Let $\mathbf{H}(s)$ denote the transfer function of the less relevant dynamics. So

$$\mathbf{H}(s) = \mathbf{A}_{12}\left(s\mathbf{I} - \mathbf{A}_{22}\right)^{-1} \mathbf{A}_{21}$$

If Eqs. (9.25) and (9.27) are now compared, it can be concluded that a choice of $\mathbf{M}$ will cause (9.25) to have λ_1 as an eigenvalue and $\mathbf{v}_{\mathbf{r}1}$ as the associated right eigenvector is

$$\mathbf{M} = \mathbf{H}(\lambda_1) = \mathbf{H}(s)|_{s=\lambda_1}$$

It should be noted that a dual result can be obtained from the definition of the left eigenvector applied to the system written in partitioned form (9.24).

9.5.2 *Basic Algorithms*

This section provides the SMA algorithms of reduced-order eigenanalysis. The starting point is the basic idea presented in the previous section. The algorithm to compute a single mode is presented first. The algorithm to compute several modes simultaneously is provided subsequently. In addition, the convergence conditions of the algorithms are detailed.

9.5.2.1 Algorithm to Compute a Single Mode

The idea of the algorithm to compute a single mode, say the one associated with eigenvalue λ_1, is very simple. Let us assume that an initial guess of the eigenvalue is available, and denote this guess by λ_1^0. A better approximation may be obtained, if the selection of relevant and less relevant variables is appropriate, by performing the eigenanalysis of the following reduced-order system

$$\dot{\mathbf{r}} = \left(\mathbf{A}_{11} + \mathbf{M}^1\right) \mathbf{r} = \mathbf{A}_R^1 \mathbf{r}$$

where $\mathbf{M}^1$ is the transfer function of the less relevant dynamics evaluated at the initial guess of the eigenvalue of interest λ_1^0

$$\mathbf{M}^1 = \mathbf{H}\left(\lambda_1^0\right) = \left[\mathbf{A}_{12}\left(\lambda_1^0 \mathbf{I} - \mathbf{A}_{22}\right)^{-1} \mathbf{A}_{21}\right]$$

The algorithm to compute a single mode can be summarized as follows:

1. Set the initial value of the iteration count to $j = 0$.
2. Provide an initial guess λ_1^0.
3. Compute $\mathbf{H}\left(\lambda_1^j\right) = \mathbf{A}_{12}\left(\lambda_1^j\mathbf{I} - \mathbf{A}_{22}\right)^{-1}\mathbf{A}_{21}$.
4. Compute $\mathbf{A}_R^{j+1} = \mathbf{A}_{11} + \mathbf{H}\left(\lambda_1^j\right)$.
5. Perform the eigenanalysis of $\mathbf{A}_R^{j+1}$ and select λ_1^{j+1} from among its eigenvalues.
6. If $\left|\lambda_1^{j+1} - \lambda_1^j\right| \leq \varepsilon$, then stop; else increment j to $j+1$ and go to Step 3.

The initial guess of the eigenvalue λ_1^0 could be obtained, in general, from the eigenanalysis of $\mathbf{A}_{11}$. A very reasonable initial guess of the eigenvalue when computing electromechanical modes of power systems is a purely imaginary value $\lambda_1^0 = j2\pi$, which corresponds to an oscillation at the typical frequency of 1 Hz.

9.5.2.2 Participation Ratio

In Sect. 9.3, we introduced the notion of subsystem participation, which was the sum of the participation factors associated with the state variables in the subsystem. The participation of the relevant subsystem in the ith mode can thus be written as $\mathbf{w}_{\mathbf{r}i}^T\mathbf{v}_{\mathbf{r}i}$, and that of the less relevant subsystem can be written as $\mathbf{w}_{\mathbf{z}i}^T\mathbf{v}_{\mathbf{z}i}$. Their ratio

$$\rho_{\mathbf{r}i} = \frac{\mathbf{w}_{\mathbf{r}i}^T\mathbf{v}_{\mathbf{r}i}}{\mathbf{w}_{\mathbf{z}i}^T\mathbf{v}_{\mathbf{z}i}}$$

will be important in describing the local convergence of the basic SMA algorithm.

9.5.2.3 Convergence Condition of Single-Mode Algorithm

We demonstrate in this section that when only one mode is of interest, the algorithm converges locally if and only if the participation ratio of the relevant variables is greater than one. Furthermore, the local convergence is linear and the convergence ratio is precisely the negative of the participation ratio of the relevant variables.

The convergence ratio of a linearly convergent algorithm is defined as

$$\lim_{\lambda_i^j \to \lambda_i} \frac{\lambda_i^j - \lambda_i}{\lambda_i^{j+1} - \lambda_i} \tag{9.28}$$

The eigenvalue λ_i in the iteration $j+1$ is obtained from the eigenanalysis of the reduced system built using the eigenvalue of the iteration j

$$\lambda_i^{j+1} = \lambda_i\left\{\mathbf{A}_R^{j+1}\left(\lambda_i^j\right)\right\} = \lambda_i\left\{\mathbf{A}_{11} + \mathbf{H}\left(\lambda_i^j\right)\right\}$$

An approximation of the eigenvalue λ_i in the iteration $j+1$ can be provided using the first-order sensitivity

$$\lambda_i^{j+1} \approx \lambda_i + \left.\frac{d\lambda_i\left\{\mathbf{A}_{11}+\mathbf{H}\left(\lambda_i\right)\right\}}{d\lambda}\right|_{\lambda=\lambda_i}\left(\lambda_i^j - \lambda_i\right) \tag{9.29}$$

Rearranging (9.29), we get

$$\lim_{\lambda_i^j \to \lambda_i} \frac{\lambda_i^j - \lambda_i}{\lambda_i^{j+1} - \lambda_i} = \left[\left.\frac{d\lambda_i\left\{\mathbf{A}_{11}+\mathbf{H}\left(\lambda\right)\right\}}{d\lambda}\right|_{\lambda=\lambda_i}\right]^{-1} \tag{9.30}$$

The sensitivity of the eigenvalue λ_i of $\mathbf{A}_{11}+\mathbf{H}\left(\lambda_i^j\right)$ can be determined as

$$\begin{aligned}
\frac{d\lambda_i\left\{\mathbf{A}_{11}+\mathbf{H}\left(\lambda\right)\right\}}{d\lambda} &= \frac{\mathbf{w}_{\mathbf{r}i}^T \left.\frac{d\{\mathbf{A}_{11}+\mathbf{H}(\lambda)\}}{d\lambda}\right|_{\lambda=\lambda_i} \mathbf{v}_{\mathbf{r}i}}{\mathbf{w}_{\mathbf{r}i}^T \mathbf{v}_{\mathbf{r}i}} \\
&= \frac{\mathbf{w}_{\mathbf{r}i}^T \left.\frac{d\mathbf{H}(\lambda)}{d\lambda}\right|_{\lambda=\lambda_i} \mathbf{v}_{\mathbf{r}i}}{\mathbf{w}_{\mathbf{r}i}^T \mathbf{v}_{\mathbf{r}i}} \\
&= -\frac{\mathbf{w}_{\mathbf{r}i}^T \mathbf{A}_{12}\left(\lambda_i \mathbf{I}-\mathbf{A}_{22}\right)^{-1}\left(\lambda_i \mathbf{I}-\mathbf{A}_{22}\right)^{-1}\mathbf{A}_{21}\mathbf{v}_{\mathbf{r}i}}{\mathbf{w}_{\mathbf{r}i}^T \mathbf{v}_{\mathbf{r}i}} \\
&= -\frac{\mathbf{w}_{\mathbf{z}i}^T \mathbf{v}_{\mathbf{z}i}}{\mathbf{w}_{\mathbf{r}i}^T \mathbf{v}_{\mathbf{r}i}} \\
&= -\frac{1}{\rho_{\mathbf{r}i}}
\end{aligned} \tag{9.31}$$

Substituting (9.31) in Eq. (9.30) shows that the convergence ratio is precisely the negative of the participation ratio of the relevant variables

$$\lim_{\lambda_i^j \to \lambda_i} \frac{\lambda_i^j - \lambda_i}{\lambda_i^{j+1} - \lambda_i} = -\rho_{\mathbf{r}i} \tag{9.32}$$

Thus the algorithm converges locally if and only if the magnitude of the participation ratio of the relevant variables is greater than one.

9.5.2.4 Performance of the Single-Mode Algorithm

The single-mode algorithm is illustrated first by computing the electromechanical eigenvalue of the model of a generator connected to an infinite bus described in Sect. 9.3.2. We have chosen the rotor angle δ and the speed deviation ω as relevant

Table 9.21 Performance of the single-mode when computing the electromechanical eigenvalue of a generator connected to an infinite bus

Iteration	λ	Residue	Convergence ratio
1	$-0.0177 + j9.6543$	6.9×10^{-2}	
2	$-0.0870 + j9.6444$	1.5×10^{-3}	46.98
3	$-0.0858 + j9.6453$	3.1×10^{-5}	46.53
4	$-0.0858 + j9.6453$	1.2×10^{-6}	46.54
5	$-0.0858 + j9.6453$	2.6×10^{-8}	46.54

variables. The magnitude of the participation ratio with this choice of relevant variables is 46.54. Table 9.21 details the iterative process. The initial guess of the eigenvalue is $j2\pi$. The convergence ratio as defined in (9.28) is provided, together with the residue of the eigenpair (the eigenvalue and the associated eigenvector) defined as $\| (\mathbf{A} - \lambda_i \mathbf{I})\, \mathbf{v}_i \|$, with the eigenvector normalized such that its length is unity.

It must be noted that the magnitude of the convergence ratio becomes exactly the magnitude of the participation ratio when the eigenvalue has almost converged. The residue of the eigenpair is also provided because it is the most accurate indicator of convergence.

9.5.2.5 Algorithm to Compute Several Modes

For the case where a single mode λ_i was of interest, we obtained our reduced-order model by adding to the core matrix $\mathbf{A}_{11}$ the correction $\mathbf{H}(\lambda_i)$, which reflected the effect of the less relevant dynamics on the mode of interest. If h modes are of interest (say the first h, i.e., $\lambda_1, \ldots, \lambda_h$), with $h \leq n$, the reduced-order system is obtained by adding to the matrix $\mathbf{A}_{11}$ a matrix $\mathbf{M}$ that reflects the effect of the less relevant dynamics on *all* modes of interest. Accordingly, in the jth iteration of the algorithm to compute $\mathbf{M}$, we find a matrix $\mathbf{M}^{j+1}$ that satisfies the following constraints

$$\mathbf{M}^{j+1}\mathbf{v}_{\mathbf{r}_i} = \mathbf{H}\left(\lambda_i^j\right)\mathbf{v}_{\mathbf{r}_i}^j, \quad i = 1, \ldots, h \tag{9.33}$$

where $\mathbf{v}_{\mathbf{r}i}^j$ is the approximation to $\mathbf{v}_{\mathbf{r}i}$ available at the j-iteration. The set of equations (9.33) can also be written in matrix form as

$$\begin{aligned} \mathbf{M}^{j+1}\left[\mathbf{v}_{\mathbf{r}_1}^j \cdots \mathbf{v}_{\mathbf{r}_h}^j\right] &= \left[\mathbf{H}\left(\lambda_1^j\right)\mathbf{v}_{\mathbf{r}_1}^j \cdots \mathbf{H}\left(\lambda_h^j\right)\mathbf{v}_{\mathbf{r}_h}^j\right] \\ \mathbf{M}^{j+1}\mathbf{V}_{\mathbf{r}h}^j &= \mathbf{F}^j \end{aligned} \tag{9.34}$$

When the number of modes of interest h is smaller than the number of relevant variables n, the solution of the linear system (9.34) is not unique. One way to choose among the solutions is to find the solution of minimum norm, which would mean that the perturbation of $\mathbf{A}_{11}$ due to $\mathbf{M}$ is minimum. Letting the superscript $+$ denoting the Moore–Penrose pseudoinverse, this minimum-norm solution is

$$\mathbf{M}^{j+1} = \mathbf{F}^{j}\left(\mathbf{V}_{\mathbf{r}h}^{j}\right)^{+}$$

In case $\mathbf{V}_{\mathbf{r}h}^{j}$ has full column rank, the pseudoinverse of $\mathbf{V}_{\mathbf{r}h}^{j}$ is determined as

$$\left(\mathbf{V}_{\mathbf{r}h}^{j}\right)^{+} = \left(\left(\mathbf{V}_{\mathbf{r}h}^{j}\right)^{T}\mathbf{V}_{\mathbf{r}h}^{j}\right)^{-1}\left(\mathbf{V}_{\mathbf{r}h}^{j}\right)^{T}$$

Another way to choose among the solutions of the linear system (9.34) is to have $\mathbf{M}^{j+1}$ be as close as possible to the set of matrices $\mathbf{H}\left(\lambda_i^j\right)$ for the h modes of interest. In other words, we can carry out the following minimization

$$\min\sum_{i=1}^{h}\beta_i^2\left|\mathbf{M}^{j+1} - \mathbf{H}\left(\lambda_i^j\right)\right|^2 \tag{9.35}$$

The solution to (9.35), subject to the constraints imposed by the linear system (9.34), is

$$\mathbf{M}^{j+1} = \mathbf{F}^{j}\left(\mathbf{V}_{\mathbf{r}h}^{j}\right)^{+} + \overline{\mathbf{H}}\left(\mathbf{I} - \mathbf{V}_{\mathbf{r}h}^{j}\left(\mathbf{V}_{\mathbf{r}h}^{j}\right)^{+}\right) \tag{9.36}$$

where

$$\overline{\mathbf{H}} = \frac{\sum_{i=1}^{h}\beta_i^2\mathbf{H}\left(\lambda_i^j\right)}{\sum_{i=1}^{h}\beta_i^2}$$

The $\mathbf{M}$ matrix can also be computed in terms of the left eigenvectors. The algorithm to compute several modes at the same time can be summarized as follows:

1. Set the initial value of the iteration count to $j = 0$.
2. Provide an initial guess $\boldsymbol{\Lambda}_h^0 = \mathrm{diag}\{\lambda_i^0\}$, $\mathbf{V}_{\mathbf{r}h}^0 = \left[\mathbf{v}_{\mathbf{r}1}^0 \cdots \mathbf{v}_{\mathbf{r}h}^0\right]$.
3. Compute $\mathbf{H}(\lambda_i^j)$ for $i = 1, \ldots, h$.
4. Solve $\mathbf{M}^{j+1}\left[\mathbf{v}_{\mathbf{r}1}^j \cdots \mathbf{v}_{\mathbf{r}h}^j\right] = \left[\mathbf{H}(\lambda_1^j)\mathbf{v}_{\mathbf{r}1}^j \cdots \mathbf{H}(\lambda_h^j)\mathbf{v}_{\mathbf{r}h}^j\right]$.
5. Compute $\mathbf{A}_R^{j+1} = \mathbf{A}_{11} + \mathbf{M}^{j+1}$.
6. Perform the eigenanalysis of $\mathbf{A}_R^{j+1}$, select $\boldsymbol{\Lambda}_h^{j+1}$ and $\mathbf{V}_{\mathbf{r}h}^{j+1}$ from among its eigenvalues and right eigenvectors.
7. If $\left|\lambda_i^{j+1} - \lambda_i^j\right| \leq \epsilon$ for $i = 1, \ldots, h$, then stop; else increment j to $j + 1$ and go to Step 3.

The initial guess of the eigenvalues $\boldsymbol{\Lambda}_h^0$ and the associated eigenvectors $\mathbf{V}_{\mathbf{r}h}^0$ could be obtained, in general, from the eigenanalysis of $\mathbf{A}_{11}$ as $\mathbf{A}_{11}\,\mathbf{v}_{\mathbf{r}i}^0 = \mathbf{v}_{\mathbf{r}i}^0\,\lambda_i^0$. The algorithm can start with a single mode of interest, and when computing the electromechanical modes of power systems, we can choose $\lambda_1^0 = j2\pi$, corresponding to an oscillatory frequency of 1 Hz.

9.5.2.6 Convergence Conditions of Multi-Mode Algorithm

Many practical examples have shown that the algorithm of reduced-order eigenanalysis exhibits good convergence if there exists a distinct dynamic pattern incorporating the modes of interest. In other words, if

$|\rho_\mathbf{r}| \gg 1$ for the n modes associated to the N relevant variables, and
$|\rho_\mathbf{r}| \ll 1$ for the remaining $N - n$ modes,

then good convergence is obtained. More precise conditions, though more difficult to apply, are given in [1].

9.5.2.7 Performance of the Multi-Mode Algorithm

The performance of the multi-mode algorithm is illustrated by computing the electromechanical eigenvalues of the three-area test system discussed in Sect. 9.3.3. We have chosen δ and ω of the six generators as relevant variables. The participation ratios of the relevant variables in all modes, except the zero mode, are shown in Tables 9.22 and 9.23. Six complex eigenvalue pairs ($\lambda_{5,6}$, $\lambda_{7,8}$, $\lambda_{9,10}$, $\lambda_{11,12}$, $\lambda_{15,16}$, and $\lambda_{17,18}$) together with the δ and ω variables of the six generators result in a distinct dynamic pattern. Only the complex pair $\lambda_{17,18}$ will not be of interest because it is a well-damped mode (its damping is 51 %) related to the system frequency dynamics. Table 9.24 details the iterative process of the multi-mode algorithm. The initial guess of the eigenvalue is $j2\pi$. Table 9.25 provides the residue of the eigenpair. The iterative process has been stopped when all residues are smaller than 10^{-4}. The algorithm converges after four iterations.

Table 9.22 Magnitude of the participation ratio of the relevant variables in the complex eigenvalues of the three-area test system

No.	λ	ρ_r
1,2	$-18.2022 \pm j20.4108$	0.0036
3,4	$-19.0475 \pm j15.5643$	0.0150
5,6	$\mathbf{-0.6543 \pm j10.7739}$	**56.7569**
7,8	$\mathbf{-0.6520 \pm j10.7411}$	**54.8053**
9,10	$\mathbf{-0.7402 \pm j10.6533}$	**36.5220**
11,12	$\mathbf{-0.1727 \pm j8.2339}$	**51.9825**
13,14	$-20.5979 \pm j7.8355$	0.0798
15,16	$\mathbf{0.2177 \pm j4.6645}$	**14.6322**
17,18	$\mathbf{-0.4591 \pm j0.7643}$	**9.6154**
19,20	$-1.4267 \pm j0.3828$	0.4504
21,22	$-21.4120 \pm j0.3112$	0.0315
23,24	$-2.1695 \pm j0.1777$	0.1196

Table 9.23 Magnitude of the participation ratio of the relevant variables in the real eigenvalues of the three-area test system

No.	λ	ρ_r	No.	λ	ρ_r
25	−0.1406	0.0159	49	−5.2669	0.2290
26	−0.1420	0.0063	50	−7.7110	0.0182
27	−0.1422	0.0047	51	−8.1064	0.0173
28	−0.1422	0.0046	52	−9.1683	0.0101
29	−0.1422	0.0046	53	−9.7709	0.0252
30	−1.4232	0.0132	54	−9.8218	0.0113
31	−1.4504	0.0129	55	−9.8684	0.0027
32	−1.5118	0.0140	56	−9.8856	0.0006
33	−1.5201	0.0087	57	−9.8931	0.0002
34	−1.5614	0.0154	58	−9.8947	0.0004
35	−1.5909	0.0053	59	−23.8707	0.1251
36	−1.9836	0.0086	60	−29.1400	0.0446
37	−1.9861	0.0072	61	−29.3942	0.0683
38	−1.9868	0.0068	62	−29.4908	0.0660
39	−2.0005	0.0003	63	−32.7496	0.0782
40	−2.7431	0.0853	64	−33.4670	0.0861
41	−2.9391	0.0534	65	−33.8097	0.0834
42	−2.9536	0.0517	66	−93.5548	0.0000
43	−2.9598	0.0509	67	−94.6309	0.0001
44	−3.8689	0.0654	68	−95.7101	0.0001
45	−3.8789	0.0660	69	−97.2405	0.0001
46	−3.9057	0.0678	70	−97.5481	0.0001
47	−4.1588	0.0709	71	−97.6598	0.0001
48	−4.6810	0.0056			

Table 9.24 Performance of the multi-mode algorithm when computing the electromechanical eigenvalues of the three-area test system: eigenvalues

λ	Iteration			
	1	2	3	4
1	$-0.5336 + j10.8572$	$-0.6491 + j10.7752$	$-0.6543 + j10.7739$	$-0.6543 + j10.7739$
2	$-0.5404 + j10.8201$	$-0.6470 + j10.7427$	$-0.6519 + j10.7412$	$-0.6520 + j10.7411$
3	$-0.6585 + j10.7197$	$-0.7358 + j10.6565$	$-0.7401 + j10.6534$	$-0.7402 + j10.6533$
4	$-0.1657 + j8.2329$	$-0.1722 + j8.2342$	$-0.1727 + j8.2339$	$-0.1727 + j8.2339$
5	$0.2283 + j4.6506$	$0.2166 + j4.6719$	$0.2177 + j4.6644$	$0.2177 + j4.6645$

9.5.3 Algorithms with Variable Transformations

The convergence conditions of the standard SMA algorithm show that its successful application depends on a good choice of the relevant variables for the modes of interest. This requires an *a priori* knowledge of the system behavior, which is not

Table 9.25 Performance of the multi-mode algorithm when computing the electromechanical eigenvalues of the three-area test system: residues

λ	Iteration 1	2	3	4
1	1.5×10^{-1}	5.4×10^{-3}	8.5×10^{-5}	1.3×10^{-6}
2	1.2×10^{-1}	5.2×10^{-3}	7.9×10^{-5}	1.2×10^{-6}
3	2.2×10^{-2}	5.4×10^{-3}	1.4×10^{-4}	3.6×10^{-6}
4	7.1×10^{-3}	5.7×10^{-4}	7.8×10^{-6}	2.2×10^{-7}
5	1.7×10^{-2}	7.4×10^{-3}	1.8×10^{-4}	2.4×10^{-5}

always available. There is an alternative to this procedure: it consists of performing a variable transformation prior to each iteration of the SMA algorithm. Starting from some selection of the relevant variables, the proposed transformation is aimed at increasing as much as possible the participation ratio of the relevant variables.

The proposed transformation can adopt one of two dual forms, depending whether it is based on the left or the right eigenvectors. The *left eigenvector transformation* affects only the relevant variables, whereas the *right eigenvector transformation* affects the less relevant ones.

9.5.3.1 The Left Eigenvector Transformation

The left eigenvector transformation is a linear transformation that only affects the relevant variables:

$$\begin{bmatrix} \bar{\mathbf{r}} \\ \bar{\mathbf{z}} \end{bmatrix} = \begin{bmatrix} \mathbf{I} & \mathbf{L}_i \\ \mathbf{0} & \mathbf{I} \end{bmatrix} \begin{bmatrix} \mathbf{r} \\ \mathbf{z} \end{bmatrix} \tag{9.37}$$

The participation ratio of the transformed relevant variables is

$$\rho_{\bar{\mathbf{r}}i} = \frac{\bar{\mathbf{w}}_{\bar{\mathbf{r}}i}^T \bar{\mathbf{v}}_{\bar{\mathbf{r}}i}}{\bar{\mathbf{w}}_{\bar{\mathbf{z}}i}^T \bar{\mathbf{v}}_{\bar{\mathbf{z}}i}} \tag{9.38}$$

If the expressions of the transformed right and left eigenvectors are substituted in (9.38), it becomes

$$\rho_{\bar{\mathbf{r}}i} = \frac{\mathbf{w}_{\mathbf{r}i}^T (\mathbf{v}_{\mathbf{r}i} + \mathbf{L}_i \mathbf{v}_{\mathbf{z}i})}{(-\mathbf{w}_{\mathbf{r}i}^T \mathbf{L}_i + \mathbf{w}_{\mathbf{z}i}^T)\mathbf{v}_{\mathbf{z}i}} \tag{9.39}$$

It should be noted that although the left eigenvector transformation only affects the relevant variables, the participation factors of both the relevant and the less relevant variables are affected.

An infinite participation ratio (9.39) can be achieved if $\mathbf{L}_i$ satisfies

$$-\mathbf{w}_{\mathbf{r}i}^T \mathbf{L}_i + \mathbf{w}_{\mathbf{z}i}^T = 0 \tag{9.40}$$

An expression for $\mathbf{L}_i$ can be obtained from the definition of the left eigenvector of the state matrix written in partitioned form

$$\begin{bmatrix} \mathbf{w}_{\mathbf{r}i}^T & \mathbf{w}_{\mathbf{z}i}^T \end{bmatrix} \begin{bmatrix} \mathbf{A}_{11} & \mathbf{A}_{12} \\ \mathbf{A}_{21} & \mathbf{A}_{22} \end{bmatrix} = \lambda_i \begin{bmatrix} \mathbf{w}_{\mathbf{r}i}^T & \mathbf{w}_{\mathbf{z}i}^T \end{bmatrix} \tag{9.41}$$

The equation corresponding to the less relevant part of the left eigenvector in (9.41) is

$$\mathbf{w}_{\mathbf{r}i}^T \mathbf{A}_{12} + \mathbf{w}_{\mathbf{z}i}^T \mathbf{A}_{22} = \lambda_i \mathbf{w}_{\mathbf{z}i}^T \tag{9.42}$$

This equation can be rewritten as

$$-\mathbf{w}_{\mathbf{r}i}^T \mathbf{A}_{12} \left(\lambda_i \mathbf{I} - \mathbf{A}_{22}\right)^{-1} + \mathbf{w}_{\mathbf{z}i}^T = \mathbf{0} \tag{9.43}$$

Comparing Eqs. (9.40) and (9.43), the matrix $\mathbf{L}_i$ can be identified as

$$\mathbf{L}_i = \mathbf{A}_{12} \left(\lambda_i \mathbf{I} - \mathbf{A}_{22}\right)^{-1} \tag{9.44}$$

When several modes (h) are of interest, a single left eigenvector transformation can be obtained by solving the linear system

$$\mathbf{L}^T \begin{bmatrix} \mathbf{w}_{\mathbf{r}1} & \cdots & \mathbf{w}_{\mathbf{r}h} \end{bmatrix} = \begin{bmatrix} \mathbf{L}_1^T \mathbf{w}_{\mathbf{r}1} & \cdots & \mathbf{L}_h^T \mathbf{w}_{\mathbf{r}h} \end{bmatrix}$$

9.5.3.2 The Right Eigenvector Transformation

The right eigenvector transformation is a linear transformation that only affects the less relevant variables

$$\begin{bmatrix} \bar{\mathbf{r}} \\ \bar{\mathbf{z}} \end{bmatrix} = \begin{bmatrix} \mathbf{I} & \mathbf{0} \\ -\mathbf{K}_i & \mathbf{I} \end{bmatrix} \begin{bmatrix} \mathbf{r} \\ \mathbf{z} \end{bmatrix} \tag{9.45}$$

The participation ratio of the transformed relevant variables is

$$\rho_{\bar{\mathbf{r}}i} = \frac{\overline{\mathbf{w}}_{\bar{\mathbf{r}}i}^T \overline{\mathbf{v}}_{\bar{\mathbf{r}}i}}{\overline{\mathbf{w}}_{\bar{\mathbf{z}}i}^T \overline{\mathbf{v}}_{\bar{\mathbf{z}}i}} \tag{9.46}$$

If the expressions of the transformed right and left eigenvectors are substituted in (9.46), it becomes

$$\rho_{\bar{\mathbf{r}}i} = \frac{\left(\mathbf{w}_{\mathbf{r}i}^T + \mathbf{w}_{\mathbf{z}i}^T \mathbf{K}_i\right) \mathbf{v}_{\mathbf{r}i}}{\mathbf{w}_{\mathbf{z}i}^T \left(-\mathbf{K}_i \mathbf{v}_{\mathbf{r}i} + \mathbf{v}_{\mathbf{z}i}\right)} \tag{9.47}$$

It should be noted that although the right eigenvector transformation only affects the less relevant variables, the participation factors of both the relevant and the less relevant variables are affected.

An infinite participation ratio (9.47) can be achieved if $\mathbf{K}_i$ satisfies

$$-\mathbf{K}_i \mathbf{v}_{\mathbf{r}i} + \mathbf{v}_{\mathbf{z}i} = 0 \tag{9.48}$$

An expression for $\mathbf{K}_i$ can be obtained from the definition of the right eigenvector of the state matrix written in partitioned form

$$\begin{bmatrix} \mathbf{A}_{11} & \mathbf{A}_{12} \\ \mathbf{A}_{21} & \mathbf{A}_{22} \end{bmatrix} \begin{bmatrix} \mathbf{v}_{\mathbf{r}i} \\ \mathbf{v}_{\mathbf{z}i} \end{bmatrix} = \begin{bmatrix} \mathbf{v}_{\mathbf{r}i} \\ \mathbf{v}_{\mathbf{z}i} \end{bmatrix} \lambda_i \tag{9.49}$$

The equation corresponding to the less relevant part of the right eigenvector (9.49) is

$$\mathbf{A}_{21} \mathbf{v}_{\mathbf{r}i} + \mathbf{A}_{22} \mathbf{v}_{\mathbf{z}i} = \mathbf{v}_{\mathbf{z}i} \lambda_i \tag{9.50}$$

This can be rewritten as

$$-\left(\lambda_i \mathbf{I} - \mathbf{A}_{22}\right)^{-1} \mathbf{A}_{21} \mathbf{v}_{\mathbf{r}i} + \mathbf{v}_{\mathbf{z}i} = \mathbf{0} \tag{9.51}$$

Comparing Eqs. (9.48) and (9.51), the matrix $\mathbf{K}_i$ can be identified as

$$\mathbf{K}_i = \left(\lambda_i \mathbf{I} - \mathbf{A}_{22}\right)^{-1} \mathbf{A}_{21}$$

In the case several modes are of interest, a single right eigenvector transformation can also be obtained by solving the linear system

$$\mathbf{K} \left[\mathbf{v}_{\mathbf{r}1} \cdots \mathbf{v}_{\mathbf{r}h} \right] = \left[\mathbf{K}_1 \mathbf{v}_{\mathbf{r}1} \cdots \mathbf{K}_h \mathbf{v}_{\mathbf{r}h} \right]$$

9.5.3.3 Algorithm with Variable Transformation to Compute Several Modes

In the case that several modes are of interest, the algorithm of reduced-order eigenanalysis that results from the application of the variable transformation can be implemented in two different forms.

The first form involves explicitly determining a variable transformation valid for the set of modes of interest. The algorithm of reduced-order eigenanalysis is then applied to the transformed system.

The second alternative avoids the computation of a variable transformation valid for all the modes of interest. Instead, for each mode of interest, the variable transformation is applied first, and the reduced-order matrix is obtained subsequently. The transfer function of the less relevant dynamics is determined from the reduced matrix. Assuming that the left eigenvectors transformation is applied, the second form of the algorithm of reduced-order eigenanalysis is as follows:

1. Set the initial value of the iteration count to $j = 0$.

2. Provide an initial guess $\boldsymbol{\Lambda}_h^0$, $\mathbf{V}_{rh}^0$.
3. For each eigenvalue of interest ($i = 1, \ldots, h$):
   (a) Compute $\mathbf{L}_i^{j+1}$ as $\mathbf{L}_i^{j+1} = \mathbf{A}_{12}\left(\lambda_i^j \mathbf{I} - \mathbf{A}_{22}\right)$.
   (b) Compute $\bar{\mathbf{A}}\left(\lambda_i^j\right)$, which is the result of applying the similarity transformation (9.37) to $\mathbf{A}$.
   (c) Compute $\mathbf{A}_R\left(\lambda_i^j\right) = \bar{\mathbf{A}}_{11} + \bar{\mathbf{A}}_{12}\left(\lambda_i^j \mathbf{I} - \bar{\mathbf{A}}_{22}\right)\bar{\mathbf{A}}_{12}$.
   (d) Compute $\mathbf{H}(\lambda_i^j) = \mathbf{A}_R\left(\lambda_i^j\right) - \mathbf{A}_{11}$.
4. Solve $\mathbf{M}^{j+1}\left[\mathbf{v}_{\mathbf{r}1}^j \ldots \mathbf{v}_{\mathbf{r}h}^j\right] = \left[\mathbf{H}(\lambda_1^j)\mathbf{v}_{\mathbf{r}1}^j \ldots \mathbf{H}(\lambda_h^j)\mathbf{v}_{\mathbf{r}h}^j\right]$.
5. Compute $\mathbf{A}_R^{j+1} = \mathbf{A}_{11} + \mathbf{M}^{j+1}$.
6. Perform the eigenanalysis of $\mathbf{A}_R^{j+1}$, and select $\boldsymbol{\Lambda}_h^{j+1}$ and $\mathbf{V}_{\mathbf{r}h}^{j+1}$ from among its eigenvalues and right eigenvectors.
7. If $\left|\lambda_i^{j+1} - \lambda_i^j\right| \leq \varepsilon$ for $i = 1, \ldots, h$, then stop; else increment j to $j+1$ and go to Step 3.

9.5.3.4 Convergence Conditions of the Algorithm with Variable Transformations

The application of variable transformations only results in infinite participation ratio of the relevant variables if the eigenvalues are exactly known. However, the eigenvalues are only known approximately. The application of a variable transformation prior to each iteration of the algorithm of reduced-order eigenanalysis makes the convergence quadratic instead of linear.

Let us assume that the left eigenvector transformation is applied. Then the participation ratio of the relevant variables (9.39) becomes

$$\rho_{\bar{\mathbf{r}}i} = \frac{1 - \mathbf{w}_{\mathbf{z}i}^T \mathbf{v}_{\mathbf{z}i} + \mathbf{w}_{\mathbf{r}i}^T \mathbf{A}_{12}\left(\lambda_i^j \mathbf{I} - \mathbf{A}_{22}\right)^{-1} \mathbf{v}_{\mathbf{z}i}}{+\mathbf{w}_{\mathbf{z}i}^T \mathbf{v}_{\mathbf{z}i} - \mathbf{w}_{\mathbf{r}i}^T \mathbf{A}_{12}\left(\lambda_i^j \mathbf{I} - \mathbf{A}_{22}\right)^{-1} \mathbf{v}_{\mathbf{z}i}} \tag{9.52}$$

Assuming that $\varepsilon_i^j = \lambda_i^j - \lambda_i$

$$\left(\lambda_i^j \mathbf{I} - \mathbf{A}_{22}\right)^{-1} = \left[(\lambda_i \mathbf{I} - \mathbf{A}_{22}) + \varepsilon_i^j \mathbf{I}\right]^{-1} \tag{9.53}$$

The first-order Taylor expansion of Eq. (9.53) is

$$\left(\lambda_i^j \mathbf{I} - \mathbf{A}_{22}\right)^{-1} \approx (\lambda_i \mathbf{I} - \mathbf{A}_{22})^{-1} - (\lambda_i \mathbf{I} - \mathbf{A}_{22})^{-2}\,\varepsilon_i^j \tag{9.54}$$

Substituting (9.54) in (9.52) shows

$$\rho_{\mathbf{r}i} = \frac{1 - \mathbf{w}_{\mathbf{z}i}^T (\lambda_i \mathbf{I} - \mathbf{A}_{22})^{-1} \mathbf{v}_{\mathbf{z}i}\, \varepsilon_i^j}{\mathbf{w}_{\mathbf{z}i} (\lambda_i \mathbf{I} - \mathbf{A}_{22})^{-1} \mathbf{v}_{\mathbf{z}i}\, \varepsilon_i^j} = \frac{1 - k_c \varepsilon_i^j}{k_c \varepsilon_i^j} \approx \frac{1}{k_c \varepsilon_i^j} \tag{9.55}$$

where

$$k_c = \mathbf{w}_{\mathbf{z}i}^T (\lambda_i \mathbf{I} - \mathbf{A}_{22})^{-1} \mathbf{v}_{\mathbf{z}i}$$

The coefficient k_c is not known *a priori*. However, an upper limit is

$$k_c \leq \left| \mathbf{w}_{\mathbf{z}i}^T \right| \cdot \min \left(\lambda_i \mathbf{I} - \lambda \left\{\mathbf{A}_{22}\right\}\right)^{-1} \cdot |\mathbf{v}_{\mathbf{z}i}|$$

The convergence ratio of the algorithm of reduced-order eigenanalysis (9.32) can also be expressed as

$$\lim_{\lambda_i^j \to \lambda_i} \frac{\lambda_i^j - \lambda_i}{\lambda_i^{j+1} - \lambda_i} = \lim_{\lambda_i^j \to \lambda_i} \frac{\varepsilon_i^j}{\varepsilon_i^{j+1}} = -\rho_{\mathbf{r}i} \tag{9.56}$$

Therefore

$$\varepsilon_i^{j+1} = \frac{-1}{\rho_{\mathbf{r}i}} \varepsilon_i^j = -k_c \left(\varepsilon_i^j\right)^2 \tag{9.57}$$

which means that the convergence ratio is quadratic.

9.5.3.5 Performance of the Algorithm with Variable Transformations

The performance of the single-mode algorithm when computing the electromechanical mode of a generator connected to an infinite bus, taking δ and ω as the relevant variables, is quite satisfactory, as shown in Table 9.21. However, if the stabilizer designed in Sect. 9.4.3 is added to the generator model, the magnitude of the participation ratio of the relevant variables changes significantly: from 46.54 to 8.51. Table 9.26 shows the corresponding convergence process assuming that the initial guess of the eigenvalue is $j2\pi$. The convergence can be improved if δ, ω, and ψ_{fd} (field flux) are taken to be the relevant variables and a variable transformation (either the left or the right eigenvector transformation) is applied prior to each iteration. Table 9.27 shows the convergence process in this more challenging case. As shown in Sect. 9.5.3.4, the convergence becomes quadratic instead of linear.

Table 9.26 Performance of the single-mode algorithm when computing the electromechanical eigenvalue of a generator connected to an infinite bus and equipped with a stabilizer

Iteration	λ	Residue	Convergence ratio
1	$-1.4095 + j9.3515$	1.6×10^{-1}	
2	$-1.5430 + j9.4117$	1.8×10^{-2}	8.68
3	$-1.5584 + j9.4050$	2.1×10^{-3}	8.53
4	$-1.5591 + j9.4032$	2.5×10^{-4}	8.51
5	$-1.5590 + j9.4030$	2.9×10^{-5}	8.51
6	$-1.5589 + j9.4030$	5.1×10^{-6}	8.51
7	$-1.5589 + j9.4030$	6.0×10^{-7}	8.51

Table 9.27 Performance of the single-mode algorithm with variable transformation when computing the electromechanical eigenvalue of a generator connected to an infinite bus and equipped with a stabilizer

Iteration	λ	Residue	Convergence ratio
1	$-1.4269 + j9.3756$	1.3×10^{-1}	
2	$-1.5588 + j9.4031$	2.0×10^{-4}	6.7×10^{2}
3	$-1.5589 + j9.4030$	6.6×10^{-10}	4.5×10^{5}

9.5.4 Composite Formulation

The composite formulation of the SMA algorithms of reduced-order eigenanalysis is an alternative that applies when the system can be expressed as a collection of dynamic subsystems interconnected through a set of static constraints, which is the case with power system models considered in stability studies. The aim of the composite formulation is to facilitate the computation of the transfer function of the less relevant dynamics $\mathbf{H}(\lambda_i^j)$ for each eigenvalue of interest in each iteration λ_i. It should be noted that the computation of $\mathbf{H}(\lambda_i^j)$ requires the computation of the inverse of $(\lambda_i^j\mathbf{I} - \mathbf{A}_{22})$. In the case of large systems whose state matrix does not exhibit any property that can be exploited, the computation of such an inverse can be a significant obstacle. Even constructing $\mathbf{A}_{22}$ itself from any initial system description can already be a challenge.

Let us consider that each dynamic subsystem is described by the following equations:

$$\dot{\mathbf{x}}_\ell = \mathbf{A}_\ell \mathbf{x}_\ell + \mathbf{B}_\ell \mathbf{u}_\ell \tag{9.58}$$

$$\mathbf{y}_\ell = \mathbf{C}_\ell \mathbf{x}_\ell + \mathbf{D}_\ell \mathbf{u}_\ell \tag{9.59}$$

where $\mathbf{x}_\ell$ is the vector of the state variables of the ℓth-subsystem, $\mathbf{u}_\ell$ is the vector of input variables of the ℓth-subsystem, and $\mathbf{y}_\ell$ is the vector of output variables of the ℓth-subsystem.

The static constraints are represented by a linear system of equations that relate the subsystem input $\mathbf{u}$ and output variables $\mathbf{y}$

$$\mathbf{J}\mathbf{u} = \mathbf{y} \tag{9.60}$$

where

$$\mathbf{u} = \left[\mathbf{u}_1^T \ldots \mathbf{u}_L^T\right]^T, \quad \mathbf{y} = \left[\mathbf{y}_1^T \ldots \mathbf{y}_L^T\right]^T \tag{9.61}$$

The original system (9.2) can be obtained from the interconnection of the dynamic subsystems (9.58–9.59) through the static constraints (9.60) as

$$\dot{\mathbf{x}}_\ell = \left[\mathbf{A}_d + \mathbf{B}_d\left(\mathbf{J} - \mathbf{D}_d\right)^{-1}\mathbf{C}_d\right]\mathbf{x} = \mathbf{A}\mathbf{x} \tag{9.62}$$

where $[\cdot]_d = \text{diag}\,\{[\cdot]_\ell\}$.

The first step of the composite formulation of the algorithms of reduced-order eigenanalysis consists in making the selection of relevant and less relevant variables at the subsystem level. Hence, (9.58–9.59) can be rewritten as

$$\begin{bmatrix} \dot{\mathbf{r}}_\ell \\ \dot{\mathbf{z}}_\ell \end{bmatrix} = \begin{bmatrix} \mathbf{A}_{11\ell} & \mathbf{A}_{12\ell} \\ \mathbf{A}_{21\ell} & \mathbf{A}_{22\ell} \end{bmatrix} \begin{bmatrix} \mathbf{r}_\ell \\ \mathbf{z}_\ell \end{bmatrix} + \begin{bmatrix} \mathbf{B}_{1\ell} \\ \mathbf{B}_{2\ell} \end{bmatrix} u_\ell \tag{9.63}$$

$$\mathbf{y}_\ell = \left[\mathbf{C}_{1\ell}\ \mathbf{C}_{2\ell}\right] \begin{bmatrix} \mathbf{r}_\ell \\ \mathbf{z}_\ell \end{bmatrix} + \mathbf{D}_\ell \mathbf{u}_\ell \tag{9.64}$$

The key idea of the composite formulation is to perform the modal reduction of the less relevant dynamics at the subsystem level, substituting $\dot{\mathbf{z}}_\ell$ by $\lambda_i^j \mathbf{z}_\ell$ and eliminating $\mathbf{z}_\ell$ subsequently. Therefore, the equations of each subsystem (9.63–9.64) become

$$\dot{\mathbf{r}}_\ell = \mathbf{A}_{R\ell}(\lambda_i^j)\mathbf{r}_\ell + \mathbf{B}_{R\ell}(\lambda_i^j)\mathbf{u}_\ell \tag{9.65}$$

$$\mathbf{y}_\ell = \mathbf{C}_{R\ell}(\lambda_i^j)\mathbf{r}_\ell + \mathbf{D}_{R\ell}(\lambda_i^j)\mathbf{u}_\ell \tag{9.66}$$

where

$$\mathbf{A}_{R\ell}(\lambda_i^j) = \mathbf{A}_{11\ell} + \mathbf{A}_{12\ell}\left(\lambda_i^j \mathbf{I} - \mathbf{A}_{22\ell}\right)^{-1}\mathbf{A}_{21\ell}$$

$$\mathbf{B}_{R\ell}(\lambda_i^j) = \mathbf{B}_{1\ell} + \mathbf{A}_{12\ell}\left(\lambda_i^j \mathbf{I} - \mathbf{A}_{22\ell}\right)^{-1}\mathbf{B}_{2\ell}$$

$$\mathbf{C}_{R\ell}(\lambda_i^j) = \mathbf{C}_{1\ell} + \mathbf{C}_{2\ell}\left(\lambda_i^j\mathbf{I} - \mathbf{A}_{22\ell}\right)^{-1}\mathbf{A}_{21\ell}$$
$$\mathbf{D}_{R\ell}(\lambda_i^j) = \mathbf{D}_{\ell} + \mathbf{C}_{2\ell}\left(\lambda_i^j\mathbf{I} - \mathbf{A}_{22\ell}\right)^{-1}\mathbf{B}_{2\ell}$$

The interconnection of the reduced-order representations of the dynamic subsystems (9.65–9.66) through the static constraints results in the state matrix of the reduced-order system

$$\dot{\mathbf{r}} = \left[\mathbf{A}_{Rd}(\lambda_i^j) + \mathbf{B}_{Rd}(\lambda_i^j)\left(\mathbf{J} - \mathbf{D}_{Rd}(\lambda_i^j)\right)^{-1}\mathbf{C}_{Rd}(\lambda_i^j)\right]\mathbf{r} = \mathbf{A}_R(\lambda_i^j)\mathbf{r} \qquad (9.67)$$

The transfer function of the less relevant dynamics (9.5.1) can be readily obtained from (9.67) as

$$\mathbf{H}(\lambda_i^j) = \mathbf{A}_R(\lambda_i^j) - \mathbf{A}_{11}$$

This approach to determining the transfer function of the less relevant dynamics has several advantages over the standard formulation:

- Only the portion of the state matrix corresponding to the relevant variables has to be built explicitly.
- In the computation of $\mathbf{H}(\lambda_i^j)$, the inverse of a matrix with a size equal to the number of less relevant variables is substituted by the inverse of a matrix with a size equal to the total number of input or output variables of the dynamic subsystems.

A particular case of composite systems is when the interconnection of the dynamic subsystems is done through a set of sparse static constraints. A set of static constraints can be represented by a graph. For a graph associated with a set of sparse static constraints, a node only has connections with a small number of other nodes. In this setting, the interconnection matrix $\mathbf{J}$ is sparse, i.e., most of the entries are 0. Dynamic subsystems are connected only to a small number of nodes of the graph in power system models arising in stability studies. In this case, Eq. (9.60) becomes

$$\begin{bmatrix} \mathbf{J}_{\alpha\alpha} & \mathbf{J}_{\alpha\beta} \\ \mathbf{J}_{\beta\alpha} & \mathbf{J}_{\beta\beta} \end{bmatrix}\begin{bmatrix} \mathbf{u} \\ \mathbf{u}_{\beta} \end{bmatrix} = \begin{bmatrix} \mathbf{y} \\ \mathbf{0} \end{bmatrix} \qquad (9.68)$$

where α denotes the subset of nodes with dynamic subsystems and β denotes the subset of nodes without dynamic subsystems.

The $\mathbf{J}$ matrix of Eq. (9.60) can be obtained by eliminating $\mathbf{u}_{\beta}$ from Eq. (9.68) to get

$$\left(\mathbf{J}_{\alpha\alpha} - \mathbf{J}_{\alpha\beta}\mathbf{J}_{\beta\beta}^{-1}\mathbf{J}_{\beta\alpha}\right)\mathbf{u} = \mathbf{y} \qquad (9.69)$$

The sparse structure of the $\mathbf{J}$ matrix can be taken advantage of using the implicit formulation of the reduced-order system (9.67)

$$\begin{bmatrix} \dot{\mathbf{r}} \\ \mathbf{0} \\ \mathbf{0} \end{bmatrix} = \begin{bmatrix} \mathbf{A}_{Rd}(\lambda_i^j) & \mathbf{B}_{Rd}(\lambda_i^j) & \mathbf{0} \\ -\mathbf{C}_{Rd}(\lambda_i^j) & \mathbf{J}_{\alpha\alpha} - \mathbf{D}_{Rd}(\lambda_i^j) & \mathbf{J}_{\alpha\beta} \\ \mathbf{0} & \mathbf{J}_{\beta\alpha} & \mathbf{J}_{\beta\beta} \end{bmatrix} \begin{bmatrix} \mathbf{r} \\ \mathbf{u} \\ \mathbf{u}_\beta \end{bmatrix} \tag{9.70}$$

which can be written in compact form as

$$\begin{bmatrix} \dot{\mathbf{r}} \\ \mathbf{0} \end{bmatrix} = \begin{bmatrix} \mathbf{A}_{Rd}(\lambda_i^j) & \mathbf{B}(\lambda_i^j) \\ \mathbf{C}(\lambda_i^j) & \mathbf{D}(\lambda_i^j) \end{bmatrix} \begin{bmatrix} \mathbf{r} \\ \mathbf{u}_\gamma \end{bmatrix} \tag{9.71}$$

where

$$\mathbf{B}(\lambda_i^j) = \begin{bmatrix} \mathbf{B}_{Rd}(\lambda_i^j) & \mathbf{0} \end{bmatrix}, \quad \mathbf{C}(\lambda_i^j) = \begin{bmatrix} -\mathbf{C}_{Rd}(\lambda_i^j) \\ \mathbf{0} \end{bmatrix}$$

$$\mathbf{D}(\lambda_i^j) = \begin{bmatrix} \mathbf{J}_{\alpha\alpha} - \mathbf{D}_{Rd}(\lambda_i^j) & \mathbf{J}_{\alpha\beta} \\ \mathbf{J}_{\beta\alpha} & \mathbf{J}_{\beta\beta} \end{bmatrix}, \quad \mathbf{u}_\gamma = \begin{bmatrix} \mathbf{u} \\ \mathbf{u}_\beta \end{bmatrix}$$

The reduced-order system can be obtained now as

$$\dot{\mathbf{r}} = \left(\mathbf{A}_{Rd}(\lambda_i^j) - \mathbf{B}(\lambda_i^j)\mathbf{D}(\lambda_i^j)^{-1}\mathbf{C}(\lambda_i^j) \right) \mathbf{r} = \mathbf{A}_R(\lambda_i^j)\mathbf{r} \tag{9.72}$$

It should be noted that the inverse of $\mathbf{D}(\lambda_i^j)$ is not computed explicitly. Rather, the LU factorization of this matrix is performed and the solution of the resulting linear system of equations for each column of $\mathbf{C}(\lambda_i^j)$ is achieved using forward-backward substitutions.

The algorithm of reduced-order eigenanalysis for composite systems is as follows:

1. Set the initial value of the iteration count to $j = 0$.
2. Provide an initial guess $\boldsymbol{\Lambda}_h^0$, $\mathbf{V}_{rh}^0$.
3. For each eigenvalue of interest (for $i = 1, \ldots, h$):
   (a) Compute $\mathbf{A}_{R\ell}(\lambda_i^j)$, $\mathbf{B}_{R\ell}(\lambda_i^j)$, $\mathbf{C}_{R\ell}(\lambda_i^j)$, $\mathbf{D}_{R\ell}(\lambda_i^j)$ for the all subsystems $\ell = 1, \ldots, L$.
   (b) Compute $\mathbf{A}_R(\lambda_i^j) = \mathbf{A}_{Rd}(\lambda_i^j) - \mathbf{B}(\lambda_i^j)\mathbf{D}(\lambda_i^j)^{-1}\mathbf{C}(\lambda_i^j)$.
   (c) Compute $\mathbf{H}(\lambda_i^j)$ as $\mathbf{H}(\lambda_i^j) = \mathbf{A}_R(\lambda_i^j) - \mathbf{A}_{11}$.
4. Solve $\mathbf{M}^{j+1}\begin{bmatrix} \mathbf{v}_{\mathbf{r}1}^j \ldots \mathbf{v}_{\mathbf{r}h}^j \end{bmatrix} = \begin{bmatrix} \mathbf{H}(\lambda_1)^j \mathbf{v}_{\mathbf{r}1}^j \ldots \mathbf{H}(\lambda_h)^j \mathbf{v}_{\mathbf{r}h}^j \end{bmatrix}$.
5. Compute $\mathbf{A}_R^{j+1} = \mathbf{A}_{11} + \mathbf{M}^{j+1}$.
6. Perform the eigenanalysis of $\mathbf{A}_R^{j+1}$, and select $\boldsymbol{\Lambda}_h^{j+1}$ and $\mathbf{V}_{\mathbf{r}h}^{j+1}$ from among its eigenvalues and right eigenvectors.
7. If $\left|\lambda_i^{j+1} - \lambda_i^j\right| \le \varepsilon$ for $i = 1, \ldots, h$, then stop; else increment j to $j+1$ and go to Step (3).

9.5.5 Application to the Power System Small-Signal Stability Problem

The composite formulation of SMA detailed in Sect. 9.5.4 shows that when the system can be expressed as a collection of dynamic subsystems interconnected through a set of static constraints, the computational burden of the basic SMA algorithms can be alleviated, thus resulting in very efficient algorithms. Section 9.5.3 has shown that variable transformations can greatly improve the performance of the basic SMA algorithms when the participation ratio of the selected relevant variables becomes attenuated by the presence of controls such as power system stabilizers. However, neither the left eigenvector transformation (9.37) nor the right eigenvector transformation (9.45) exploits the properties of the power system model that arises in small-signal stability studies.

It has been found that if δ, ω, and ψ_{fd} of each generator are the relevant variables, the submatrix $\mathbf{L}_i$ of (9.37) becomes a diagonal matrix, which means that the left eigenvector transformation can be applied at generator level

$$\begin{bmatrix} \bar{\mathbf{r}} \\ \bar{\mathbf{z}} \end{bmatrix} = \begin{bmatrix} \mathbf{I} & \mathbf{L}_{di} \\ \mathbf{0} & \mathbf{I} \end{bmatrix} \begin{bmatrix} \mathbf{r} \\ \mathbf{z} \end{bmatrix}$$

where

$$\mathbf{L}_{di} = \mathrm{diag}\left\{\mathbf{L}_{\ell i}\right\}, \quad \mathbf{L}_{\ell i} = \mathbf{A}_{12\ell}\left(\lambda_i \mathbf{I} - \mathbf{A}_{22\ell}\right)^{-1}$$

The SMA algorithm with variable transformation exhibits quadratic convergence in the case that only one mode is of interest. We explore now the convergence properties when computing several modes at the same time. Figure 9.19 compares the performance of the SMA algorithm using the left eigenvector transformation with the basic SMA algorithm in the case of the three-area test system discussed in Sect. 9.3.3, incorporating the stabilizers designed in Sect. 9.4.5. The variation of the eigenpair residue (corresponding to the electromechanical eigenvalues) with the iteration number is displayed. Assuming a common eigenpair residue threshold (10^{-4}) to reach convergence, the SMA algorithm with the left eigenvector variable transformation exhibits better performance than the basic SMA algorithm.

The application of the SMA algorithm with the left eigenvector transformation is limited to medium size systems (300 generators) because the SMA algorithm involves the eigenanalysis of a matrix whose size is three times the number of generators.

Small-signal stability analysis of very large power systems concerns only a small fraction of the electromechanical modes: the inter-area modes. Inter-area modes correspond to low-frequency oscillations. Inter-area modes are poorly damped, in contrast to local modes because most controls (power system stabilizers) are aimed at damping local oscillations.

A divide-and-conquer approach has been designed to overcome the size limitations of the SMA algorithm with the left eigenvector transformation. Instead of computing all electromechanical modes at the same time assuming a *single-area*

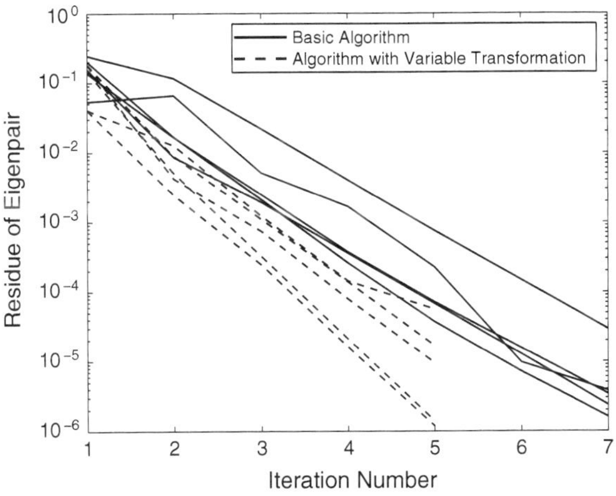

Fig. 9.19 Comparison of the performance of the basic algorithm and the algorithm left eigenvector transformation in the computation of electromechanical eigenvalues of the three-area test system

problem, a *multi-area* approach assumes the existence of areas of coherent generators and calculates separately the inter-area and local (intra-area) modes of each area. Inter-area modes correspond to the coherent oscillations among groups of generators whereas the intra-area modes correspond to the local oscillations within an area.

The determination of inter-area modes is facilitated by the application of a right eigenvector transformation that exploits the coherent nature of the inter-area oscillations and allows one to represent each area by just one generating unit.

A *grouping matrix* that associates each generator to one area ($\mathbf{K}_g(i, j) = 1$ if the ith generator belongs to the area whose reference generator is the jth one; otherwise, $\mathbf{K}_g(i, j) = 0$) is used as the matrix of a right eigenvector transformation (9.45) for a set of modes that can be applied at the area level

$$-\mathbf{K}_g\mathbf{V_r} + \mathbf{V_z} = \mathbf{0}$$

The grouping matrix can be determined using either the *slow coherency* method [36, 37] or the *synchrony* approach [38, 39].

Figure 9.20 shows the performance of the algorithm to compute the two inter-area modes of the three-area test system discussed earlier. Three areas are assumed. The generators are grouped according to: (1,2), (3,4), and (5,6). The reference generators are generator numbers 1, 3, and 5, respectively. The variation of the eigenpair residue (corresponding to the inter-area electromechanical eigenvalues) with the iteration number is displayed. Although the algorithm requires more iterations than the single-area approach, it proves the feasibility of exploiting through a right eigenvector transformation the fact that inter-area modes correspond to coherent areas.

The procedure for the computation of the local (intra-area) modes assumes the local character of these oscillations (i.e., only the generators of the area of interest

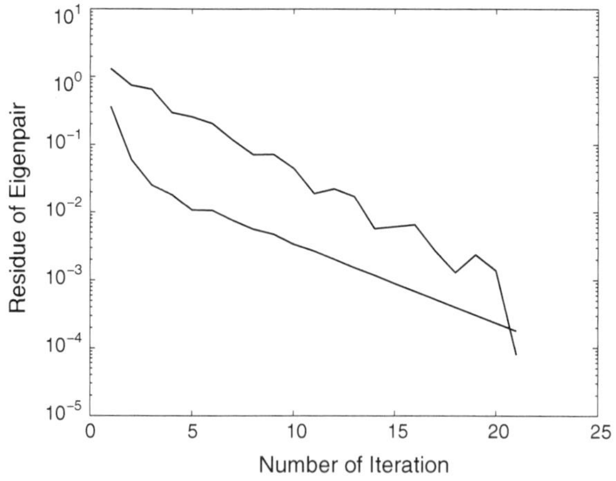

Fig. 9.20 Performance of the algorithm to compute the two inter-area modes of the three-area test system

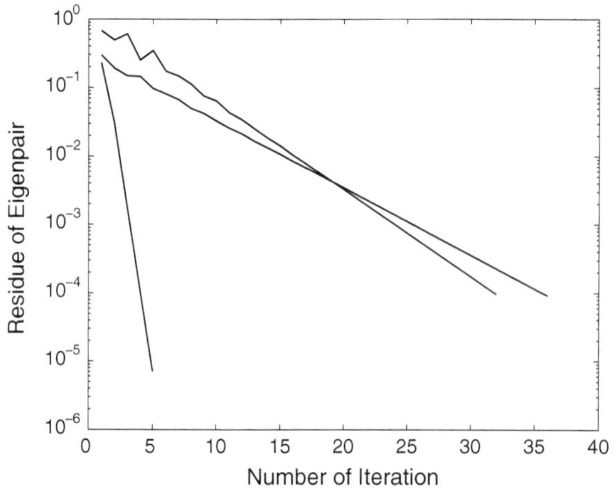

Fig. 9.21 Performance of the algorithm to compute the three local modes of the three-area test system

are involved) which in SMA parlance means that the dynamics of the remaining generators are treated as less relevant.

Figure 9.21 shows the performance of the algorithm to compute the three local modes in the three-area test system discussed in this section. The variation of the eigenpair residue (corresponding to the local electromechanical eigenvalues) with the iteration number is displayed. The number of iterations required to converge greatly depends on the eigenvalue. Nevertheless, the feasibility of the divide-and-conquer strategy to compute electromechanical modes in large power systems is illustrated.

Acknowledgments Several other individuals have also contributed to the development of the SMA approach in different ways. We are grateful for the contributions of Drs. José Luis Sancha, Ángel Saíz-Chicharro, Damián Laloux, Ganesh Ramaswamy (deceased), Julián Barquín, Hermann Vargas, and Bogdan Marinescu. We have also been fortunate to have the support of our own institutions and a number of companies: Iberduero, Iberdrola, Red Eléctrica de España, Electricité de France, Réseau de Transport d'Electricité de France, EPRI, Ontario Hydro, and CAMMESA.

Appendix: Power System Data

Single-Machine Infinite Bus System

Table 9.28 provides the synchronous machine operating point of the single-machine infinite bus system. Tables 9.29–9.32 contain the synchronous machine, excitation system, governor-turbine, and transformer-transmission line data. Figures 9.22 and 9.23 depict the excitation system and the governor-turbine models.

Table 9.28 Synchronous machine operating point

P(pu)	Q(pu)	V(pu)
0.9	0.4359	1.0

Table 9.29 Synchronous machine data

H(s)	T'_{d0}(s)	T''_{d0}(s)	T'_{q0}(s)	T''_{q0}(s)	X_d(pu)	X_q(pu)	X'_d(pu)	X''_d(pu)	X'_q(pu)	X''_q(pu)	R_a(pu)
3.558	8.0	0.03	0.4	0.05	1.8	1.7	0.3	0.55	0.25	0.25	0.003

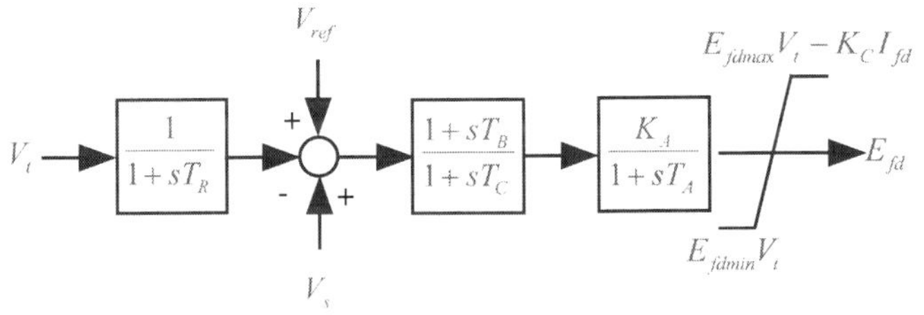

Fig. 9.22 Excitation system model

Table 9.33 provides the transformer, transmission line, and series capacitor data of the single-machine connected to an infinite bus via a series-compensated transmission line. It should be noted that the total reactance of the transmission line and the series capacitor is equal to the line reactance of Table 9.32. Tables 9.34, 9.35 contain the inertia and stiffness constants of the rotor masses. It is also worth noting that the total inertia of the masses of Table 9.34 is identical to the rotor inertia provided in Table 9.29. Figure 9.24 shows the rotor multi-mass model.

Table 9.30 Excitation data

K_A(pu)	T_A(s)	T_B(s)	T_C(s)	T_R(s)
200	0	0	0	0.01

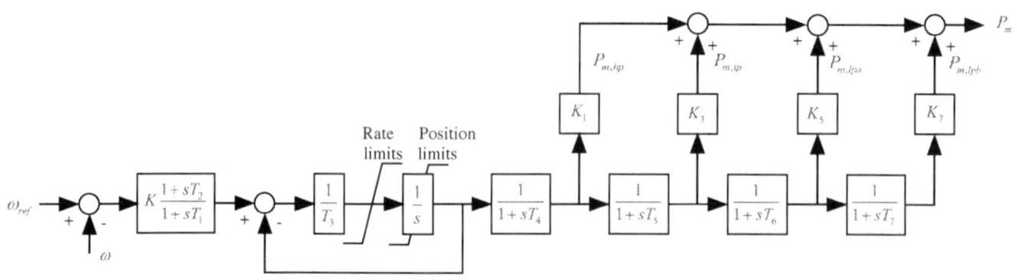

Fig. 9.23 Governor-turbine model

Table 9.31 Governor-turbine data

K(pu)	T_1(s)	T_2(s)	T_3(s)	T_4(s)	T_5(s)	T_6(s)	T_7(s)	K_1(pu)	K_3(pu)	K_5(pu)	K_7(pu)
20	0.1	0	0.3	0.3	7	0.5	0	0.3	0.3	0.2	0.2

Table 9.32 Transformer and transmission line data in per unit on synchronous machine base

X_t	X_ℓ
0.15	0.1

Fig. 9.24 Rotor multi-mass model

Table 9.33 Transformer, transmission line, and series capacitor data in pu on synchronous machine base

X_t	X_ℓ	X_C
0.15	0.2	−0.1

Table 9.34 Inertia of rotor masses in seconds on machine base

H_{hp}	H_{ip}	H_{lpa}	H_{lpb}	H_g
0.124	0.232	1.155	1.192	0.855

Table 9.35 Stiffness constants of rotor masses in pu

K_{hpip}	K_{iplpa}	K_{lpalpb}	K_{lpbg}
21.8	48.4	75.6	62.3

9.5.6 Three-Area System

Tables 9.36–9.41 contain the generator, load, transmission line, transformer, series capacitor, and shunt capacitor data of the three-area system.

Table 9.36 Generator data

Generator	MVA	P_G(MW)	V(pu)
1	375	300	1.05
2	125	100	1.03
3 (swing bus)	750	658.43	1.05
4	250	200	1.03
5	562.5	450	1.05
6	187.5	150	1.03

Table 9.37 Load data

Bus	P_L(MW)	Q_L(Mvar)
9	800	160
11	500	100
16	200	40
19	300	60

Table 9.38 Line data (230 kV, 100 MVA, $r = 0.0001$ pu/km, $x = 0.001$ pu/km, $b = 0.00175$ pu/km)

From Bus	To Bus	Length (km)
7	8	10
8	9	25
9	10	110
10	11	110
11	12	25
12	13	10
14	15	10
15	16	25
16	17	110
18	19	110
16	20	55
20	19	55

Table 9.39 Transformer data ($x = 0.15$ pu)

From Bus	To Bus	MVA
1	7	375
2	8	125
3	13	750
4	12	250
5	14	562.5
6	5	187.5

Table 9.40 Series capacitor (100 MVA)

From Bus	To Bus	x(pu)
17	18	−0.11

Table 9.41 Shunt capacitor

Bus	Mvar
10	200
20	100

9.5.7 Wind Generator Model

Tables 9.42 and 9.43 contain the induction machine and DC link and network converter filter data. Table 9.44 details the wind generator operating point.

Table 9.42 Induction machine data

H(s)	R_s(pu)	X_s(pu)	X_m(pu)	R_r(pu)	X_r(pu)
3	0.01	0.15	5	0.01	0.15

Table 9.43 DC link and network converter filter data

C(s)	R_a(pu)	L_a(pu)
0.05	0.06	0.6

Table 9.44 Wind generator operating point

v_{sd}(pu)	v_{sq}(pu)	i_{sd}(pu)	i_{sq}(pu)	$i_{rd}{}^{\psi_s}$(pu)	$i_{rd}{}^{\psi_s}$(pu)
1	0	−1	0	0	1

References

1. I.J. Pérez-Arriaga, selective modal analysis with applications to electric power systems. Ph.D. Thesis, Department of Electrical Engineering and Computer Science, Massachusetts Institute of Technology, June 1981
2. I.J. Pérez-Arriaga, G.C. Verghese, F.C. Schweppe, Selective modal analysis with applications to electric power systems. Part I: Heuristic introduction. Part II: The dynamic stability problem. IEEE Trans. Power Apparatus Syst. **PAS–101**(9), 3117–3134 (1982)
3. G. Cedersund, L. Rouco, G.C. Verghese, Structured reduction of dynamic biochemical models. In Proceedings of 10th International Conference on Systems Biology, Poster 2.091, Stanford, California, USA, 2009
4. P. Kundur, *Power System Stability and Control* (McGraw Hill, New York, 1994)
5. G. Rogers, *Power System Oscillations* (Kluwer Academic Publishers, New York, 2000)
6. A.M.A. Hamdan, Coupling measures between modes and state variables in power system dynamics. Int. J. Control **43**(3), 1029–1041 (1986)
7. W.A. Hashlamoun, M.A. Hassouneh, E.H. Abed, New results on modal participation factors: Revealing a previously unknown Dichotomy. IEEE Trans. Autom. Control **54**(7), 1439–1449 (2009)

8. A. Saíz-Chicharro, I.J. Pérez-Arriaga, Selective modal analysis of subsynchronous resonance. In Proceedings of the 9th Power Systems Computation Conference, Casçais, Portugal, Sept 1987, pp. 902–906
9. L. Rouco, J.L. Zamora, Dynamic patterns and model order reduction in small-signal models of doubly fed induction generators for wind power applications. In Proceedings of IEEE Power Engineering Society General Meeting, paper no. 06GM1126, Montreal, Canada, June 2006
10. J. Tabernero, L. Rouco, Dynamic patterns in small-signal models of multipole synchronous generators for wind power applications. In Proccedings of IEEE Power Tech, 2007, Lausanne, Switzerland, July, 2007
11. F.L. Pagola, I.J. Pérez-Arriaga, G.C. Verghese, On sensitivities, residues, and participations: Applications to oscillatory analysis and control. IEEE Trans. Power Syst. **PWRS–4**(1), 278–285 (1989)
12. F.L. Pagola, I.J. Pérez-Arriaga, Design of power system stabilizers using eigenvalue sensitivities and SMA techniques. In Proceedings of the 8th Power Systems Computation Conference, Helsinki, Finland, August 1984, pp. 997–1001
13. F.L. Pagola, L. Rouco, I.J. Pérez-Arriaga, Analysis and control of small signal stability in electric power systems by selective modal analysis. Eigenanalysis and Frequency Domain Methods for System Dynamic Performance, IEEE, Publication 90TH0292-3-PWR, 1990, pp. 77–96
14. L. Rouco, F.L. Pagola, A.García-Cerrada, J.M. Rodríguez, R.M. Sanz, Damping electromechanical oscillations in power systems with superconducting magnetic energy storage systems: Location and controller design. In Proceedings of the 12th Power Systems Computations Conference, Dresden, Germany, August 1996, pp. 1097–1104
15. L. Rouco, F.L. Pagola, An eigenvalue sensitivity approach to location and controller design controllable series capacitors for damping power systems electromechanical oscillations. IEEE Trans. Power Syst. **PWRS–12**(4), 1660–1666 (1997)
16. L. Rouco, Coordinated design of multiple controllers for damping power system oscillations. Int. J. Electr. Power Energy Syst. **23**(7), 517–530 (2001)
17. L. Sigrist, L. Rouco, Design of Damping Controllers for Doubly Fed Induction Generators Using Eigenvalue Sensitivities. Proceedings of IEEE Power Systems Conference and Exposition Meeting, Seattle, Washington, USA, March, 2009
18. J.L. Sancha, I.J. Pérez-Arriaga, Selective modal analysis of power system oscillatory instability. IEEE Trans. Power Syst. **PWRS–3**(2), 429–438 (1988)
19. L. Rouco, I.J. Pérez-Arriaga, Multi-area analysis of small signal stability in large electric power systems by SMA. IEEE Trans. Power Syst. **PWRS–8**(3), 1257–1265 (1993)
20. B. Marinescu, L. Rouco, A Unified Framwork for Nonlinear Dynamic Simulatiuon and Modal Analysis. In Proceedings of the 15th Power Systems Computation Conference, Paper No. 175, Liège, Belgium, August 2005
21. J.H. Wilkinson, *The Algebraic Eigenvalue Problem* (Oxford University Press, Oxford, 1988)
22. G.H. Golub, C.F. Van Loan, *Matrix Computations* (John Hopkins University Press, Baltimore, 1988)
23. F.P. de Mello, P.J. Nolan, T.F. Laskowski, J.M. Undrill, Coordinated application of stabilizers in multimachine power systems. IEEE Trans. Power Apparatus Syst. **PAS–99**(3), 892–901 (1980)
24. B.E. Eliasson, D.J. Hill, Damping structure and sensitivity in the NORDEL power system. IEEE Trans. Power Syst. **7**(1), 97–105 (1992)
25. IEEE Working Group on Prime Mover and Energy Supply Models for System Dynamic Performance Studies, Dynamic models for Fossil Fueled Steam Units in power system studies. IEEE Trans. Power Syst. **6**(2), 753–761 (1991)
26. IEEE Recommended Practice for Excitation System Models for Power System Stability Studies, IEEE Std 421.5-2005, 2005
27. A. Elices, L. Rouco, H. Bourlès, T. Margotin, Design of robust controllers for damping interarea oscillations: Application to the European power system. IEEE Trans. Power Syst. **19**(2), 1058–1067 (2004)

28. P.M. Anderson, B.L. Agrawal, J.E. Van Ness, *Subsynchronous Resonance in Power Systems* (IEEE Press, New York, 1989)
29. R. Pena, J.C. Clare, G.M. Asher, Doubly fed induction generator using back-to-back PWM converters and its application to variable-speed Wind-energy generation. IEE Proc. Electr. Power Appl. **143**(3), 231–241 (1996)
30. B.C. Lesieutre, A.M. Stanković, J.R. Lacalle-Melero, A study of state variable participation in nonlinear limit-cycle behavior. In Proceedings of the 4th IEEE Conference on Control Applications, Albany, New York, USA, September 1995, pp. 79–84
31. N. Martins, L.T.G. Lima, Eigenvalue and frequency domain analysis of small-signal electromechanical stability problems. Eigenanalysis and Frequency Domain Methods for System Dynamic Performance, IEEE, Publication 90TH0292-3-PWR, 1990, pp. 17–33
32. N. Martins, L.T.G. Lima, Determinations of suitable locations for power system stabilizers and static VAR compensators for damping electromechanical oscillations in large power systems. IEEE Trans. Power Syst. **PWRS–5**(4), 1455–1469 (1990)
33. N. Martins, H.J.C.P. Pinto, A. Bianco, N.J.P. Macedo, TCSC Control Structures for Line Power Scheduling and Methods to Determine their Location and Tuning to Damp Power System Oscillations, Proceedings of IV SEPOPE, Paper SP-36, Foz do Iguaçu, Brazil, May, 1994
34. J.E. Van Ness, F.P. Imad, J.M. Boyle, Sensitivities of large, multi-loop control systems. IEEE Trans. Autom. Control **AC–10**(3), 308–315 (1965)
35. G.F. Franklin, J.D. Powell, A. Emami-Naeini, *Feedback Control of Dynamic Systems*, 4th edn. (Prentice Hall, New York, 2002)
36. J.R. Wilkenman, J.H. Chow, B.C. Bowler, B. Avramović, P.V. Kokotović, An analysis of interarea dynamics of multi-machine systems. IEEE Trans. Power Apparatus Syst. **PAS–100**(2), 754–763 (1981)
37. J.H. Chow, *Time-Scale Modeling of of Dynamic Networks with Applications to Power Systems, Lecture Notes in Control and Information Sciences*, vol. 46 (Springer-Verlag, New York, 1982)
38. G.N. Ramaswamy, G.C. Verghese, L. Rouco, C. Vialas, C.L. DeMarco, Synchrony, aggregation, and multi-area eigenanalysis. IEEE Trans. Power Syst. **PWRS–10**(4), 1986–1993 (1995)
39. G.N. Ramaswamy, L. Rouco, O. Fillatre, G.C. Verghese, P. Panciatici, B.C. Leisutre, D. Peltier, Synchronic modal equivalencing (SME) for structure-preserving dynamic equivalents. IEEE Trans. Power Syst. **PWRS–11**(1), 19–29 (1996)

Chapter 10
Interarea Mode Analysis for Large Power Systems Using Synchrophasor Data

Luigi Vanfretti, Yuwa Chompoobutrgool and Joe H. Chow

Abstract Interarea oscillations are predominantly governed by the slower electromechanical modes which, in turn, are determined by the coherent machine rotor angles and speeds. The issue is that, although these rotor angles and speeds provide the best visibility of such modes, currently they are not available from phasor measurement units (PMU). As such, the aim of this chapter is to demonstrate that interarea oscillations are observable in the network variables, such as voltages and line currents, which are measured by PMU. By analyzing the electromechanical modes in the network variables, we can trace *how* electromechanical oscillations are spread through the power network following a disturbance. Applying eigenvalue and sensitivity analysis, we provide an analytical framework to understand the nature of these network oscillations through a relationship termed *network modeshapes*. Using this relationship, a novel concept, "dominant interarea oscillation paths," is developed to identify the passageways *where* the interarea modes of concern travel the most. We demonstrate the concept of the dominant path with an equivalent two-area system. We propose an algorithm for identification of the dominant paths and illustrate with a reduced model of a large-scale network. Finally, we end this chapter with an important application of the concept: feedback input signal selection for damping controller design.

L. Vanfretti (✉) · Y. Chompoobutrgool
KTH Royal Institute of Technology, Stockholm, Sweden
e-mail: luigiv@kth.se

J. H. Chow
Rensselaer Polytechnic Institute, Troy, New York, USA
e-mail: chowj@rpi.edu

J. H. Chow (ed.), *Power System Coherency and Model Reduction*,
Power Electronics and Power Systems 94, DOI: 10.1007/978-1-4614-1803-0_10,

10.1 Introduction

Electromechanical modes are predominantly determined by the machine rotor angles and speeds [1–5], and as a result, they provide the best visibility of such modes. Electromechanical modes are also observable in the network variables, such as voltages and line current phasors, which are measured by Phasor Measurement Units (PMUs) [6] and observed through simulations [7]. In this study, by computing the electromechanical modes in the network variables, we can trace *how* electromechanical oscillations are spread through the power network following a disturbance. In particular, interarea modes can travel throughout a large network. These network variables have received less attention because they are algebraic variables which can change abruptly. Recently, an analytical framework to study interarea oscillations present in the bus voltage and frequency variables has been proposed [8]. This technique uses the linearized model of a power system with bus voltages and frequencies as output variables.

In the first part of this chapter, we aim to generalize the results in [8] to analyze power system oscillations which may be present in any of the network variables. To motivate, we first analyze the oscillations observed during a major system disturbance that took place on 2/26/2008 in the Florida Reliability Coordination Council (FRCC) service area [9, 10]. The modal components from distributed PMU measurements have a common feature: they do not peak at the same time instants. Therefore, there exists a time difference (or delay) between the oscillations at different points of the network. This time difference is effectively a phase shift between the network oscillations.

With the goal of explaining the origin of the phase shifts in the network variables, we use the multi-machine linearized power system electromechanical model [11] and extend the mode shapes to the network variables. We show that with no damping and constant impedance loads, all electromechanical oscillations are in phase. However, when damping or control equipment such as voltage regulator models are included, the eigenvector matrix will indicate phase shifts. The time delays related to these phase shifts show a strong resemblance to those observed in PMU data.

To obtain the mode shapes, we perform a detailed sensitivity analysis of the network variables and provide analytical expressions [12]. This analysis provides a theoretical understanding of oscillations as directly measured by PMUs on high-voltage transmission systems. Finally, network sensitivities and eigenvectors are used to obtain the modal components in the network variables, thus providing a rationale explaining the phase shifts observed in the modal components of the synchrophasor data.

In the second part of this chapter, we aim to present two different applications of the concept of dominant interarea oscillation paths: an algorithm for tracing the spread of interarea modes (i.e., identification of dominant paths) and an application for feedback control input signals selection. We demonstrate the algorithm with a reduced Nordic grid model while implementing a fundamental study on damping control design using different input signals, PMU's and non-PMU's, with a conceptualized two-area system. The results are useful for determining not only proper feedback input signals but also PMU siting.

The remainder of this chapter is organized as follows: In Sect. 10.2, we analyze the network oscillations originating from the 2/26/2008 FRCC disturbance. In Sect. 10.3, we determine the origin of the phase shifts in the electromechanical mode shape. In Sect. 10.4, we perform sensitivity analysis of the network variables and provide closed-form expressions of the network sensitivities. Afterward, in Sect. 10.5 we map the network sensitivities onto the electromechanical mode shapes and discuss the nature of the modal components in the network variables, illustrating this with a four-generator, two-area system [12, 13]. Applications for interarea mode tracing and feedback control input signal selection are demonstrated in Sect. 10.6. Conclusions are given in Sect. 10.7.

10.2 Nature of Network Oscillations Observed from Phasor Measurement Data

We study the power system interarea oscillations excited by the FRCC system disturbance by analyzing archived phasor measurement data. The PMUs considered in this analysis are: Manitoba, near the city of Winnipeg, Canada; Maine near Bangor; Florida near Jacksonville; West Tennessee (W. Tenn.) near Memphis; and East Tennessee (E. Tenn.) near Knoxville, as shown in Fig. 10.1a. In Fig. 10.1b we show the bus frequency measured during the disturbance by the PMUs, showing the wide-area impact of the disturbance, the electromechanical swing that propagated from Florida, to E. Tenn. and W. Tenn., and subsequently to Manitoba, and finally to Maine.

We aim to analyze the oscillatory components and characteristics contained in these measurements. We use the Eigensystem Realization Algorithm (ERA) [14] to

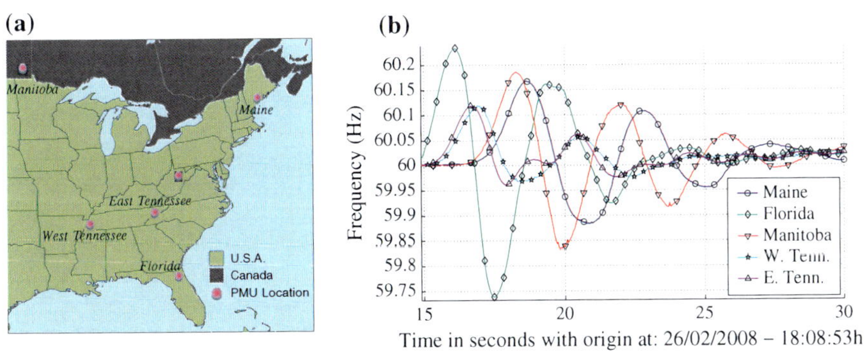

Fig. 10.1 PMU locations and bus frequency traces obtained during the 2008 Florida disturbance. **a** PMU locations. **b** Bus frequency traces

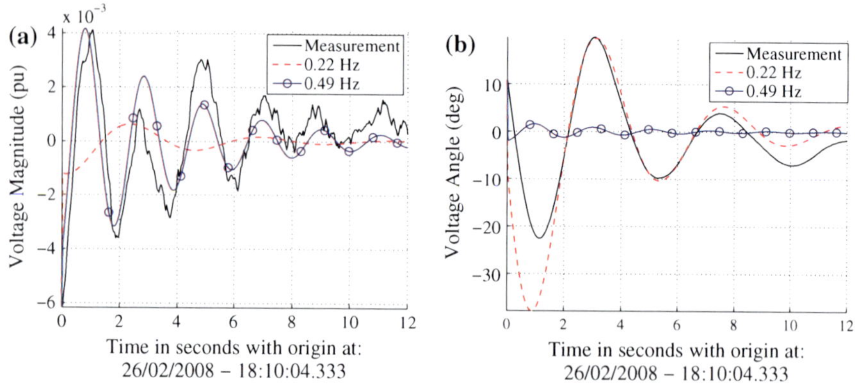

Fig. 10.2 Identified oscillatory components in the voltage phasor at W. Tenn. **a** Voltage magnitude. **b** Voltage angle

identify the individual modal components in the voltage magnitude, voltage angle, and active power flow measurements available from each PMU [9]. The identified low-frequency interarea modes include frequencies of 0.22 and 0.49 Hz. In Fig. 10.2 the voltage phasor measurements of the PMU at W. Tenn. are shown along with their ERA approximation for the 0.22 and 0.49 Hz components of the signals.

All identified components for the 0.22 Hz mode in the voltage angle are shown in Fig. 10.3a, and those for the 0.49 Hz mode are shown in Fig. 10.3b. The starting time $t = 0$ s corresponds to 18:10:04.333 h. From the voltage angles of the 0.22 Hz mode, we note that Florida oscillates against Maine and Manitoba: that is, it is a North versus South mode.

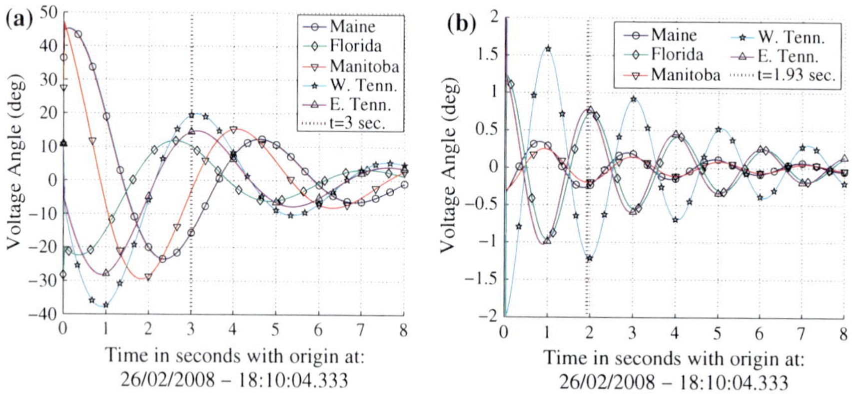

Fig. 10.3 Voltage angle interarea oscillations and projections. **a** $\Delta\theta_i$ 0.22 Hz components. **b** $\Delta\theta_i$ 0.49 Hz components

The 0.49 Hz mode is more difficult to analyze from this limited data set. However, it can be noted that the voltage angle at W. Tenn., E. Tenn., and Florida have the largest oscillations while Manitoba and Maine have a less significant contribution. More important, E. Tenn. and W. Tenn. are in anti-phase suggesting that the pivot node of the oscillation is located somewhere between them.

The most important observation that could be made about the oscillations discussed above is the following: *for all of the network variables, the individual modal components do not peak at the same instants*. There is in fact a time shift (or delay) between the modal components. This time delay can be viewed as a phase shift between the modal components in the frequency domain. In the next section, we will investigate this phase shift by analyzing the mode shapes of a test network.

10.3 The Electromechanical Mode Shape

10.3.1 Multi-Machine Systems Electromechanical Model

In this section, we investigate the origin of electromechanical mode phase shifts by performing eigenanalysis on different linearized models of multi-machine power systems. We start with the electromechanical model [1]. For an N-machine power system, the linearized electromechanical model in state-space form is given by

$$\underbrace{\begin{bmatrix} \Delta\dot{\delta} \\ \Delta\dot{\omega} \end{bmatrix}}_{\dot{x}} = \underbrace{\begin{bmatrix} 0 & \Omega I_{(N\times N)} \\ M^{-1}K & M^{-1}D \end{bmatrix}}_{\bar{A}} \underbrace{\begin{bmatrix} \Delta\delta \\ \Delta\omega \end{bmatrix}}_{x} \tag{10.1}$$

where

$$\begin{aligned} \Delta\delta &= [\Delta\delta_1 \ \cdots \ \Delta\delta_N]^T, \ (N\times 1) \\ \Delta\omega &= [\Delta\omega_1 \ \cdots \ \Delta\omega_N]^T, (N\times 1) \\ M^{-1} &= \operatorname{diag}\left(\tfrac{1}{2H_i}\right), \ (N\times N) \\ D &= \operatorname{diag}(D_i), \ (N\times N); \quad K = \left[K_{ij}\right], (N\times N) \end{aligned} \tag{10.2}$$

where $i, j = 1, \ldots, N$. We will refer to this model (10.1) as the electromechanical model with damping, and to matrix $\bar{A}$ as the state matrix with damping. If the damping terms are neglected ($D_i \approx 0$), the model becomes

$$\underbrace{\begin{bmatrix} \Delta\dot{\delta} \\ \Delta\dot{\omega} \end{bmatrix}}_{\dot{x}} = \underbrace{\begin{bmatrix} 0 & \Omega I_{(N\times N)} \\ A_\omega & 0 \end{bmatrix}}_{A} \underbrace{\begin{bmatrix} \Delta\delta \\ \Delta\omega \end{bmatrix}}_{x} \tag{10.3}$$
$$A_\omega = M^{-1}K$$

We will refer to A as the state matrix without damping. Many properties of $\bar{A}$ and A have been studied [11, 15]. In this investigation, some characteristics of the eigenvectors of $\bar{A}$ and A are analyzed. These characteristics have strong effects on the phase shift of network variables as discussed later.

10.3.2 Eigenvectors of the Electromechanical Models

Consider the two-area, four-machine power system in Fig. 10.4 [12, 13] modeled using (10.3). In the resulting state matrix A, all elements corresponding to the machine damping $M^{-1}D$ are zero. Eigenanalysis is performed on A resulting in the eigenvector matrix

$$W(A) = \begin{bmatrix} 0.5 & -0.5930 & 0.7871 & -0.0547 \\ 0.5 & -0.5997 & -0.6167 & -0.0512 \\ 0.5 & -0.4539 & 0.0155 & 0.9972 \\ 0.5 & 0.2876 & -0.0015 & -0.0041 \end{bmatrix} \tag{10.4}$$

Note that only the components corresponding to the machine angles are shown. All the components of $W(A)$ are real. Column 2, the interarea mode mode shape,

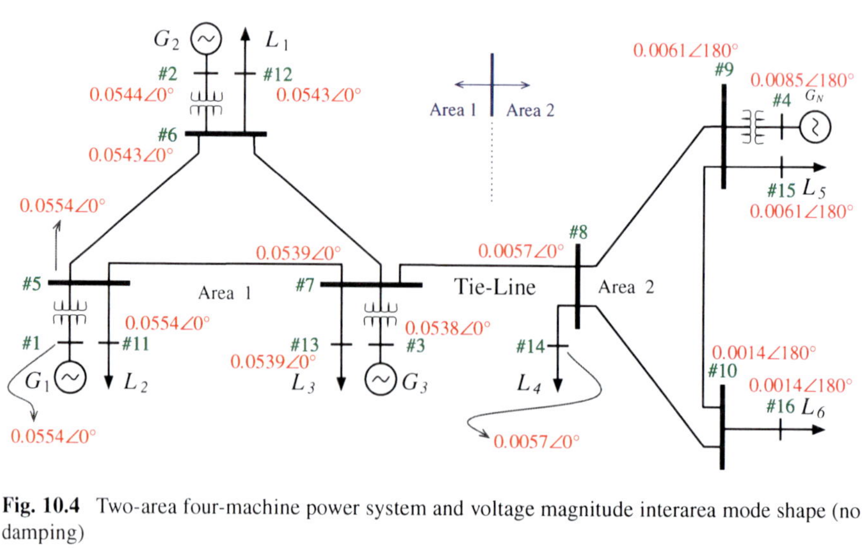

Fig. 10.4 Two-area four-machine power system and voltage magnitude interarea mode shape (no damping)

shows that G_1, G_2, and G_3 are oscillating against G_4: that is, Area 1 oscillates against Area 2.

Column 3, the mode shape for Local Mode 1, indicates that G_1 and G_2 are mostly oscillating against each other. Finally, Local Mode 2 (Column 4) is mostly confined within Area 1 with G_3 oscillating against G_1 and G_2. The most important characteristic to note is that the oscillations are either completely in-phase or anti-phase.

Using (10.1), we analyze the effect that machine damping has on the eigenvectors. The elements corresponding to the machine damping $M^{-1}D$ in $\bar{A}$ are given by

$$M^{-1}D = \text{diag}\left(\begin{bmatrix} -0.001 & -0.3 & -0.015 & -0.2 \end{bmatrix}\right) \tag{10.5}$$

and all other elements remain unchanged with respect to A.

Eigenanalysis is performed on $\bar{A}$, yielding the eigenvector matrix

$$W(\bar{A}) = \begin{bmatrix} 0.5\angle 0^\circ & 0.5882\angle 172.6196^\circ & 0.7878\angle 0^\circ & 0.0548\angle -178.5539^\circ \\ 0.5\angle 0^\circ & 0.6066\angle -180^\circ & 0.6084\angle 165.8682^\circ & 0.0507\angle 172.6111^\circ \\ 0.5\angle 0^\circ & 0.4548\angle 176.1869^\circ & 0.0951\angle 68.6262^\circ & 0.9972\angle 0^\circ \\ 0.5\angle 0^\circ & 0.2814\angle -6.3759^\circ & 0.0072\angle -116.0791^\circ & 0.0041\angle 176.6467^\circ \end{bmatrix} \tag{10.6}$$

Note that the eigenvector matrix is now complex. For convenience, it is shown in polar form. The main oscillatory characteristics discussed for the mode shapes from A are maintained for the mode shapes of $\bar{A}$. However, the components of each mode shape now present a phase shift due to the inclusion of machine damping. This phase shift is readily observed in the phasor diagrams in Fig. 10.5a, b. A more important observation is that in the time domain these phase shifts translate to time delays, making the oscillations in each mode peak at different instants as shown in Fig. 10.5c, d. It is interesting to note that the time-shifts shown in Fig. 10.5c, d strongly resembles those of the measurement data presented in the previous section.

Why do the network measurements exhibit phase shifts similar to those in the machine angle mode shapes of $\bar{A}$? In the following section, we answer this question by building upon the understanding gained from the mode shapes from $\bar{A}$, and generalize the results obtained in [8].

10.4 Sensitivity Analysis of Network Variables

Network sensitivities are used to provide linear relationships that predict changes in the network variables given a small perturbation in the power system. Here, we provide closed-form expressions for the network sensitivities [12]. This analysis gives a rationale about the nature of the network oscillations directly measured by PMU data.

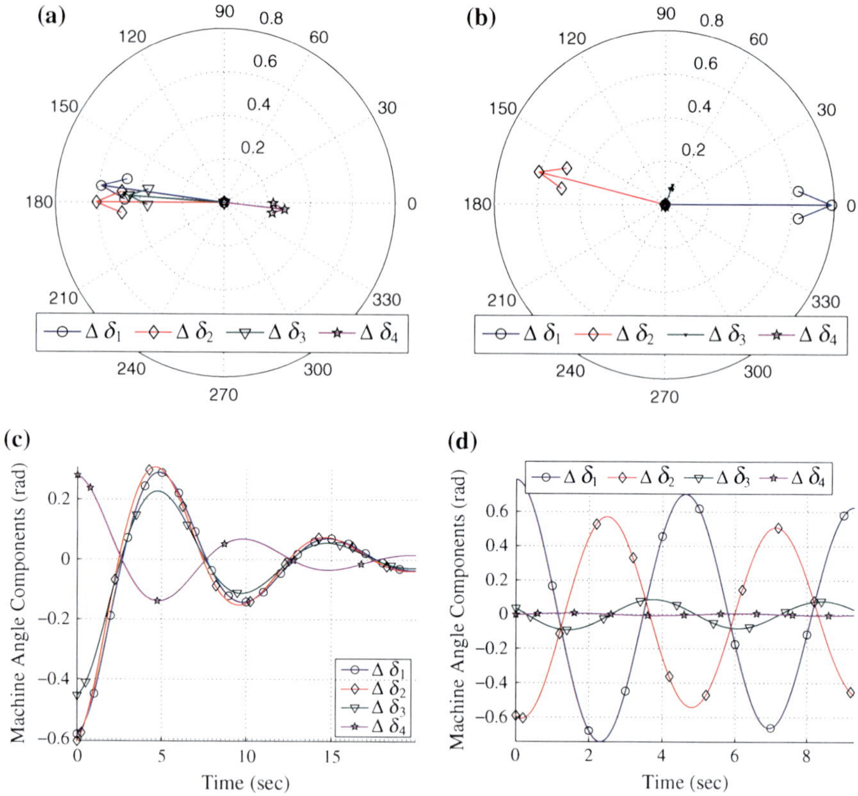

Fig. 10.5 Machine angle mode shapes with damping—phasor diagram (**a**, **b**), and time response (**c**, **d**). **a** Interarea mode. **b** Local mode 1. **c** Interarea mode. **d** Local mode 1

10.4.1 Network Sensitivities

By computing the sensitivities of the entire network, it is possible to predict the incremental behavior of the network variables when a small perturbation occurs in the power system. This is of particular interest because PMUs are capable of measuring synchronized network variable changes across wide-areas of the power system that emerge from small perturbations. Thus, an understanding of network sensitivities can provide insight for PMU data analysis of small signal oscillations occurring in large-scale power networks.

To illustrate, consider the n-bus, N-machine power system shown in Fig. 10.6. Buses i to $(i + 4)$ are transmission and load buses remotely connected to generators $j = 1, \ldots, N$. Regardless of their distance from the generators, these buses will be affected by changes in the internal machine angle of any generator. For example, a change in the internal angle of Machine j, $\Delta\delta_j$, will be reflected in the voltage magnitude at Bus i, V_i, in a proportion dictated by the network sensitivity

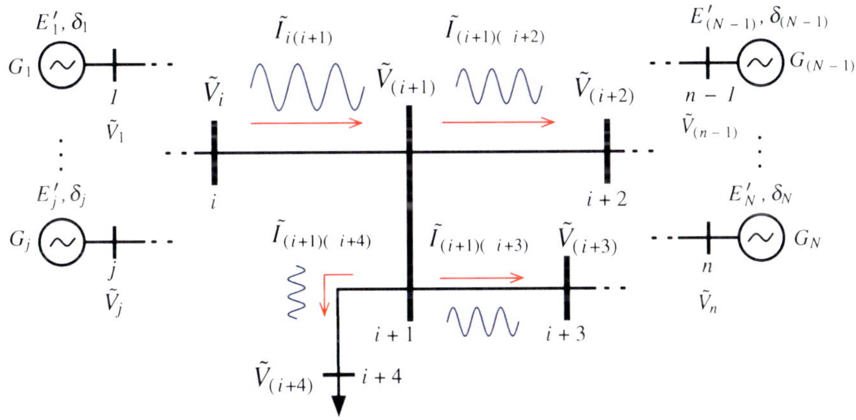

Fig. 10.6 Changes in the network variables with respect to changes in δ_j's

$(\partial V_i/\partial \delta_j)|_{(0)}$.[1] The total change in the voltage magnitude at Bus i, ΔV_i, will be the sum of all the changes in the machine internal angles scaled by their corresponding network sensitivity.

Similarly, if the current flow among Buses i to $(i + 4)$ is as indicated in Fig. 10.6, the change in the current flow will also be affected by the change in machine internal angles. The complex line current flow changes will be proportionally distributed satisfying Kirchhoff's current law, i.e., $\Delta\tilde{I}_{i(i+1)} = \Delta\tilde{I}_{(i+1)(i+2)} + \Delta\tilde{I}_{(i+1)(i+3)} + \Delta\tilde{I}_{(i+1)(i+4)}$. Thus, it becomes possible to trace how the current oscillations are being divided among multiple lines/paths and propagated across the entire network.

To obtain the change of the voltage magnitude (ΔV_i) and angle ($\Delta\theta_i$) at Bus i of the network with respect to the change of machine internal angles, we obtain the Taylor series expansion of V_i and θ_i about an equilibrium. Ignoring the higher order terms, the change of voltage magnitude at Bus i due to the change in the machine internal angles is given by

$$\Delta V_i = \left(\frac{\partial V_i}{\partial \delta_1}\right)\bigg|_0 \Delta\delta_1 + \left(\frac{\partial V_i}{\partial \delta_2}\right)\bigg|_0 \Delta\delta_2 + \cdots + \left(\frac{\partial V_i}{\partial \delta_N}\right)\bigg|_0 \Delta\delta_N \tag{10.7}$$

where $\left(\partial V_i/\partial \delta_j\right)|_0$ is the sensitivity of the i-th bus voltage magnitude to the j-th machine angle at the equilibrium point. In matrix form, the voltage magnitude changes at Buses $i = 1, \ldots, n$ due to the change in the machine internal angles are given by

[1] The subscript (0) denotes the evaluation of the sensitivity at a stable equilibrium point.

$$\underbrace{\begin{bmatrix} \Delta V_1 \\ \vdots \\ \Delta V_n \end{bmatrix}}_{\Delta V} = \underbrace{\begin{bmatrix} \left(\frac{\partial V_1}{\partial \delta_1}\right)\Big|_{(0)} & \cdots & \left(\frac{\partial V_1}{\partial \delta_N}\right)\Big|_{(0)} \\ \vdots & \ddots & \vdots \\ \left(\frac{\partial V_n}{\partial \delta_1}\right)\Big|_{(0)} & \cdots & \left(\frac{\partial V_n}{\partial \delta_N}\right)\Big|_{(0)} \end{bmatrix}}_{C_{V\delta}} \underbrace{\begin{bmatrix} \Delta \delta_1 \\ \vdots \\ \Delta \delta_N \end{bmatrix}}_{\Delta \delta} \tag{10.8}$$

where ΔV is the vector of bus voltage magnitude changes of size $n \times 1$, and $C_{V\delta}$ is the bus voltage magnitude sensitivity matrix of size $n \times N$.

Similarly, sensitivities may also be obtained for any other network variable. For the bus voltage angle changes, we have

$$\Delta \theta = C_{\theta\delta} \Delta \delta \tag{10.9}$$

where $C_{\theta\delta}$ is the bus voltage angle sensitivity matrix of size $n \times N$.

In a power system with a total number of ℓ transmission lines connecting the sending end buses, f, to the receiving end buses, t, the current magnitude and angle changes with respect to the machine angle changes are given by

$$\Delta I_{ft} = C_{I_{ft}\delta} \Delta \delta \tag{10.10}$$

$$\Delta \phi_{ft} = C_{\phi_{ft}\delta} \Delta \delta \tag{10.11}$$

where ΔI_{ft} and $\Delta \phi_{ft}$ are the current magnitude and angle changes, respectively. The matrix $C_{I_{ft}\delta}$ is the current magnitude sensitivity matrix while $C_{\phi_{ft}\delta}$ is the current angle sensitivity matrix, both of size $\ell \times N$.

In general, the sensitivity matrices $C_{V\delta}$ and $C_{\theta\delta}$ are obtained by numerical perturbation using simulation software such as PST [16]. Analytical expressions may provide further insight about their properties and the parameters affecting them. In the next section, we provide closed-form expressions of different network sensitivities.

10.4.2 Analytical Derivation of the Network Sensitivities

The first step to develop closed-form expressions of the different sensitivities is to obtain a general expression of the network variables as a function of the machine internal nodes. To this aim, it is possible to write nodal voltage equations that extend to the machine internal nodes. To include E_j', the voltage behind transient reactances x_{dj}', we add N buses to the n-bus power system network, thus extending the admittance matrix to the machine internal nodes.

The internal machine buses are denoted by $n+1, \ldots, n+j, \ldots, n+N$, which are the buses behind the transient reactances, x_{dj}'. The resulting admittance matrix differs from the admittance matrix used in load flow analysis in that the additional internal machine buses are included to account for the machine internal voltages, $\tilde{E}'_j$. In addition, loads are modeled as constant admittances and included in the diagonal elements of the admittances of the extended $\tilde{Y}$ matrix. As a result, injected currents in all nodes other than the generator internal nodes are zero, i.e., $\tilde{I}_i = 0, \quad i = 1, \ldots, n$.

Denoting all the generator current injections as $\tilde{I}_N$, generator voltages as $\tilde{E}'_N$, and bus voltages as $\tilde{V}_n$, the node voltage equations are

$$\begin{bmatrix} 0 \\ \tilde{I}_N \end{bmatrix} = \begin{bmatrix} \tilde{Y}_{nn} & \tilde{Y}_{nN} \\ \tilde{Y}_{nN}^T & \tilde{Y}_{NN} \end{bmatrix} \begin{bmatrix} \tilde{V}_n \\ \tilde{E}'_N \end{bmatrix} \tag{10.12}$$

From (10.12) an expression for the bus voltage phasors as a function of the machine internal voltages is obtained as

$$\tilde{V}_n = -\tilde{Y}_{nn}^{-1} \tilde{Y}_{nN} \tilde{E}'_N \tag{10.13}$$

where $\tilde{Y}_{nn}^{-1} \tilde{Y}_{nN}$ is referred to as the bus voltage reconstruction matrix and has a size of $(n \times N)$. The bus voltage reconstruction coefficient matrix is given by

$$\tilde{\kappa} = -\tilde{Y}_{nn}^{-1} \tilde{Y}_{nN} = \kappa \angle \gamma \tag{10.14}$$

where κ and γ are the magnitude and angle of $\tilde{\kappa}$. We can now obtain a generalized expression relating the voltage phasor at Bus i with the machine internal voltages using (10.14)

$$\tilde{V}_i = \tilde{\kappa}_{i1} \tilde{E}'_1 + \tilde{\kappa}_{i2} \tilde{E}'_2 + \cdots + \tilde{\kappa}_{iN} \tilde{E}'_N = \sum_{j=1}^{N} \tilde{\kappa}_{ij} \tilde{E}'_j \tag{10.15}$$

Hence, the voltage at the i-th bus is a function of the machine internal angles δ_j. In this expression, $\tilde{V}_i$ depends on the value of the machine internal voltage magnitudes at the equilibrium, E'_j, and the admittances in the voltage reconstruction coefficient matrix $\tilde{\kappa}_{ij}$, with $j = 1, \ldots, N$.

It is also possible to develop a generalized expression for the complex line current flow through any line of the power network. Consider the π-equivalent of a transmission line: the current from Bus f to Bus t is given by

$$\tilde{I}_{ft} = \left(\tilde{y}_{ft} + \tilde{y}_{f0}\right) \tilde{V}_f - \tilde{y}_{ft} \tilde{V}_t \tag{10.16}$$

Letting $\tilde{Y}_{ft0} = \tilde{y}_{ft} + \tilde{y}_{f0}$ and $\tilde{Y}_{ft} = \tilde{y}_{ft}$, and writing the voltage for Buses f and j in terms of the internal machine voltages using (10.15), we obtain

$$\tilde{I}_{ft} = \sum_{j=1}^{N} \left(\tilde{Y}_{ft0} \tilde{\kappa}_{fj} - \tilde{Y}_{ft} \tilde{\kappa}_{tj} \right) \tilde{E}'_j = \sum_{j=1}^{N} \tilde{\Psi}_{FTj} \tilde{E}'_j \tag{10.17}$$

where $\tilde{\Psi}_{FTj} = \Psi_{FTj}\angle\psi_{FTj}$, and

$$\Psi_{FTj} = \left|\tilde{Y}_{ft0}\tilde{\kappa}_{fj} - \tilde{Y}_{ft}\tilde{\kappa}_{tj}\right|, \ \psi_{FTj} = \angle\left(\tilde{Y}_{ft0}\tilde{\kappa}_{fj} - \tilde{Y}_{ft}\tilde{\kappa}_{tj}\right) \tag{10.18}$$

The complex line current flow in Line f-t is a function of the machine angles δ_j. The phasor $\tilde{I}_{ft}$ depends on the value of the machine internal voltage magnitudes at the equilibrium, E'_j, the admittances in the voltage reconstruction coefficient matrix $\tilde{\kappa}_{ij}$, and the admittances and shunts between Buses f and t. Expressions (10.15) and (10.17) are used to derive closed-form expressions of the network sensitivities. A complete derivation can be found in [12].

10.4.2.1 Voltage Sensitivities

Given an N-machine power system, the bus voltage magnitude sensitivities for Bus i with respect to the j-th machine angle are given by

$$\frac{\partial V_i}{\partial \delta_j} = \frac{1}{|\tilde{V}_i|}(\alpha), \ j = p; \quad \frac{\partial V_i}{\partial \delta_j} = \frac{1}{|\tilde{V}_i|}(-\alpha), \ j \neq p \tag{10.19}$$

where

$$|\tilde{V}_i| = \sqrt{\left(\sum\nolimits_{j=1}^{N} \tilde{\kappa}_{ij}\tilde{E}'_j\right)^2} \tag{10.20}$$

$$\begin{aligned}\alpha = &-\sum\nolimits_{p=1}^{N-1}\sum\nolimits_{q=p+1}^{N} \kappa_{ip}\kappa_{iq}E'_pE'_q \sin\left(\delta_p + \gamma_{ip} - \delta_q - \gamma_{iq}\right)\\ &+\sum\nolimits_{\substack{p=1\\p\neq j}}^{N-1}\sum\nolimits_{\substack{q=p+1\\q\neq j}}^{N} \kappa_{ip}\kappa_{iq}E'_pE'_q \sin\left(\delta_p + \gamma_{ip} - \delta_q - \gamma_{iq}\right)\end{aligned} \tag{10.21}$$

For the bus voltage angle

$$\frac{\partial \theta_i}{\partial \delta_j} = \frac{1}{|\tilde{V}_i|^2}\left[\kappa_{ij}^2E_j'^2 + \sum_{\substack{q=1\\q\neq j}}^{N} \kappa_{ij}\kappa_{iq}E'_jE'_q \cos\left(\delta_j + \gamma_{ij} - \delta_q - \gamma_{iq}\right)\right] \tag{10.22}$$

is the closed-form formula to obtain the sensitivity of the i-th bus voltage angle to the j-th machine angle.

10.4.2.2 Current Sensitivities

The line current magnitude sensitivities in any line from Bus f to Bus t with respect to the j-th machine angle are given by

$$\frac{\partial I_{ft}}{\partial \delta_j} = \frac{1}{|\tilde{I}_{ft}|}(\beta),\ j = p;\quad \frac{\partial I_{ft}}{\partial \delta_j} = \frac{1}{|\tilde{I}_{ft}|}(-\beta),j \neq p \tag{10.23}$$

where

$$|\tilde{I}_{ft}| = \sqrt{\left(\sum_{j=1}^{N} \tilde{\Psi}_{FTj}\tilde{E}_j'\right)^2} \tag{10.24}$$

is the value of the line current magnitude at equilibrium, and β is given by (10.25)

$$\begin{aligned}\beta = &-\sum_{p=1}^{N-1}\sum_{q=p+1}^{N} \Psi_{FTp}\Psi_{FTq}E_p'E_q' \sin\left(\delta_p + \psi_{FTp} - \delta_q - \psi_{FTq}\right) \\ &+\sum_{\substack{p=1\\ p\neq j}}^{N-1}\sum_{\substack{q=p+1\\ q\neq j}}^{N} \Psi_{FTp}\Psi_{FTq}E_p'E_q' \sin\left(\delta_p + \psi_{FTp} - \delta_q - \psi_{FTq}\right)\end{aligned} \tag{10.25}$$

For the line current angle

$$\frac{\partial \phi_{ft}}{\partial \delta_j} = \frac{1}{|\tilde{I}_{ft}|^2}\left[\Psi_{FTj}^2 E_j'^2 + \sum_{\substack{q=1\\ q\neq j}}^{N} \Psi_{FTj}\Psi_{FTq}E_j'E_q' \cos\left(\delta_j + \psi_{FTj} - \delta_q - \psi_{FTq}\right)\right] \tag{10.26}$$

is the closed-form formula to obtain the sensitivity of ϕ_{ft} with respect to the j-th machine angle.

Finally, the sensitivities of the real and imaginary part of the line current (10.16) are given by

$$\frac{\partial \Re\{\tilde{I}_{ft}\}}{\partial \delta_j} = -\sum_{j=1}^{N} \Psi_{FTj}E_j' \sin\left(\delta_j + \psi_{FTj}\right) \tag{10.27}$$

$$\frac{\partial \Im\{\tilde{I}_{ft}\}}{\partial \delta_j} = \sum_{j=1}^{N} \Psi_{FTj}E_j' \cos\left(\delta_j + \psi_{FTj}\right) \tag{10.28}$$

Similar expressions can be obtained for the current from Bus t to Bus f by substituting the subscripts ft and FT with tf and TF, and using the appropriate coefficients and admittances.

10.4.3 Properties of the Network Sensitivities

We finalize this discussion by providing some intrinsic properties of the sensitivities discussed above. For the bus voltage sensitivities, the properties are[2]

$$\sum_{j=1}^{N}\left(\frac{\partial V_i}{\partial \delta_j}\right)=0,\quad \sum_{j=1}^{N}\left(\frac{\partial \theta_i}{\partial \delta_j}\right)=1 \tag{10.29}$$

and for the line currents the sensitivity properties are

$$\sum_{j=1}^{N}\left(\frac{\partial I_{ft}}{\partial \delta_j}\right)=0,\quad \sum_{j=1}^{N}\left(\frac{\partial \phi_{ft}}{\partial \delta_j}\right)=1 \tag{10.30}$$

Both voltage and current sensitivities have the same property: *the sum of sensitivities of the magnitude of the phasor w.r.t. all the machine angles is zero, while the sum of the sensitivities of the angle of the phasor w.r.t. machine angles is equal to unity.*

These sensitivity properties can be explained by using the phasor diagram in Fig. 10.7. The voltage at the i-th bus of a two-machine system, $\tilde{V}_i = V_i\varepsilon^{j\theta_i}$, is perturbed by introducing small changes to the machine angles, $\Delta\delta_1$ and $\Delta\delta_2$, resulting in the perturbed voltage $\tilde{V}_i^* = V_i^*\varepsilon^{j\theta_i^*}$. When the perturbations introduced to the machine angles are identical, the resulting perturbation to the bus angle will be equal to the value used to perturb the machines. The perturbation of each machine is appropriately scaled by the corresponding sensitivity. Hence, the sum of all sensitivities is one. Note that if only one of the machines is perturbed, the bus angle will only be changed in the proportion dictated by the sensitivity and the perturbation value.

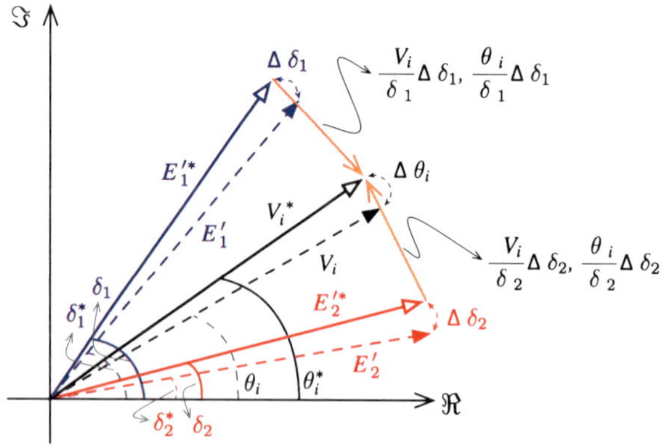

Fig. 10.7 Illustration of the sensitivity properties

[2] A formal proof for each property can be found in [12].

In addition, observe that the magnitude before the perturbation is the same as the magnitude after the perturbation, i.e., $V_i = V_i^*$. Note from the phasor diagram that $\Delta\delta_1$ scaled by $\partial V_i/\partial\delta_1$ will oppose the change in the voltage magnitude from $\Delta\delta_2$ scaled by $\partial V_i/\partial\delta_2$. Observe that the value of the bus voltage sensitivities for this case is identical. Because both sensitivities are equal with opposing signs, the change in the bus voltage magnitude, $\Delta V = 0$.

10.5 Network Modeshapes: Sensitivity Mapping onto the Electromechanical Mode Shapes

In the previous section, we investigated the origin of network sensitivities, and how they can be used to compute the total change in the network variables resulting from small perturbations in the machine internal angles. These changes include all $(N - 1)$ modal components of the power system. Measurements from PMUs are able to capture these changes in the network variables, and separation of their modal components can be performed with various techniques such as ERA or Prony analysis. In this section, we investigate how the eigenvector matrices discussed in Sect. 10.3, can be used to separate the components of each oscillatory mode contained in the total change of the network variables. This result was previously exploited in [8], where we showed that it is possible to compute the bus voltage magnitude and frequency mode shapes. Here we extend this concept to include any type of network variable, and to understand how by mapping the network sensitivities to a particular mode shape from $\bar{A}$, it is possible to replicate the phase shift observed in the modal components of PMU measurements.

By mapping the network sensitivities onto the right eigenvector, we obtain a *network modeshape*, which indicates the observability of a particular mode in a specific network variable. The mode shapes as observed in the bus voltage magnitudes and angles for all the network buses are given by

$$S_V = C_{V\delta} W \tag{10.31}$$

$$S_\theta = C_{\theta\delta} W \tag{10.32}$$

where W is of size $N \times N$.

Similarly, for the line current magnitude and angle we have

$$S_{I_{ft}} = C_{I_{ft}\delta} W \tag{10.33}$$

$$S_{\phi_{ft}} = C_{\phi_{ft}\delta} W \tag{10.34}$$

where the subscript *ft* indicates that these are the modal components from Bus f to Bus t. Modal components for any other network variable can be obtained in similar fashion.

When the eigenvectors are computed from $\bar{A}$, the phase shifts due to damping will be mapped onto the bus voltages in a proportion dictated by the sensitivities, reproducing the phase shifts observed in the network variables measured by PMUs.

To illustrate, consider the two-area four-generator system discussed in Fig. 10.4, Sect. 10.3.2. The bus voltage magnitude sensitivities of Bus 7 w.r.t. all machine angles are given by

$$C_{V\delta_{(7,k)}} = [-0.0463 \quad -0.0270 \quad 0.0146 \quad 0.0587] \tag{10.35}$$

Multiplying by the eigenvector column corresponding to the interarea mode, the interarea mode shape viewed from the bus voltage magnitude at Bus 7 is

$$\begin{aligned} S_{V_{(7,2)}} &= \frac{\partial V_7}{\partial \delta_1} W_{(1,2)} + \frac{\partial V_7}{\partial \delta_2} W_{(2,2)} + \frac{\partial V_7}{\partial \delta_3} W_{(3,2)} + \frac{\partial V_7}{\partial \delta_4} W_{(4,2)} \\ S_{V_{(7,2)}} &= -0.0463\ W_{(1,2)} - 0.0270\ W_{(2,2)} + 0.0146\ W_{(3,2)} + 0.0587\ W_{(4,2)} \end{aligned} \tag{10.36}$$

Note that each sensitivity scales its corresponding element of interarea mode shape.

The eigenvector matrix computed from (10.3) has the interarea mode shape given by Column 2 of (10.4)

$$W_{(j,2)}(A) = [-0.5930 \quad -0.5997 \quad -0.4539 \quad 0.2876]^T \tag{10.37}$$

Using this mode shape, it is possible to compute the interarea component of the bus voltage magnitude at Bus 7 (10.36)

$$\begin{aligned} S_{V_{(7,2)}}(A) &= (-0.0463)(-0.5930) + (-0.0270)(-0.5997) \\ &\quad + (0.0146)(-0.4539) + (0.0587)(0.2876) = 0.05391 \end{aligned} \tag{10.38}$$

Similarly, if the eigenvector matrix is computed from model (10.1), the interarea mode shape is given by Column 2 of (10.6), and the interarea component of the bus voltage magnitude becomes

$$S_{V_{(7,2)}}(\bar{A}) = 0.0534\angle -5.25^\circ \tag{10.39}$$

Comparing (10.39) to (10.5) helps in illustrating the origin of the phase shifts observed from phasor measurement data: that is, the presence of damping.

10.5.1 Illustration with the Two-Area Four-Machine System

We extend our discussion to consider the different network variables across the entire power system. The bus voltage magnitude mode shapes (the mapping of the sensitivities onto the electromechanical mode shapes) for all network buses are shown in the one-line diagrams in Figs. 10.4 and 10.8 for the interarea mode, for both the case without damping and with damping, respectively, and the phasor diagrams in Fig. 10.9. These figures clearly show that the damping in $\bar{A}$ will give rise to the phase shifts across all bus voltage magnitudes in the power system. Consider the phase shift introduced to the voltage magnitude at Bus 1: when no damping is included $S_{V_{(1,2)}}(A) = 0.0544\angle 0°$ and, with the effect of damping $S_{V_{(1,2)}}(\bar{A}) = 0.0547\angle -7.14°$. Thus the phase shift $-7.14°$ is a result of damping. Similarly for Bus 4 $S_{V_{(4,2)}}(A) = 0.0085\angle 180°$, while $S_{V_{(4,2)}}(\bar{A}) = 0.0085\angle 176.14°$. The 3.86° of difference between these two last quantities are the result of including damping. A more interesting case is Bus 8, which lies at the right end of the tie-line. Note that $S_{V_{(8,2)}}(A) = 0.0057\angle 0°$, while $S_{V_{(4,2)}}(\bar{A}) = 0.0056\angle -6.05°$, the $-6.05°$ being a result of the inclusion of damping.

It is worthwhile to note that this holds for any mode. For example, consider the bus voltage magnitude mode shape for Local Mode 1 shown in Fig. 10.10. In this case, there is a more prominent phase shift between Buses 1 and 12 than in Fig. 10.9 because Local Mode 1 mostly involves G_1 and G_2. This is also reflected in the voltage mode shapes. It is also important to highlight that the observations above also hold for any network variable. Consider the bus voltage angle mode shapes in Fig. 10.11. In the case without damping in Fig. 10.11a, it is observed that the bus voltage angles are either completely in phase or anti-phase. In contrast, Fig. 10.11b shows that the result of including damping is to have phase shifts in all the mode shapes. Another important feature of this mode shape is that when comparing Figs. 10.11b to 10.5b, *the machine angles outline a boundary within which the bus voltage angles exist*. In other words, *a bus voltage angle close to a generator will have a smaller relative phase angle difference with the generator than other generators farther apart from it*.

The relative time delay between the voltage magnitudes at Buses p and q can be calculated from the oscillatory frequency of the k-th mode and the phase shift at each

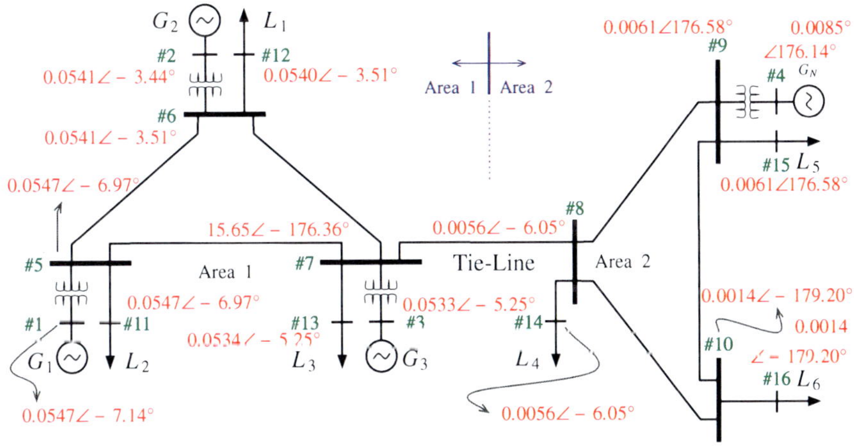

Fig. 10.8 Two-area four-machine power system and voltage magnitude interarea mode shape (damping)

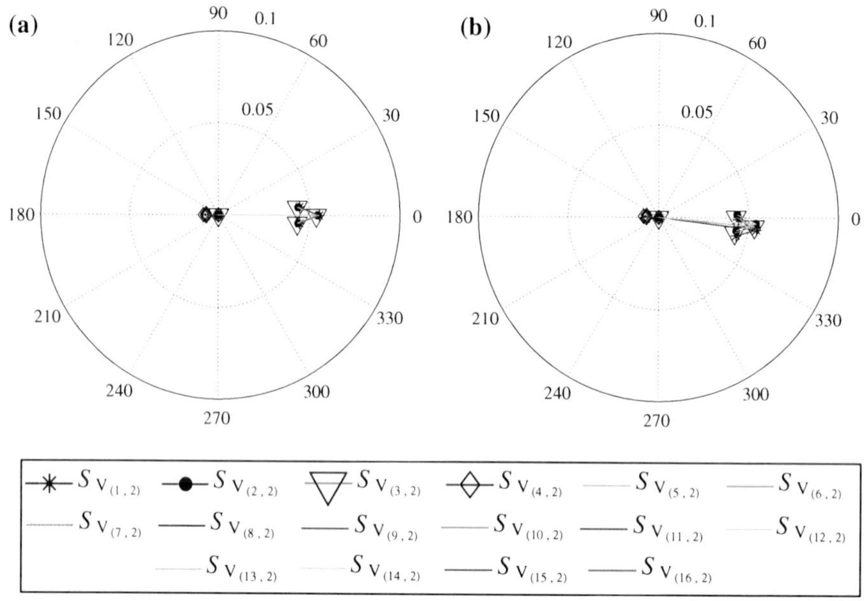

Fig. 10.9 V_i interarea mode shape with and without damping. **a** Without damping. **b** With damping

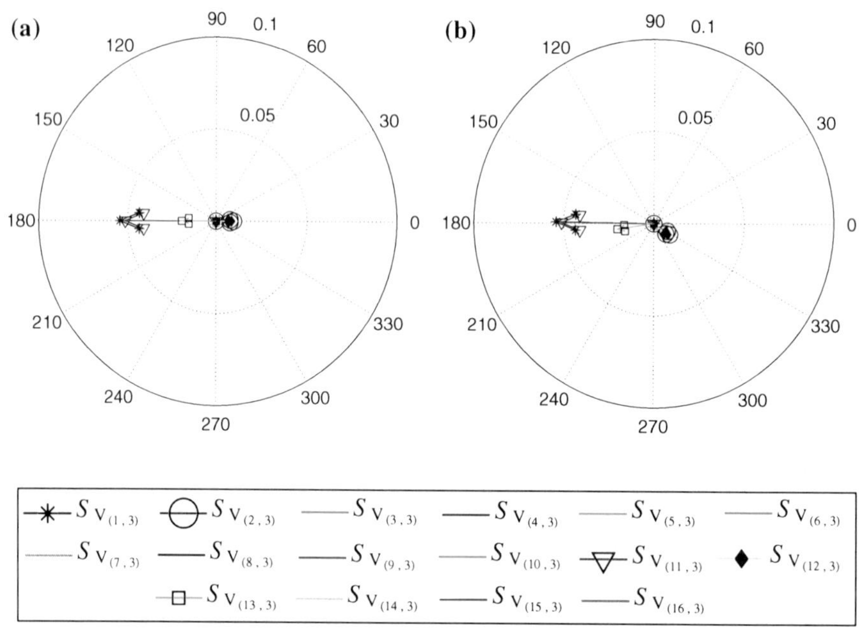

Fig. 10.10 V_i Local 1 mode shape with and without damping. **a** Without damping. **b** With damping

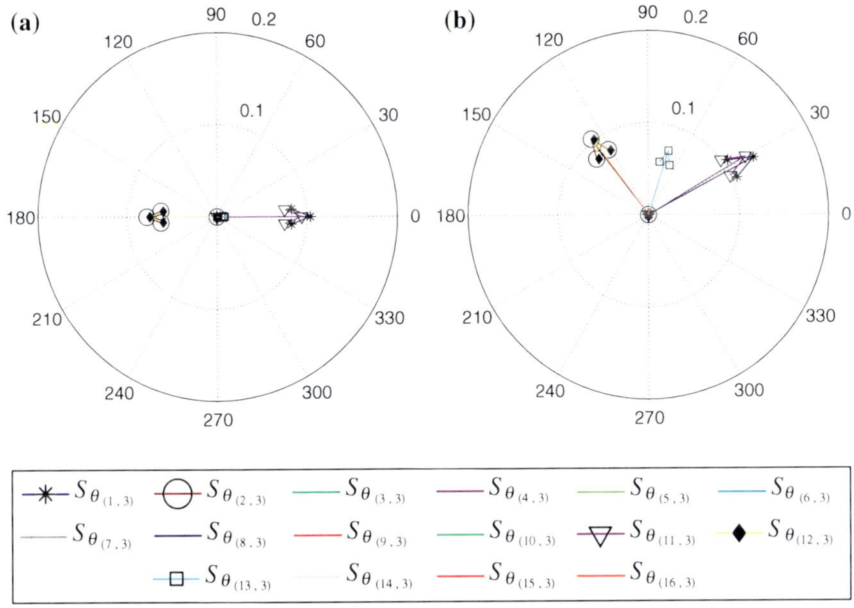

Fig. 10.11 θ_i Local 1 mode shapes with and without damping. **a** Without damping. **b** With damping

bus by computing

$$\tau_{(p-q,k)} = \frac{\theta_{(p,k)} - \theta_{(q,k)}}{2\pi} \times \frac{1}{f_k} \tag{10.40}$$

where f_k is the frequency of the k-th mode of the system, and

$$\theta_{(p,k)} = \angle\left(S_{V_{(p,k)}}\right), \quad \theta_{(q,k)} = \angle\left(S_{V_{(q,k)}}\right) \tag{10.41}$$

are the k-th mode phase shifts of the voltage magnitude oscillation in rad. at Buses p and q, respectively, and $j = 1, \ldots, N$.

As an example, consider the oscillations from Local Mode 1 at Bus 12, $S_{V_{(12,3)}}$, and Bus 1, $S_{V_{(1,3)}}$, as shown in Fig. 10.10. The relative time delay is given by

$$\tau_{(12-1,3)} = \frac{(5.6295) - (3.1291)}{2\pi \times 0.21516} = 1.8495\,\text{s}$$

This time delay is shown in Fig. 10.12, where we compare oscillations for Local Mode 1 in the bus voltage magnitude at Bus 1 and Bus 12.

Equation (10.40) can be used to compute the time delays for any network mode shape by selecting the sensitivity and mode of interest. Conversely, the relative phase shift between the buses can also be computed by knowledge of the time response of

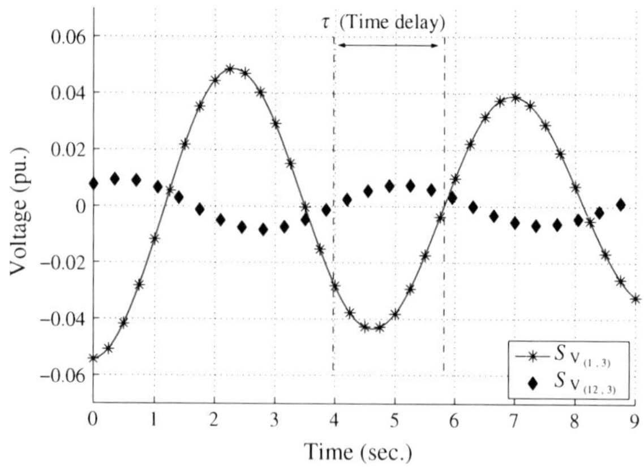

Fig. 10.12 Time delay between $S_{V_{(12,3)}}$ and $S_{V_{(1,3)}}$

the network (or ERA results from processing PMU data) from

$$\theta_{(p,k)} - \theta_{(q,k)} = 2\pi f_k \tau_{(p-q,k)} \tag{10.42}$$

where f_k is the frequency of the k-th mode of the system. Note that this frequency is different for each system mode, i.e., the interarea mode frequency is different from the Local Mode 1 frequency. Using the expression above, and selecting a reference voltage magnitude, we obtain the mode shapes from the time response, or more important, from PMU data.

Also of interest is the mapping for the line complex current flow mode shape, shown in Fig. 10.13 for $W(\bar{A})$. Note that the phase shifts are also mapped onto these network variables. The phase lags shown in the phasor diagram in Fig. 10.13a will translate to the different time delays that can be observed in Fig. 10.13b. More important, it is interesting to observe how the interarea mode distributes in different lines as shown in Fig. 10.13c. Note that it is possible to compute the current balance for the real and imaginary part, and that we can also determine how the interarea mode is distributed in each line. This new capability to "track" the oscillations has important implications as discussed next.

10.6 Applications for Interarea Mode Tracing: Identification of Dominant Interarea Oscillation Paths

One goal of a Wide Area Measurement System (WAMS) is to have tracking tools for oscillatory dynamics in an interconnected power grid, particularly those which are critical to operational reliability, i.e., interarea oscillations [17]. Insufficient damping of low-frequency interarea oscillations arises as weak interconnected power systems

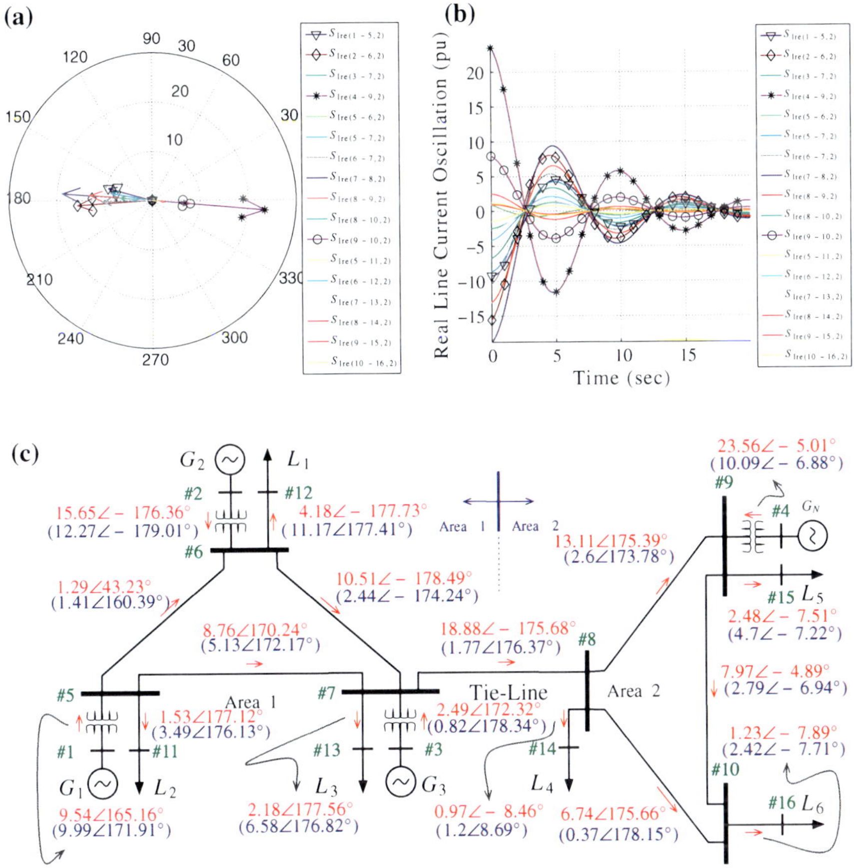

Fig. 10.13 Interarea mode $\tilde{I}_{ft}$ *mode shapes* (model with damping, imaginary quantities are shown inside *parenthesis*). **a** Phasor diagram. **b** Time response. **c** One-line diagram

are stressed to meet up with increasing demand [18]. This inadequacy may lead to oscillatory instability, resulting in system collapse.

"Interaction paths," defined in [17] as the group of transmission lines, buses, and controllers which the generators in a system use for exchanging energy during swings, are one important source of dynamic information necessary for WAMS. If the interaction paths of interarea swings can be identified, monitored, and tracked, proper preventive measures or control actions can be carried out to enhance the system's transfer capacity while maintaining high security.

A characteristic of power oscillations is that, for every mode of oscillation, there exists a series of connecting corridors in which the highest content of the mode would propagate. For a particular case of interarea modes, the path is termed "dominant interarea oscillation path" [19], a concept based on the notion of interaction paths. These dominant interarea oscillation paths are deterministic [20]. Furthermore,

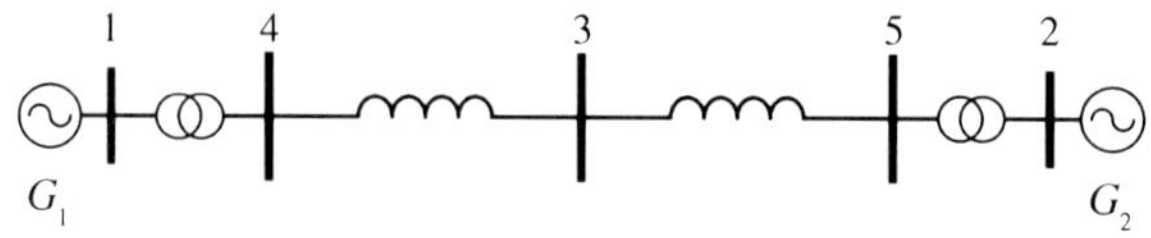

Fig. 10.14 Simplified dominant interarea path represented with a two-area, two-machine system

signals from the dominant path are the most observable and have the highest content of interarea modes. Results from the study suggest that by using well-selected dominant path signals for wide-area control, adequate damping performance can be achieved.

10.6.1 Dominant Interarea Oscillation Paths

Dominant interarea oscillation paths are defined as the passageway containing the highest content of the interarea oscillations. Consider a simplified dominant interarea path, represented by the two-area system shown in Fig. 10.14, where G_1 and G_2 represent coherent groups of machines involved in the interarea swing while transformers and line impedances represent elements of the dominant path connecting the two areas.

The characteristics of dominant interarea paths[3] can be demonstrated using the computed bus voltage magnitude (S_V) and angle modeshapes (S_θ) as illustrated in Fig. 10.15a, b, respectively. Fig. 10.15c, d illustrate the corresponding magnitude ($|S_\theta|$) and phase ($\angle S_\theta$) of the voltage angle modeshapes, S_θ. Two transfer levels, Case A and Case B, are compared in this figure. The labels "1,100" and "900" MW represent the amount of power transfer over the dominant path. The x-axis represents the bus number in the dominant path; the distance between buses is proportional to the line impedance magnitude. According to Fig. 10.15, important features of the dominant path are summarized below.

- The smallest $|S_\theta|$ element(s) (Fig. 10.15c) or the largest S_V (Fig. 10.15a) indicates the center of the path. This center can be considered as the "interarea mode center of inertia" or the "interarea pivot" for each of the system's interarea modes.
- The pivot divides the path into two groups where their respective phases are opposing each other (Fig. 10.15d).
- The difference between the S_θ elements of two edges of the path (Fig. 10.15b) are the largest among any other pair within the same path. In other words, the oscillations are the most positive at one end while being the most negative at the other end. Hence, they can be considered as the "tails" for each interarea mode.

[3] These are similar to the characteristics of voltage change and angle change of the first swing mode in Fig. 13 [21] where the mode is described by a single wave equation with one spatial dimension.

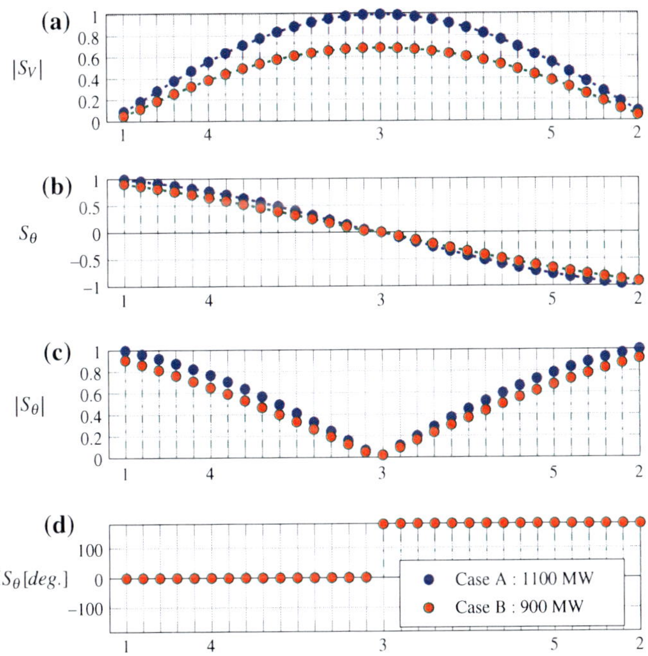

Fig. 10.15 Voltage magnitude and angle modeshapes of the dominant path in the two-area system. **a** Voltage magnitude modeshape, **b** Voltage angle modeshape, **c** Magnitude of voltage angle modeshape, **d** Phase of voltage angle modeshape

- The S_V elements of the edges (Fig. 10.15a) are the smallest or one of the smallest within the path.
- The interarea contents of the voltage magnitude modeshapes are more observable in a highly stressed system.

10.6.2 Dominant Interarea Oscillation Paths Identification Algorithm

As illustrated in Sect. 10.5.1, the current magnitude modeshapes ($S_{I_{ft}}$) indicate how much contents of the interarea modes are distributed among the transfer corridors. Thus, corridors having the highest content of current magnitude modeshapes signify the paths where the interarea oscillations will travel the most; hence the term "dominant interarea oscillation paths." On the other hand, the magnitude of voltage magnitude and angle modeshapes (S_V and S_θ) indicates the modal observability of a signal. The larger in magnitude the modeshape is, the more observable the signal measured (from the dominant path) becomes. This will be helpful when selecting feedback signals having high interarea modal content.

An algorithm for identification of dominant interarea oscillation paths is proposed here. Note that this algorithm takes synchrophasor measurements into consideration by exploiting ambient data.[4]

Ambient measurements are synthesized by simulating the time response of the power system with random noise and small step inputs at all loads [23]. The algorithm for handling ambient measurements is described next.

Dominant Path Identification Algorithm using Ambient PMU data

Step 1. Pre-process a parcel of ambient measurements, ΔI_{ft}, ΔV, and $\Delta\theta$ by filtering all the measurements such that only the interarea modal content of interest is preserved in the signals.

Step 2. Compute the power spectral densities (PSD) of ΔI_{ft}.

Step 3. Select an appropriate window. Find the peak PSD for each signal within the selected window, and sort the contents in descending order.

Step 4. The dominant interarea oscillation path is determined from the signals having the largest PSD contents. Compare to the schematic diagram of the system of study and identify such paths.

Step 5. Using one of the edges of the path in *Step 4* as a reference, the cross power spectral densities (CPSD) of the preprocessed ΔV and $\Delta\theta$ of the dominant path are computed.

Step 6. Select appropriate windows for each type of signal. Find the corresponding largest CPSD magnitude for each measurement within the selected window.

Step 7. Verify the characteristics of the dominant path using its corresponding peak CPSD of voltage magnitude and angle. The results should resemble the dominant interarea oscillation path's features shown in Fig. 10.15.

10.6.3 Algorithm Illustration with the KTH-NORDIC32 System

The system under study, namely KTH-NORDIC32 [24] (illustrated in Fig. 10.16), is a simplified model of the Swedish power system and its neighbors [25]. It has 20 generators and 52 transmission lines. Small signal stability analysis [25] reveals that the system has two lightly damped low-frequency interarea oscillations: 0.50 and 0.74 Hz. Note that to develop a fundamental understanding, we consider a case where controls such as exciters and turbine governors are disabled.[5] Two scenarios, namely Case 1 and Case 2, are used to illustrate the algorithm throughout this study.

[4] For algorithms dealing with other set of information such as known system model or transient measurements, refer to [22].

[5] The model with controls is studied in [24].

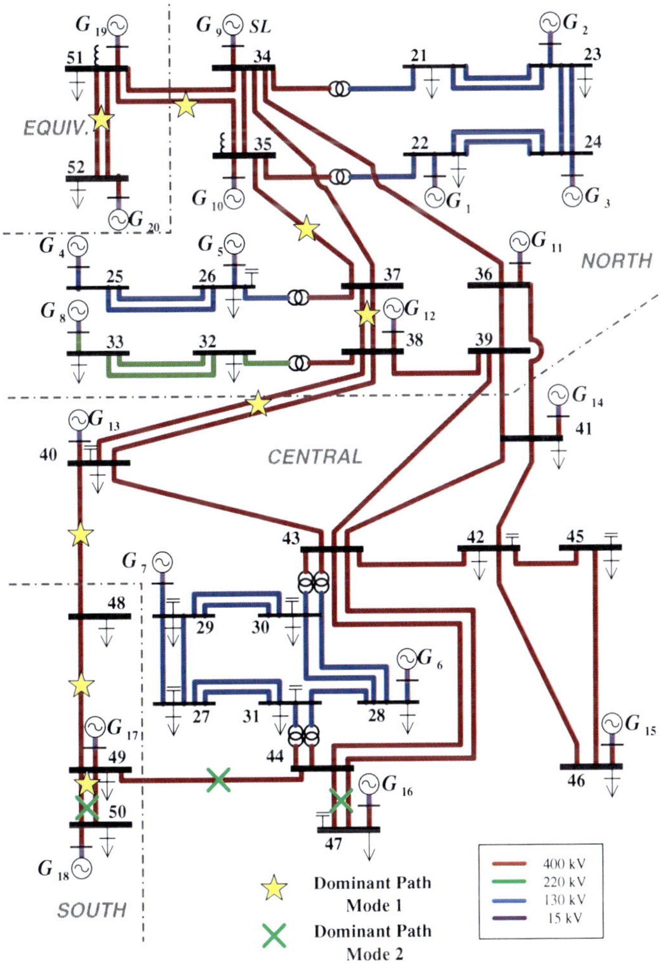

Fig. 10.16 KTH-NORDIC32 system

Power flow from the northern to the southern regions for each respective scenario are 3,134, and 2,933 MW. Note that the step-by-step demonstration will be described only for Case 1.

Step 1. ΔI_{ft} measurements before and after pre-processing are illustrated in Fig. 10.17.

Step 2. The computed PSD of ΔI_{ft} are illustrated in Fig. 10.18. The dashed (red) lines indicate the cutoff frequencies used in the preprocessing.

Step 3. The window of the selected data corresponds to the dashed lines in Fig. 10.18. The peak PSD for each ΔI_{ft} are sorted in descending order; the ten largest values and their corresponding sending and receiving buses are summarized in Table 10.1.

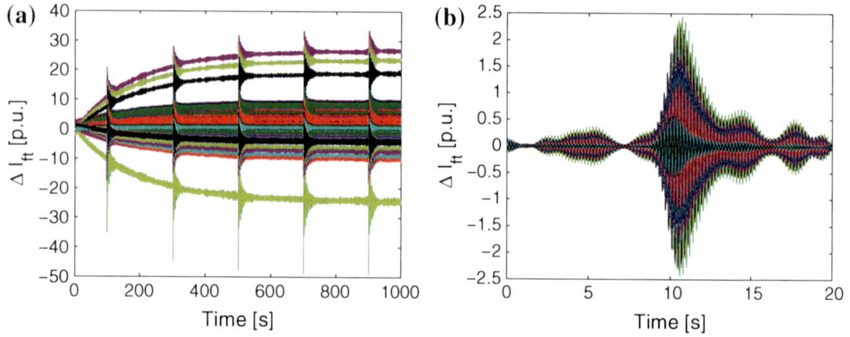

Fig. 10.17 ΔI_{ft} signals: before and after pre-processing. **a** Before pre-processing. **b** After pre-processing, (from 0 to 20 s)

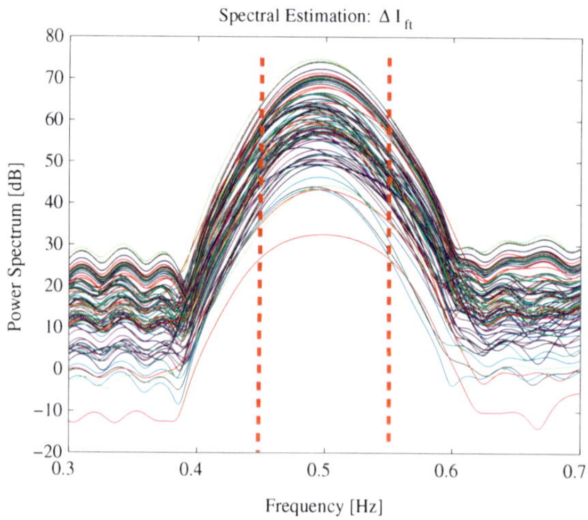

Fig. 10.18 PSD of ΔI_{ft}, mode 1

Step 4. The dominant interarea oscillation path for Mode 1 is identified to be 52-51-35-37-38-40-48-49-50[6] as marked by the (yellow) stars in Fig. 10.16.

Step 5. ΔV_{52} and $\Delta\theta_{52}$ are used as references for the CPSD computation of voltage magnitude and voltage angle measurements of the dominant path, respectively. The corresponding computed CPSD are illustrated in Fig. 10.19.

Step 6. After selecting the appropriate windows, the largest (absolute) CPSD values for each measurement within the dominant path are used to reconstruct the

[6] Observe from Table 10.1 that the path 42-43-44-49-50 has considerably high content of Mode 1; however, it has lesser content than that of the specified dominant path. This second path is thus termed "secondary dominant interarea oscillation path." For details regarding multiple interaction paths, refer to [26].

Table 10.1 Ten largest PSD of ΔI_{ft}

Sending bus, f	Receiving bus t	PSD [dB]
18	50	74.69
40	48	74.14
48	49	71.35
44	49	70.81
49	50	69.77
49	50	69.77
42	43	67.68
43	44	66.60
43	44	66.60
38	40	66.46

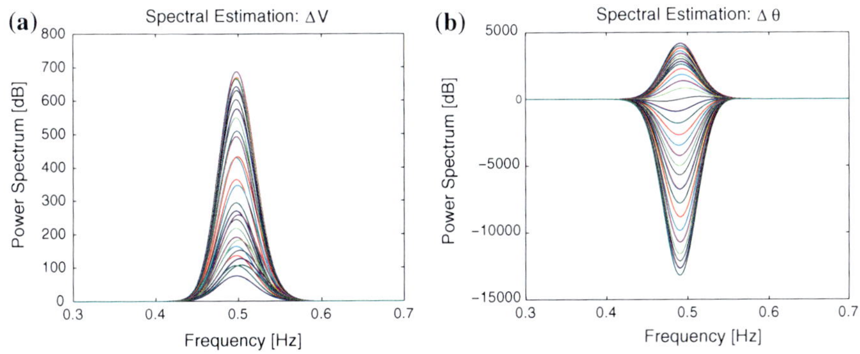

Fig. 10.19 Computed CSD of the dominant path. **a** Voltage magnitude. **b** Voltage angle

path. Using the characteristics of the dominant path from Fig. 10.15c and d, the bus having the smallest magnitude of voltage angle deviations, $|\Delta\theta|$, is the pivot of the path and thus used as the reference. The reconstructed dominant path of Mode 1 is as shown in Fig. 10.20a with blue dots for Case 1.

Step 7. The characteristics of the obtained path is verified by comparing to that of Fig. 10.15. The main features of the dominant paths remain preserved, and, thus, the path is justified.

Comparing the two case studies in Fig. 10.20a, Case 1 has an overall larger modal content in ΔV's while $\Delta\theta$'s are about the same in both cases. The differences between the two edges of both cases are comparable.

Repeating *Step* 1–7 for Mode 2, the dominant interarea oscillation path for Mode 2 is identified to be 50-49-44-47. Voltage magnitude and angle deviations of the path are illustrated in Fig. 10.20b. Similar to the result from Mode 1, Case 1 has larger modal content in ΔV's.

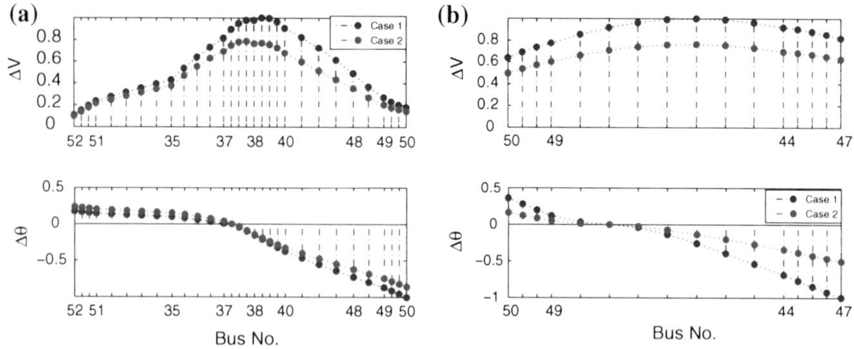

Fig. 10.20 Dominant interarea oscillation paths for both interarea modes, ambient measurements. **a** Mode 1. **b** Mode 2

As a final remark, note that Fig. 10.19b shows clear indications of the coherency groups: buses with positive CPSD peaks correspond to the group on the left and those with negative peaks to the group on the right of the dominant path. This is the characteristic shown in Fig. 10.15b for the two-area system. A similar analogy is valid for the CPSDs in Figs. 10.19a and 10.15a. In Fig. 10.19a, the CPSD peaks are all positive and with magnitudes according to the location of the buses within the dominant path where the signal is taken (see Fig. 10.15a). Observe that according to the voltage magnitude mode shape in Fig. 10.15a, those buses close to the "tails" of the dominant path have the lowest voltage oscillation (lowest CPSD magnitude) while those within the middle of the dominant path experience the highest voltage oscillations (highest CPSD magnitude). The fact that all CPSDs from the dominant path are positive corroborates that all voltage magnitude oscillations in a dominant path are either in phase or showing only a small phase shift between each other. Note that the shift in the CPSDs peak corresponds also to the phase shift illustrated in Sect. 10.5.1.

10.6.4 Damping Controller Design: Feedback Signals Selection

One major challenge in damping control design is the selection of feedback input signals. Conventionally, power system stabilizers (PSSs) use local measurements for input signals, such as active power in the outgoing transmission line, generator speed, and frequency at the terminal bus. With the availability of signals from PMUs, choices of inputs are not only limited to those local but now include wide-area signals. Several studies suggest that wide-area signals are preferable to local signals [27, 28]; therefore, the exploitation of PMU signals is desirable. However, the main issue is which signal, among all the available signals, would give satisfactory damping performance.

Results from previous studies on the concept of dominant interarea oscillation paths [19, 26] have suggested that effective damping control can be achieved by using signals from the dominant paths. The aim of this section is thus to carry out a fundamental study on feedback control using PMU signals from a dominant path. As such, the two-area system introduced in Sect. 10.6.1 is used to illustrate PSS control design for damping enhancement. Two types of signals, namely voltage angle and generator rotor speed, are used as inputs for a PSS controller. The first type represents the signals available from PMU while the latter represents one of the most commonly used signals in PSS damping design.[7] Their corresponding performances are analyzed and compared. The results of this study offer promising and feasible choices of signals to be used in feedback control. Although only PSS is considered, the concepts are applicable for any other damping controllers.

In this study, the impact of different feedback input signals on the damping of the two-area test system will be evaluated. Each signal requires different controller (PSS) parameters, as well as different structures. The purpose of the design is to achieve a specific damping performance. As such, for each input signal, PSS parameters will be tuned such that the system achieves $\zeta = 15\,\%$ damping ratio.

Voltage angle and generator speeds will be used as feedback input signals. Controller performance is evaluated considering the following factors: (1) effective gain (the cumulative gain of the PSSs which can be computed from $\alpha^n K_d$), (2) overshoot (M_p), and (3) rise time (t_r).

The monitored signal is the bus voltage terminal at G_1, V_1, which will be used to evaluate the control performance.

10.6.4.1 Controller Structure

The objective of the design is to improve damping of the interarea mode by installing a PSS at G_1 modulating the AVR error signal. Following the design in [29], the structure of the PSS includes lead/lag compensators in the form

$$PSS = K_d \left[\alpha \frac{s+z}{s+p} \right]^n \left[\frac{T_w s}{1 + T_w s} \right]^m \tag{10.43}$$

where n and m are the number of compensator stages and T_w is the washout filter time constant having the value of 10 s. The exponent $m = 1$ when the input signal is a generator rotor speed while $m = 2$ when the input signal is a voltage angle difference.

Note that generator speed, as well as angle difference, has high components of torsional modes [4]. Therefore, a torsional filter is added to the PSS structure *when generator speed and bus voltage angle differences are used as feedback input signals*. The impact of the torsional filter on PSS design is represented by a lower-order transfer function

[7] Note that the generator speed is not currently available from PMUs.

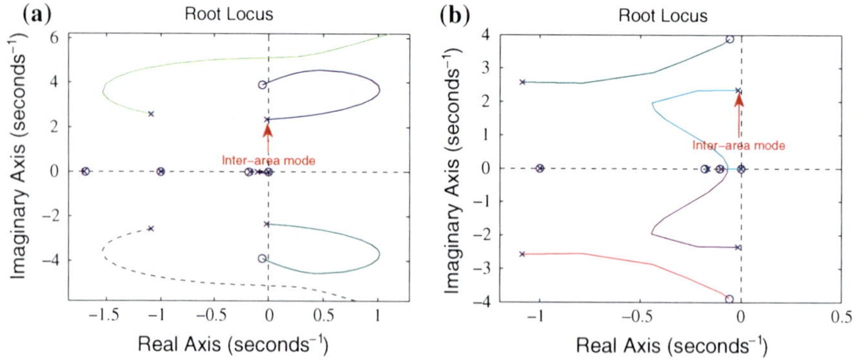

Fig. 10.21 Root-locus plots of the system with $\Delta\theta_{14}$ as feedback input signal. **a** No phase compensation. **b** With phase-lead compensation

$$G_{tor}(s) = \frac{1}{0.0027s^2 + 0.0762s + 1}. \tag{10.44}$$

The controller parameters α, poles (p), and zeros (z) can be computed from the following equations

$$\phi_m = \frac{180° - \theta_{\text{dep}}}{n} \tag{10.45}$$

$$\alpha = \frac{1 + sin(\phi_m)}{1 - sin(\phi_m)} \tag{10.46}$$

$$p = \sqrt{\alpha}\omega_c, \quad z = p/\alpha \tag{10.47}$$

where θ_{dep}, ϕ_m, and ω_c represent the angle of departure of the interarea mode, the phase compensation required, and the frequency of the mode in rad/s, respectively.

10.6.5 Controller Design Illustration

In this illustration, the signal $\Delta\theta_{14} = \theta_1 - \theta_4$, the voltage angle difference between Bus 1 and Bus 4 (see Fig. 10.14), is used as the feedback input signal. A root-locus plot of the open-loop system (no phase compensation) including two washout filters is shown in Fig. 10.21a. The angle of departure (θ_{dep}) of the interarea mode is 16.167°. Using this angle, the PSS parameters are computed using Eqs. (10.43–10.47). Applying the designed controller to the system, the root-locus plot is shown in Fig. 10.21b, which shows a stabilizing direction of the interarea mode; i.e., the damping of interarea mode is improved. The gain K_d is obtained when moving along the branch of the root loci of the interarea mode until the desired damping ratio (15 %) is reached. Finally, for the signal used, the obtained PSS has the form

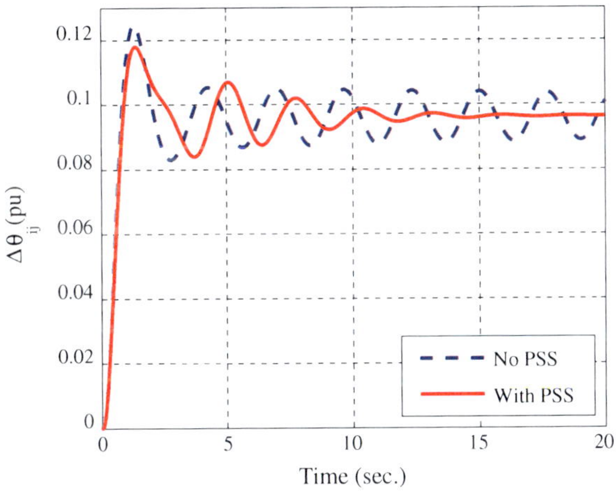

Fig. 10.22 Damping control performance using $\Delta\theta_{14}$ as feedback input signal

Table 10.2 System performance using voltage angle differences as feedback input signals.

Signals	Effective gain	M_p (%)	t_r (s)
$\Delta\theta_{14}$	130.51	18.02	0.60
$\Delta\theta_{43}$	55.87	16.58	0.59
$\Delta\theta_{45}$	34.85	15.61	0.58
$\Delta\theta_{42}$	33.47	15.14	0.57

Note The effective gain is defined as the cumulative gain of the PSSs which can be computed from $\alpha^n K_d$ in (10.43)

$$PSS = 0.00325\left[200.284\frac{s+0.1659}{s+33.2357}\right]^2\left[\frac{10s}{1+10s}\right]^2. \tag{10.48}$$

The responses of the terminal voltage at Bus 1 with and without PSS are compared in Fig. 10.22.

10.6.5.1 Voltage Angle Differences as Input Signals

Using the same design process, the system performance using voltage angle differences as feedback input signals is summarized in Table 10.2. All angle differences in Table 10.2 use a 2-stage lead compensator. Note that θ_4 is used as a reference, and the order of representation corresponds to the location of the buses in Fig. 10.14.

According to the results in Table 10.2, it can be concluded that the overall performance corresponds to the voltage angle modeshape (S_θ) relationship. That is, the larger the angle modeshape (difference), the lesser the gain is required, the smaller the overshoot is, and the faster the rise time becomes. In other words, the signal $\Delta\theta_{42}$, having the largest S_θ difference, requires the smallest gain and has the

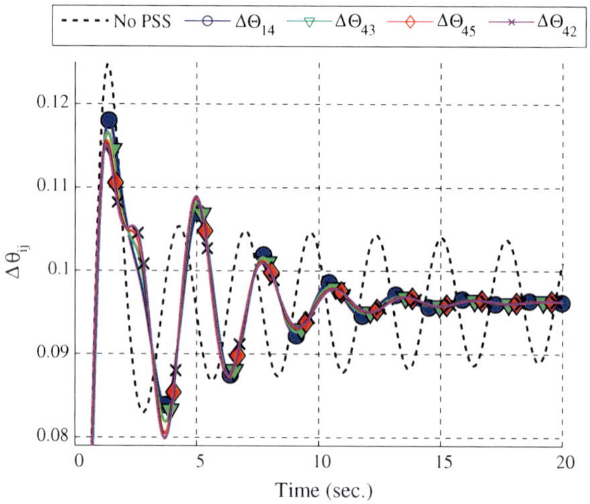

Fig. 10.23 Damping control performance using $\Delta\theta_{ij}$ as feedback input signals

Table 10.3 System performance using signal combinations of voltage angle differences as feedback input signals

Signals	Effective gain	M_p (%)	t_r (s)
$\Delta\theta_{12}$	26.94	15.64	0.58
$\Delta\theta_{avg,1}$	15.19	15.62	0.58
$\Delta\theta_{avg,2}$	22.44	15.60	0.58

smallest overshoot and rise time. This is because the signals have modal contents proportional to the voltage angle difference dictated by the voltage angle modeshape (see Fig. 10.15b).

Table 10.2's corresponding responses of the terminal voltage at Bus 1 are illustrated in Fig. 10.23. Note that, in order to have the same sign, $\Delta\theta_{14}$ is used instead of $\Delta\theta_{41}$.

The possible signal combinations using angle differences between the two areas are shown in Table 10.3 where the average angle differences $\Delta\theta_{\text{avg},1}$ represents $(\theta_1 + \theta_4) - (\theta_2 + \theta_5)$, and $\Delta\theta_{\text{avg},2}$ represents $(\theta_a + \theta_4) - (\theta_b + \theta_5)$. Note that Bus a is a bus in the middle between Bus 4 and Bus 3, whereas Bus b is a bus in the middle between Bus 3 and Bus 5. The aim of using the average angle differences in the two areas is to increase the interarea mode content of the resulting feedback signal, in larger power networks it can be used to reduce the excitation of the local oscillations [30]. All two-area angle difference combinations use a 2-stage lead compensator.

According to the results in Table 10.3, by combining signals from both areas, the effective gains are significantly reduced compared to the results in Table 10.2. In addition, overshoots are slightly reduced, while rise times are similar. Overall, $\Delta\theta_{\text{avg},1}$ requires the least amount of gain with comparable overshoot and rise time

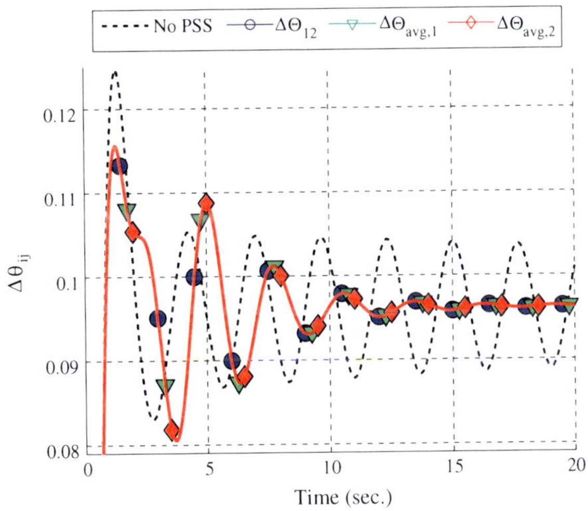

Fig. 10.24 Damping control performance using signal combinations of $\Delta\theta_{ij}$ as feedback input signals

performance. However, in practice, θ_1 and θ_2 are generator buses and thus not usually available from PMUs (see [31]). Hence, for any practical implementation, the most feasible combination is $\Delta\theta_{avg,2}$, for which the desired damping performance can be achieved while requiring a smaller gain than those from individual signals.

The time responses of the terminal voltage at Bus 1 are illustrated in Fig. 10.24.

10.6.5.2 Rotor Speeds as Input Signals

For comparison purposes, we consider speed signals from generators, although only available locally and not commonly available from PMUs [31]. The system performance using generator speeds as feedback input signals is summarized in Table 10.4. All signals in Table 10.4 use a 2-stage lead compensator except for ω_2 which requires a 3-stage lead compensator.[8]

According to the results in Table 10.4, overall, using speeds as feedback input signals not only results in larger overshoots and longer rise times but also requires considerably larger gain than using bus voltage angle difference signals, particularly using ω_2.[9]

[8] ω_2 is in anti-phase to any signals starting from ω_1 to the interarea pivot (middle point in Fig. 10.15b). Thus, two compensators are required for a rotation of 180° and an additional stage for the needed phase compensation.

[9] Although not shown here, the angle of departure of ω_2 is negative while the others are positive. This is the reason why not only an additional stage compensator is required but also a very high gain.

Table 10.4 System performance using generator speed as feedback input signals.

Signals	Effective gain	M_p (%)	t_r (s)
ω_1	582.27	35.50	0.72
ω_2	136,713	20.33	0.49
$\omega_1 - \omega_2$	297.32	35.50	0.72

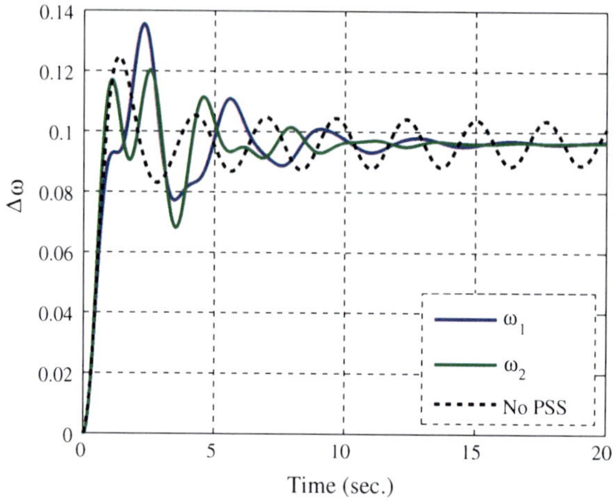

Fig. 10.25 Damping control performance using ω_i as feedback input signal

The time responses of the terminal voltage at Bus 1 are illustrated in Fig. 10.25. Note that because responses of ω_1 and $\omega_1 - \omega_2$ are exactly the same, only the response using ω_1 is presented in the figure.

10.6.5.3 Input Signal Comparison

A comparison of Tables 10.2, 10.3, and 10.4 reveals that using bus voltage angle differences as feedback input signals for damping control is as effective (or even better) as using generator speeds, for interarea mode damping. From Tables 10.2, 10.3 and 10.4, it can be observed that using $\Delta\theta_{ij}$ as input signals, similar damping performance can be obtained while having much lower overshoots and, more important, using much lower effective gain.[10] Comparing $\Delta\theta_{12}$ to $\omega_1 - \omega_2$, the angle difference outperforms the speed signals in effective gain required, overshoot, and rise time.[11]

Results from Sects. 10.6.5.1 and 10.6.5.2 indicate that angle difference is the most effective feedback input signal with superior overall performance compared to

[10] The effective gain for the generator speed signals is naturally higher comparing to the angles due to the scaling introduced by the system frequency, $2\pi f_b = 314.16$.

[11] However, it is noted that none of the signals are actually available from PMUs: $\Delta\theta_{12}$ due to placement practice [31] and $\omega_1 - \omega_2$ due to PMU characteristics [30].

generator speed. Future work will investigate the use of relative generator speeds and bus frequencies for use as feedback input signals.

The overall performance of each signal of the voltage angle differences is in accordance with their corresponding network modeshapes (see Fig. 10.15). That is, the performance of signals having high network modeshape is superior to that of those with lower network modeshape.

Different loading effects are not yet considered in this study. This is relevant because, for different loading scenarios, the open-loop observability of the dominant path signals shifts depending on the loading level. Further work is necessary to determine if the closed-loop observability on different loading levels maintains the same properties as those revealed in this study.

The selection of the "right" input signals from PMUs is critical for effective damping control. However, in the case of signal loss (due to communication failures), the controller must be adjusted even if new signals are used to replace a lost signal so that the highest damping can be obtained. These adjustments must occur adaptively and must be initiated by an adequate switch-over logic that guarantees the continued operation of the damping controller. Depending on the types of signals as well as the signal combination, the controller structure must be adapted accordingly to achieve optimal damping. As such, "adaptive" controllers, which can automatically adjust their parameters for each input signal feeding in, are promising and desirable.

10.7 Conclusions

In this chapter, the fundamental results presented in the analysis of network modeshapes provide a novel understanding of power system oscillations as viewed from network variables. Furthermore, they have several applications for interarea mode tracing and monitoring, PMU placement, and damping of power system oscillations. An algorithm for identifying dominant paths using ambient PMU data was presented, and the use of signals from the dominant path for feedback control input signals was investigated for the case of PSS damping control. The demonstration of these applications corroborates the fact that interarea modes are visible in network variables measured by PMUs which are readily exploitable.

References

1. J.H. Chow, G. Peponides, P.V. Kokotović, B. Avramović, J.R. Winkelman, *Time-Scale Modeling of Dynamic Networks with Applications to Power System* (Springer, New York, 1982)
2. M. Klein, G.J. Rogers, P. Kundur, A fundamental study of inter-area oscillations in power systems. IEEE Trans. Power Sys. **6**, 914–921 (1991)
3. G.C. Verghese, I.J. Pérez-Arriaga, F.C. Schweppe, Selective modal analysis with application to electric power systems, part I: heuristic introduction, part II: the dynamic stability problem. IEEE Trans. Power Apparatus Sys. **PAS-101**, 3117–3134 (1982)

4. R.A. Lawson, D.A. Swann, G.F. Wright, Minimization of power system stabilizer torsional interaction on large steam Turbine-generators. IEEE Trans. Power Apparatus Sys. **PAS-97**(1), 183–190 (1978)
5. G. Rogers, *Power System Oscillations* (Kluwer, 1999)
6. J.H. Chow, L. Vanfretti, A. Armenia, S. Ghiocel, S. Sarawgi, N. Bhatt, D. Bertagnolli, M. Shukla, X. Luo, D. Ellis, D. Fan, M. Patel, A.M. Hunter, D.E. Barber, G.L. Kobet, Preliminary Synchronized Phasor Data Analysis of Disturbance Events in the US Eastern Interconnection, in *Proceedings of IEEE PES Power Systems Conference and Exposition*, March 2009
7. L. Vanfretti, *Notions of Phasor Measurement-Based Power System Model Reduction of Large Power Systems* M.Sc. Thesis, Rensselaer Polytechnic Institute, Troy, New York, 2007
8. L. Vanfretti, J.H. Chow, Computation and Analysis of Power System Voltage Oscillations from Interarea Modes, *in Proceedings of IEEE PES General Meeting*, July 2009
9. L. Vanfretti, T.M.L. Assis, J.H. Chow, L. Dosiek, J. Pierre, D. Trudnowski, Y. Liu, Data analysis of the 2/26/08 Florida disturbance. NASPI Work Group Meeting, Sacramento, CA, 2009, http://www.naspi.org/meetings/workgroup/workgroup.stm
10. FRCC Event Analysis Team, FRCC system disturbance and underfrequency load shedding event report, 26 Feb 2008 at 1:09 pm. Florida Reliability Coordinating Council, 30 Oct 2008
11. P.W. Sauer, M.A. Pai, *Power System Dynamics and Stability* (Prentice Hall, Englewood Cliffs NJ, 1998)
12. L. Vanfretti, *Phasor Measurement-Based State Estimation of Electric Power Systems and Linearized Analysis of Power System Network Oscillations*, Ph.D Thesis, Rensselaer Polytechnic Institute, Troy, New York, Dec 2009
13. E.Z. Zhou, Power oscillation flow study of electric power systems. Electri. Power Energy Sys. **17**(2), 143–150 (1985)
14. J. Juang, *Applied System Identification* (Prentice-Hall, New Jersey, 1994)
15. J.H. Chow, P.V. Kokotović, Time scale modeling of sparse dynamic networks. IEEE Trans. Autom. Control **AC–30**, 714–722 (1985)
16. J.H. Chow, K.W. Cheung, A toolbox for power system dynamics and control engineering education. IEEE Trans. Power Sys. **7**, 1559–1564 (1992)
17. J.F. Hauer, D.J. Trudnowski, J.G. DeSteese, A Perspective on WAMS Analysis Tools for Tracking of Oscillatory Dynamics, in *Proceedings of IEEE PES General Meeting*, July 2007
18. G. Rogers, Demystifying power system oscillations. IEEE Comput. Appl. Power, pp. 30–35 (1996)
19. Y. Chompoobutrgool, L. Vanfretti, On the Persistence of Dominant Inter-Area Oscillation Paths in Large-Scale Power Networks, in *Proceedings of 8th IFAC Symposium on Power Plants and Power Systems* (Toulouse, France, Sep, 2012), pp. 2–5
20. J.F. Hauer, W.A. Mittelstadt, K.E. Martin, J.W. Burns, and H. Lee, Integrated Dynamic Information for the Western Power System: WAMS Analysis in 2005, Chapter 15 in Power System Stability and Control: The Electric Power Engineering Handbook, ed. by L.L. Grigsby (CRC Press, Boca Raton, 2007)
21. R.L. Cresap, J.F. Hauer, Emergence of a new swing mode in the western power system. IEEE Trans. Power Apparatus Sys. **PAS–100**(4), 2037–2045 (1981)
22. Y. Chompoobutrgool, L. Vanfretti, Identification of power system dominant inter-area oscillation paths. IEEE Trans. Power Sys. (2012)
23. L. Vanfretti, S. Bengtsson, V.H. Aarstrand, J.O. Gjerde, Applications of Spectral Analysis Techniques for Estimating the Nordic Grid's Low Frequency Electromechanical Oscillations, in *Proceedings of 16th IFAC Symposium on System Identification* (Brussel, Belgium, July 2012), pp. 11–13
24. Y. Chompoobutrgool, W. Li, L. Vanfretti, Development and Implementation of a Nordic Grid Model for Power System Small-Signal and Transient Stability Studies in a Free and Open Source Software, *in proceedings of IEEE PES General Meeting*, July, 2012
25. Y. Chompoobutrgool, L. Vanfretti, Linear Analysis of the KTH-NORDIC32 System, Report Smarts-Lab-2011-001, KTH Royal Institute of Technology (2011), http://urn.kb.se/resolve?urn=urn:nbn:se:kth:diva-51255

26. Y. Chompoobutrgool, L. Vanfretti, Persistence of Multiple Interaction Paths for Individual Inter-Area Modes, in *Proceedings of 8th IFAC Symposium on Power Plants and Power Systems* (Toulouse, France, Sep 2012), pp. 2–5
27. M.E. Aboul-Ela, A.A. Sallam, J.D. McCalley, A.A. Fouad, Damping controller design for power system oscillations using global signals. IEEE Trans. Power Sys. **11**(2), 763–773 (1996)
28. J.H. Chow, J.J. Sanchez-Gasca, H. Ren, S. Wang, Power system damping controller design using multiple input signals. IEEE Control Sys. Mag. pp. 82–90 (2000)
29. J.H. Chow, G.E. Boukarim, A. Murdoch, Power system stabilizers as undergraduate control design projects. IEEE Trans. Power Sys. **19**(1), 144–151 (2004)
30. J.H. Chow, S. Ghiocel, *An Adaptive Wide-Area Power System Damping Controller using Synchrophasor Data* (Control and Optimization Methods for Electric Smart Grids, Springer Series in Power Electronics and Power Systems, 2012)
31. J.H. Chow, L. Beard, M. Patel, P. Quinn, A. Silverstein, D. Sobajic, L. Vanfretti, Guildelines for siting phasor measurement units, North American synchroPhasor initiative (2011), http://tinyurl.com/naspi-placement-guide

Index

J. H. Chow (ed.), *Power System Coherency and Model Reduction*, Power Electronics and Power Systems 94, DOI: 10.1007/978-1-4614-1803-0,

Printed by Publishers' Graphics LLC
MLSI130603.11.24.25